Pollution, Politics, and International Law

Science, Technology, and the Changing World Order
edited by Ernst B. Haas and John Gerard Ruggie

Scientists and World Order:
The Uses of Technical Knowledge
in International Organizations,
by Ernst B. Haas, Mary Pat Williams and Don Babai

Pollution, Politics, and International Law:
Tankers at Sea,
by R. Michael M'Gonigle and Mark W. Zacher

Plutonium, Power, and Politics:
International Arrangements for the Disposition of
Spent Nuclear Fuel
by Gene I. Rochlin

Pollution, Politics,

and International Law

TANKERS AT SEA

R. Michael M'Gonigle
and
Mark W. Zacher

University of California Press
Berkeley • Los Angeles • London

University of California Press
Berkeley and Los Angeles, California

University of California Press, Ltd.
London, England

First Paperback Printing 1981

ISBN 0-520-04513-0
Library of Congress Catalog Card Number: 78-54799
Printed in the United States of America

1 2 3 4 5 6 7 8 9

The 337,000 ton Danish supertanker *Karama Maersk* is a vision of size and power. The ship was recently constructed by Odense Steel Shipyards of Denmark. (Photo: Odense Steel Shipyards.)

To our families

Contents

APPENDIXES

Tables, Maps, and Charts

TABLES

MAPS

CHARTS

APPENDIX TABLES

Glossary of Acronyms

Information on organizations is provided in the text; consult index for location.

ACMRR	Advisory Committee on Marine Resource Research
ACOPS	Advisory Committee on Oil Pollution of the Sea
ARAMCO	Arabian American Oil Company
ASEAN	Association of South-East Asian Nations
BIMC	Baltic International Maritime Conference
BP	British Petroleum
CACOPS	Coordinating Advisory Committee on Oil Pollution of the Sea
CCMS	Committee on the Challenges of Modern Society (NATO)
CLC	Civil Liability Convention
CMI	Comité Maritime Internationale
COW	Crude-oil-washing
CRISTAL	Contract Regarding an Interim Settlement to Tanker Liability for Oil Pollution
DWT	Dead Weight Tonnage
ECOSOC	United Nations Economic and Social Council
E & P	Exploration and Production Forum
EPTA	Expanded Program of Technical Assistance
FAO	Food and Agricultural Organization
FOE	Friends of the Earth
GEMS	Global Environmental Monitoring System
GESAMP	Joint Group of Experts on the Scientific Aspects of Marine Pollution
GIPME	Global Investigation of Pollution in the Marine Environment
GRT	Gross registered tonnage
IACS	International Association of Classification Societies
ICAO	International Civil Aviation Organization
ICNT	Informal Composite Negotiating Text
ICS	International Chamber of Shipping
ICSPRO	Inter-Secretariat Committee on Scientific Programs
IDOE	International Decade of Ocean Exploration
IGO	Intergovernmental organization
IGOSS	Integrated Global Ocean Station System

ILO	International Labor Organization
IMCO	Intergovernmental Maritime Consultative Organization
IMO	International Maritime Organization
IOC	International Oceanographic Commission
IPIECA	International Petroleum Industry Environmental Conservation Association
IRS	International Referral System
ITIA	International Tanker Indemnity Association
IUCN	International Union for the Conservation of Nature and Natural Resources
IUMI	International Union of Marine Insurers
IAEA	International Atomic Energy Agency
LEPOR	Long-Term Expanded Program of Ocean Research
LOT	Load-on-top
MARPOL	International Convention for the Prevention of Pollution from Ships
MEPC	Marine Environment Protection Committee
MSC	Maritime Safety Committee
NAS	National Academy of Sciences
NATO	North Atlantic Treaty Organization
NGO	Nongovernmental organization
OCIMF	Oil Companies International Marine Forum
ODAS	Ocean Data Acquisition System
OECD	Organization for Economic Cooperation and Development
P & I	Protection and Indemnity Clubs
RSNT	Revised Single Negotiating Text
SCMP	Subcommittee on Marine Pollution
SCOP	Subcommittee on Oil Pollution
SNT	Single Negotiating Text
SOLAS	Safety of Life at Sea Convention
TCC	Technical Cooperation Committee
TOVALOP	Tanker Owners' Voluntary Agreement on Liability for Oil Pollution
TSPP	Tanker Safety and Pollution Prevention Conference
ULCC	Ultra-large crude carrier
UMCC	United Maritime Consultative Council
UNCITRAL	United Nations Commission on International Trade Law
UNCLOS	United Nations Conference on the Law of the Sea
UNCTAD	United Nations Conference on Trade and Development
UNDP	United Nations Development Program
UNEP	United Nations Environment Program
UNESCO	United Nations Education and Scientific Organization
WEO	West European and Others
WMO	World Meterological Organization

Foreword

By Maurice F. Strong*

The history of civilization has been primarily concerned with man's attempts to assert his sovereignty, his management, and the rule of his laws over some 30 percent of the earth's surface—the land masses and contiguous waters. Now in a single generation we face the tasks of resolving the legal status of and establishing managerial regimes for the other 70 percent of our planet—the oceans.

The struggle for the oceans has now been joined. Few battles have had a more decisive impact on the course of human history than this one will have, for its results will profoundly affect the political and economic shape of our world for centuries to come. The struggle engages the interests, the rivalries, and the aspirations of all nation-states as well as the interests and ambitions of powerful corporations.

It has thus far been a silent war. The majority of the peoples of the world whose future security and well-being are at stake in this struggle are largely unaware of it. It takes place in the assemblies and conference rooms of the United Nations Conference on the Law of the Sea (UNCLOS), in meetings of committees and working parties of International Maritime Consultative Organization (IMCO), in the quiet diplomacy of bilateral negotiations between nations, and in the sheltered sanctuaries of corporate board rooms. But the rivalries are just as intense, the potential for conflict just as acute. The use of raw power, coercion, intrigue, and undercover operations, however subtly they may be invoked, are as intense and as relentless as in any conventional war. And only the knowledge that in the nuclear age all parties could be losers in an armed conflict, together with the realization that ultimately the very nature of the oceans requires cooperation among nations for effective management, has thus far prevented nations from militarizing the struggle.

Most of the actors involved in this struggle are interested in the ocean primarily for its economic resources—oil and minerals from the seabed,

*Maurice F. Strong was formerly the Secretary-General of the 1972 U.N. Conference on the Human Environment and the Executive Director of the U.N. Environment Program. He is presently Chairman of the Board of the International Development Research Center in Ottawa and Chairman of the International Union for the Conservation of Nature and Natural Resources.

fisheries, and the use of the surface for transport. Only such organizations as the United Nations Environment Program and the International Union for the Conservation of Nature and Natural Resources are exclusively concerned with the conservation of ocean resources and the preservation of the marine environment. It is in these issues that the short-term economic interests of many nations and corporations come into conflict with the noncommercial and longer term interests of the human community as a whole. And it is here that the main focus is on oil. Oil spills have triggered public awareness of the dangers of ocean pollution and have provided the political impetus for governments to act. Yet effective action must contend with powerful economic realities. Oil is the principal commodity transported over the world's oceans and the key factor in the world's economy. The oceans provide the energy lifeline of many nations, including some of the strongest, and many of the most powerful corporations are producers or users of oil which must be transported by sea.

The great increase in ocean traffic and the dramatic accidents of the *Torrey Canyon*, *Argo Merchant*, and *Amoco Cadiz* have spurred international efforts to agree on the control of pollution of the oceans by oil. This book documents the story of these efforts and relates them to the larger context of protecting the global environment as well as creating a viable legal and managerial regime for the oceans. By focusing clearly and specifically on oil pollution by vessels, the authors paint a vivid and authoritative picture of the real nature of the struggle that is taking place for the oceans as a whole. They correctly identify the fundamental political character of the struggle and with remarkable perception analyze thoroughly the interests of the various actors—governmental, intergovernmental, and corporate. They trace the complex interplay of the forces which these actors have brought to bear in the negotiations at UNCLOS and IMCO as well as in their unilateral initiatives.

This book is one of the finest pieces of work in the environmental field I have seen. It is a rich source of the kind of historical and technical information that makes it an essential reference book for the expert. Yet the authors have succeeded in presenting their material in a clear and interesting manner in which the sense of high drama in the realization of what is at stake pervades the entire book and is present in even the most technical of chapters.

This is a timely and important book and in producing it the authors have done a great service to the entire environmental cause as well as to the cause of establishing a viable regime for the oceans. By their brilliant analysis of the political process affecting the control of oil pollution they have cogently pointed up both the difficulties and the possibilities of turning the struggle for the oceans into a victory for all mankind.

Acknowledgments

During the research and writing of this book, many individuals and organizations have assisted us. We interviewed over two hundred government, international, and industry officials, and more than twenty individuals read and commented on our drafts. We have seldom been able to attribute information or opinions in the text to these people, since most of them still hold official positions. To all of them we would like to express our appreciation for their kindness and assistance. We are only sorry that a limitation on space prevents us from thanking each of them in this preface.

Three individuals were exceedingly generous in helping us to understand the political and technical dimensions of oil pollution control. They are Claude Walder, executive secretary of the Oil Companies International Marine Forum; Rear Admiral Sidney Wallace, U.S. Coast Guard; and Gordon Victory (Ret.), Marine Division, U.K. Department of Trade. The Secretariat of the Intergovernmental Maritime Consultative Organization (IMCO) assisted us in a variety of ways. We greatly benefited from insights of Thomas Mensah, Yoshio Sasamura, and Secretary-General C. P. Srivastava. And Anthea Meldrum and Charlotte Hornstein greatly facilitated our research at IMCO. Members of the Canadian government helped us in many ways. In particular, we would like to thank Gareth Davies, Ron Macgillivray, and Reg Parsons of the Ministry of Transport and Terry Bacon, Robert Hage, Barry Mawhinney, and Erik Wang of the Department of External Affairs.

Information and critiques were also kindly provided by James Barnes, Stephen Gibbs, Sir Colin Goad, Clancy Hallberg, Maurice Holdsworth, Christopher Horrocks, Kitty Gillman, Eileen Gladman, Robert McManus, Donald McRae, Dr. A. C. R. M'Gonigle, Sonia Pritchard, Joseph Rubin, Harvey Silverstein, H. Steyn, and Robert Stockman. We also greatly benefited from the advice and assistance of the series co-editors, Ernst Haas and John Ruggie, and Grant Barnes, Nancy Jarvis, and Sheila Levine of the University of California Press. We were able to attend the Tanker Safety and Pollution Prevention Conference in February 1978 as delegates of Friends of the Earth, and we would like to thank that organization and its delegates, Richard Sandbrook and Victor Sebek, for providing us this opportunity.

James McConnell was an excellent research assistant and friend during his two summers of work with us. We are particularly grateful to him for his thorough research on Chapter 2. Dean Peter Larkin of the Faculty of Graduate Studies at the University of British Columbia helped us through his support for the Institute of International Relations' project "Canada and the International Management of the Oceans." In London the Centre of International Studies at the London School of Economics facilitated our library research. Also, to the Donner Canadian Foundation, the Canada Council, and Imperial Oil Limited (Canada), we are indebted for generous financial assistance. We have, in addition, been most fortunate in having excellent secretarial assistance. In London Cassie Palamar helped us during our research, and in Vancouver Betty Greig heroically and efficiently typed more drafts than she would like to remember.

A special and unique expression of gratitude goes to the Madhumati Restaurant in London. On many evenings during our work in London its curries enlivened both the palate and the mind. Finally, we would like to thank Wendy, Carol, Nicole, and Glenn, whose interest in things other than oil pollution and IMCO provided us with a balanced perspective on life.

Vancouver
Summer 1978

R. MICHAEL M'GONIGLE
MARK W. ZACHER

PART ONE

The Issues and the International Organization Context

Chapter I

The Nature of the Challenge

The protection of the global environment has only recently become an issue of international concern. Indeed, despite the massive, worldwide industrialization of the last century, the health of the environment was long assumed or ignored. With the mounting pollution and ecological disruption brought on by such industrialization, attitudes have begun to change, but slowly. Even today, the seriousness of the problem and the size of the changes needed to deal with it are still uncertain. Whether, as some would argue, man's present patterns put him "on a collision course with the laws of nature"[1] is probably debatable. Yet the desirability of preserving a clean, attractive and healthy environment is widely accepted and the need to achieve a long-term environmental balance, undeniable. The point at issue is not the need for environmental protection. It is how much and how.

Our present way of life is premised on technological expansion, economic growth, and rising expectations of wealth—processes that have been subject to little control. And today it is these very processes that are most widely cited as the ultimate sources of the environmental "crisis."[2] In addition,

1. Paul R. Ehrlich, Anne H. Ehrlich, and John P. Holden, *Human Ecology: Problems and Solutions* (San Francisco: W. H. Freeman, 1973), p. 3. The list of offenses against nature is seemingly endless, and no attempt is made here to elucidate them. However, for a recent bibliography of environmental problems, see William Ophuls, *Ecology and the Politics of Scarcity* (San Francisco: W. H. Freeman, 1977), pp. 249–284. Many respectable scientists have argued that mankind is indeed "at the turning point." Thirty-three leading British scientists concurred earlier in the decade, "If current trends are allowed to persist, the breakdown of society and the irreversible disruption of our life support systems on this planet . . . within the lifetimes of our children are inevitable." "Blueprint for Survival," *The Ecologist* 2 (January 1972), p. 1. For an even more dire prognosis, see "Report of the MIT Study of Critical Environmental Problems," *Man's Impact on the Global Environment: Assessment and Recommendations for Action* (Cambridge, Mass.: MIT Press, 1970).

2. The most important early statement on the side effects of technology was Jacques Ellul's *La Technique* in 1954 (translated as *The Technological Society*, New York: Knopf, 1964). See also *inter alia* Ezra Mishan, *Technology and Growth: The Price We Pay* (New York: Praeger, 1969); David Hamilton, *Technology, Man and the Environment* (New York: Scribners, 1973); William R. Kinter, *Technology and International Politics: The Crisis of Wishing* (Lexington, Mass.: Lexington Books, 1975), and books subsequently referred to. There has been an even more vigorous

the political and legal systems that have legitimized and encouraged these processes in the past are now being asked to control them or, at least, to minimize their unwanted consequences. To accomplish such a transformation is clearly a difficult undertaking.

The task of environmental protection is made even more difficult by its global dimensions. Pollution often ignores boundaries, but the political solutions necessary to control it cannot. Nowhere is this divergence between environmental and political reality more apparent than in the oceans. As Barbara Ward and René Dubos have noted eloquently in *Only One Earth*, "It is, above all, at the edge of the sea that the pretensions of sovereignty cease and the fact of a shared biosphere begins more strongly with each passing decade to assert its inescapable reality."[3]

The protection of our global heritage has become a significant concern at the international level only in recent years. During the 1920s and 1930s, there were attempts by a few governments and the League of Nations to conclude an international convention to control ship pollution—the only major intergovernmental meetings to discuss questions of international pollution before the Second World War—and these came to naught. Until the 1950s, in fact, almost all conservation treaties dealt solely with the protection of migratory wildlife.[4] In the late 1940s and the 1950s the United Nations Transport and Communications Commission promoted renewed discussions on a possible treaty to control oil pollution of the oceans, and its activities were partially responsible for Britain's finally convening such a conference in 1954. This conference and the convention it produced, while they attracted little public attention, were a long overdue beginning to international environmental regulation.

It was another decade, however, before a general concern for the environment and a recognition of the need to reorient some of our political and economic behavior finally began to develop in earnest. Long dismissed as the ravings of "eccentric bird-watchers, butterfly chasers, and overstrenuous hikers,"[5] calls for an end to the reckless degradation of the natural environment increased with the mounting evidence of man's ecological folly. In particular, the publication of Rachel Carson's *Silent Spring* in 1963

---

debate on the economic nature of the environmental crisis. This has been reflected in the writings of the "limits to growth" and the "steady-state" economic theorists. See *inter alia* Donella H. Meadows et al., *The Limits to Growth* (New York: Universe, 1972); Herman E. Daly, ed., *Toward a Steady State Economy* (San Francisco: W. H. Freeman, 1973); Wilfred Beckerman, *In Defence of Economic Growth* (London: Cape, 1974); and Ophuls, *Ecology*.

3. (New York: W. W. Norton, 1972), p. 203.

4. And often even these have been disastrously unsuccessful as the history of, for example, the *International Convention for the Regulation of Whaling* (1946) would show.

5. Richard Falk, *This Endangered Planet: Prospects and Proposals for Human Survival* (New York: Random House, 1971), p. 183.

sounded an alarm and became "the sort of rallying point of the movement to protect the environment that the anti-slavery book *Uncle Tom's Cabin* had been for the movement to abolish slavery in the 1850s."[6] From the industrialized countries, particularly the United States, came a rising chorus of protest.

This growing awareness in the developed countries spilled over to the international level as well. In 1965, for example, it was learned that excessive high seas exploitation had finally completed the commercial extinction of the blue whale, the largest animal ever to inhabit the planet. Two years later, visual confirmation of our environmental carelessness was dramatically portrayed before a shocked world audience as the grounded Liberian supertanker *Torrey Canyon* poured 120,000 tons of heavy crude oil onto a hundred miles of British and French coastlines. That vision was repeated two years later when, in 1969, an oil well ran out of control off the coast of Santa Barbara, California. Meanwhile, with the expansion of man's ambitions in *outer* space came a new awareness of *inner* space. Photographs of the "Spaceship Earth" soon confirmed "what previously could only be grasped intellectually: that earth is indeed small, lovely, unitary, finite and vulnerable."[7]

The issue was now beginning to command worldwide attention. In 1968, UNESCO convened its Biosphere Conference, and scientists from around the world gathered to warn of the need to develop a coordinated and rational approach to the use and conservation of the earth's resources. In 1970, the Food and Agricultural Organization called together hundreds of fisheries experts in Rome to consider the effects of growing marine pollution on the global fish resource. Even NATO entered the scene with conferences and programs organized by its Committee on the Challenges to Modern Society. In 1972, world concern crystallized with the holding of the United Nations (Stockholm) Conference on the Human Environment. This meeting stimulated such vast public interest that governments were forced to consider a wide range of environmental issues that had long been neglected. It brought together hundreds of delegates from almost every country and, under the scrutiny of the world press, approved numerous principles to guide future international conduct. To see that these principles were implemented, the conference initiated a new international environmental agency, the United Nations Environment Program (UNEP). "The growth of all great social movements has been marked at certain critical

6. Charles R. Ersendrath, "Environmental Protection: A State of Mind," *Horizons U.S.A.* 20 (1977), p. 26.

7. Thomas W. Wilson, *International Environmental Action: Global Survey* (Cambridge: Dunellen, 1971), p. 6.

points by catalyzing or consolidating events. The United Nations Conference on the Human Environment [was] such an event."[8]

Yet appearance belies reality: progress has been painfully slow. Certainly some minor adjustments have been made, but the basic problems, although identified, remain untackled and unresolved. This is especially true of those issues which, at the international level, can be resolved only by negotiation and compromise. As a report to the United Nations Secretary General noted in 1969, "Technological solutions exist to most industrial pollution problems but are not applied usually for economic and political reasons."[9] Only an understanding of the constraints on political action will allow observers to prescribe realistic and constructive avenues for change.

To illuminate the nature of the international political challenge posed by environmental pollution is the purpose of this book. The environmental issue is a broad one and cannot easily be considered in its entirety. We present here a study of the political processes leading to the creation and application of international environmental law for one pollutant, oil. The negotiations to control this marine pollutant have involved numerous actors over many years, but the focus of debate has been in one international organization, the Intergovernmental Maritime Consultative Organization (IMCO). Oil pollution was one of the first environmental problems to have been confronted internationally (a convention was concluded on it in 1954), and IMCO is a U.N. agency with a twenty-year history in the field of international environment regulation. A study of oil and of IMCO presents, therefore, an ideal opportunity to see political—and economic—interests in operation when confronted by an environmental problem of international complexity. To understand the many policies and powers which must combine to protect against this contaminant is our first and central objective.

8. Lynton K. Caldwell, *In Defense of Earth: International Protection of the Biosphere* (Bloomington, Ind.: University of Indiana Press, 1972), p. 5. A recent commentator noted in looking back at the Stockholm Conference that the environmental issue has become even more complex in the few years since. The conference, he says, "could not have foreseen . . . the linkages of general environment concerns with the sudden world crisis in energy, in food production, in massive mid-African droughts with starvation for millions, and the near breakdown of world pricing systems and of commodity, trade and monetary arrangements." Maxwell Cohen, "International Law and the Environment," *Queen's Quarterly* 81 (1974), p. 444.

9. Cited in J. E. S. Fawcett, *Priorities in Conservation* (London: David Davies Memorial Lecture, Royal Institute of International Affairs, June 1970), p. 8. Professor Fawcett also commented that "it is an illusion to suppose that problems created by technological advance —the many forms of pollution, the loss of amenities that go with industrialization and the growth of urban networks—are capable merely of technological solutions. Far too much of the literature and debate on conservation is buried under scientific and pseudo-scientific technicalities and jargon—pesticide persistency, food chains, eutrophication, toxicity, levels of tolerance, the mega-mouse experiment, the population explosion—treated with varying degrees of alarm and complacency, while *the political roots of conservation are not grasped.*"

The regulations IMCO has produced can be classified into three basic categories. First, IMCO has created rules to *prevent* pollution. These encompass standards to control both intentional (or operational) discharges and, in recent years, accidental spillages as well. Here, technological solutions often have been thwarted for "economic and political reasons." Of course, pollution prevention does not always work, and negotiations have focused, second, on the provision of *remedies* when prevention fails. Most important in this area are the negotiations that have created new rights for coastal states both to act against threats of pollution and to receive compensation when oil damage does occur. With such dramatic accidents as the *Torrey Canyon*, the *Argo Merchant*, and the *Amoco Cadiz*, both these issues have been of much concern to the public. Finally, one must look at how the *jurisdictional powers* to prescribe and enforce particular pollution control standards have developed for all concerned states. With the increasing threat that pollution from foreign ships has posed to the coastal states, the issue of who has what jurisdiction over vessels has been a significant source of contention at IMCO and, in the last few years, at the United Nations Law of the Sea Conference (UNCLOS).

International negotiations on these issues have focused on the creation of treaties or conventions. Since 1954, therefore, the history of international oil pollution control has been replete with committee meetings and conference diplomacy. For the present study, a review has been completed of all the submissions and debates at IMCO since its creation in 1958 and, before that time, at the London Conference of 1954. Several committee meetings and one conference were attended by the authors. In reviewing the documentation of IMCO's committees, only those issues specifically concerned with pollution prevention were studied. Those environmental improvements occurring as a spin-off from negotiations on related maritime topics (such as safety of life at sea or crew regulations) were excluded, and this has unavoidably restricted the scope of our treatment of pollution resulting from accidents.[10] Analysis of the documentary material was augmented by extensive interviews with a large number of industry, government, and international officials. There is clearly a limit on the extent to which one can pursue a study of each participant in the negotiations, but the key state and industry actors have been investigated.[11]

10. Several proposals for controlling accidental spills were considered in 1971 and 1973. However, it was not until 1977 that the prevention of accidental pollution finally was treated as a major separate issue and not simply as an adjunct of the many more general regulations for the safety of all ships. That this is so is itself an important observation, but it does limit our treatment of this source of pollution. Given the vast scope and complexity of maritime regulations, only those specifically created for environmental purposes will be analyzed in detail.

11. For two studies of a variety of national environmental politics, see Cynthia H. Enloe, *The Politics of Pollution in a Comparative Perspective: Ecology and Power in Four Nations* (New York:

The giant supertanker, *Globtik Tokyo*, was until recently the largest ship in the world. At 477,000 tons (dwt), it stretches a quarter of a mile and has a surface area that could accommodate two and a half soccer fields. This ship has been superseded by the 542,000 ton French tanker *Batilus* as the world's largest. (Photo: Miller Services/Camera Press.)

Finally, information obtained from the documentation and interviews has been supplemented by evidence available from a wide variety of secondary sources.

Oil pollution is a political problem and, by examining the areas in which specific international environmental laws have been created, a fairly complete picture of the many factors determining changes in the international legal regime can be identified. Moreover, when combined with an understanding of the problems encountered in actually achieving compliance with the formal legal regime, the entire landscape of international environmental control—from formulation to implementation—emerges for this one substance. To portray this landscape is the first task of this study.

David McKay, 1975), and Donald R. Kelley et al., *The Economic Superpowers and the Environment: The United States, the Soviet Union and Japan* (San Francisco: W. H. Freeman, 1976).

A second and closely linked goal is to understand what the politics in this one issue reveal about the character of the changing international system of which they are a part. After all, oil is only one of many pollutants, and pollution is itself but one of the many new problems to be found on the international political agenda. Indeed, as modern society expands to nudge the very edge of the biosphere, the entire setting of political activity has begun to change. Recent literature on international law and politics has been much concerned with these changes, particularly with the many new actors, issues, and strategies that have started to appear on the world stage. These changes are reflected in the environmental sphere. Through analysis of IMCO's work on oil pollution, we can highlight the shadows.

With the recent and rapid changes in the international political arena, accepted perceptions of the operation of the political system itself have been challenged. In the face of new global challenges, some have been led to question the very ability of the traditional international structure of independent nation-states to respond to them.[12] A wholesale reevaluation of this structure is beyond the scope of this work but, by watching its operation in one field over many years, some of its shortcomings are visible. At the same time as the limitations and parochialism of nation-states are being decried, many more "states" have emerged. Today, in addition to the United States and the Soviet Union, other centers of power—Western Europe, Japan, China—have achieved greater prominence. The developing countries of the "Third World," a few years ago only a setting for great power rivalries, are now increasingly restless and self-assertive. With their special interests, the developing nations cannot help but have important future effects on international politics in general and on the environmental issue in particular.

In addition to the advent of these new *state* actors, the salience of nongovernmental organizations (whether they be associations of national interest groups or multinational corporations) has received increasing attention in recent years.[13] In the past, industrial actors were seen largely as national organizations which affected only the policies of their local governments. Now it is recognized that they are also independent "transnational" actors

12. As one book has noted recently about the entire nation-state system: "The human population of the earth may be found together in a common fate, but parochial tribalism continues to sustain a fragmented international order more relevant to the seventeenth than to the late-twentieth century." Harold and Margaret Sprout, *Toward a Politics of the Planet Earth* (Princeton: Van Nostrand, 1971), p. 401. For a similar appraisal, see Richard Falk, "The Logic of State Sovereignty Versus the Requirements of World Order," *1973 Yearbook of World Affairs* (London: Royal Institute of International Affairs), p. 7.

13. See particularly Robert O. Keohane and Joseph S. Nye, eds., "Transnational Relations and World Politics," *International Organization* 35 (1971), pp. 329–748. See also Raymond Vernon, *Sovereignty at Bay: The Multinational Spread of U.S. Enterprises* (New York: Basic Books, 1971); Jeffrey Harrod, "Transnational Power," *1976 Yearbook of World Affairs*, p. 97; and Susan Strange, "Transnational Relations," *International Affairs* 52 (1976), p. 333.

(often with interests in many states), who participate in international decision-making and offer their various home states new forms of leverage on the policies of other countries. These new perspectives are explored in this volume, particularly with respect to the oil, shipping, and maritime insurance industries and their international spokesmen. Indeed, it is impossible to understand the international politics of oil pollution control without a consideration of the interests and activities of these private actors. Public interest environmental organizations also will be analyzed, but to date their contribution in this area has largely been through national and not "transnational" channels.

New conceptions of old actors are also emerging. The study of "transnational" relations has provided a new perspective of just what constitutes international politics. This has undermined the old vision of the state itself, which often is not a "unitary actor."[14] Instead, with the diversification of actors and issues, various competing bureaucratic interests act across national boundaries, carrying on their business subnationally (or "transgovernmentally"), building foreign alliances in much the same way as do the new nongovernmental actors. Although it is not possible to examine in detail the bureaucratic processes of every participant state,[15] some general patterns are evident and important at IMCO.

Finally, with these new insights into the operation of national and non-national actors, our conception of the nature of the international organization itself is changing: "We need to think of international organizations less as institutions than as clusters of intergovernmental and transgovernmental networks associated with the formal institutions."[16] That is, within these organizations, there is such a continuous flow of officials dealing with such a variety of issues that the function of the agency may be as much a place to "activate potential coalitions" as to engage in more formalized undertakings. It is, for example, important to understand just when an issue will come into IMCO in order to see the organization in its larger informal context. However, within the agency, neither the formal nor the informal processes can be ignored in analyzing the development of international environmental regulations. It is certainly true that there is such a variety of interactions occurring within the organization that one cannot refer to *an* international organization without some qualification.[17] Yet it is, at the

14. Keohane and Nye, "Transnational Relations." See also (by the same authors) "Transgovernmental Relations and International Organizations," *World Politics* 27 (1974), p. 39, and works cited in fns. 16 and 18.

15. In fact, throughout most of the book, states are referred to as monolithic actors. Certain trends in transgovernmental relations will be evident, however, and these will be brought out more explicitly in the concluding chapters.

16. Robert O. Keohane and Joseph S. Nye, *Power and Interdependence: World Politics in Transition* (Boston: Little Brown, 1977), p. 240.

17. Probably the classic study on international organizations within the global political process is Robert Cox and Harold Jacobson, eds., *The Anatomy of Influence: Decision-Making in*

same time, the organization as formally conceived that is able to create specific international laws through the diplomatic conference and the international convention. In a global political environment and within an organization where activities often are irregular, if not haphazard, rule creation is a strictly formalized procedure.

In addition to the influx of new actors and the transformation of the old, we also have observed a diversification of the political agenda with which they deal. Many and more varied issues have emerged in international politics as matters of important concern. The "bipolar" system has fragmented and the "Cold War" receded. The nuclear threat has by no means disappeared, but the endless jockeying between the two superpowers that dominated the politics of the two decades following the Second World War has achieved a relative stability in recent times. Meanwhile, issues of economic and ecological interdependence have begun to predominate over those of war. The future of the world economy, the possibilities for economic development and a more equitable economic distribution, the uncertainties of uncontrolled resource exploitation and use, and the deterioration of the global environment have all emerged as pressing and difficult concerns. The vision of catastrophe has not left us; it has become more diversified.

It is, therefore, hardly surprising to observe a fragmentation in the treatment of the issues themselves: "Analysts of world politics have begun to talk less about *the* international system, and to realize that there are significant variations among systems in different issue-areas."[18] No longer is the military balance of power or the Cold War axis of conflict assumed to underlie all political issues. The bases of policy, power, and influence are, in a variety of "issue-areas," a matter for empirical analysis. Such analysis has led to a questioning of the very nature of political power itself. Many regard it as transformed in the new, more fragmented system. For example, one important study of international organizations has argued that there must now be recognized a distinction between power in the "general environment" of world politics as traditionally conceived and power in the "specific environment" of a particular issue.[19] If this were so, it could herald the advent of a genuine international pluralism that would transform the nature of international political change. This has been an important topic of recent debate.[20]

International Organizations (New Haven, Conn.: Yale University Press, 1973). They define such an organization as "a system of interaction including all of those who directly participate in decisions taken within the framework of the organization, and in addition all officials and individuals who in various ways determine the positions of the direct participants" (p. 16).

18. Robert O. Keohane and Joseph S. Nye, "Transgovernmental Relations and International Organizations," *World Politics* 32 (1974), p. 55.

19. Cox and Jacobson, *Anatomy of Influence*, especially pp. 15–36 and 409–423.

20. Keohane and Nye (*Power and Interdependence*) have constructed a comprehensive overview of the process of "regime change" in the global political system. They have identified four basic sources for this change: (1) the normal, uncontrolled impact of economic and

Admittedly, the need for such analysis is not new, but it does seem to have acquired more urgency of late. Many are moved by the belief that the international system may really be changing in an historic way. Others argue that, in its essentials, the political system is not changing at all but that it must if we are to survive the demands of the changes occurring around it. Whatever the perspective, there is certainly today a new momentum behind those inquiring about the future of "world order."

These are not minor issues and, if we are to pursue fruitful paths in search of a just and peaceful world order, they are issues which must be confronted. They are certainly not remote from the challenge posed by the ecological crisis or, specifically, by the need to control marine oil pollution. Indeed, our investigation into the factors affecting the policies and influences of states in this one area should tell us much not only about the development of international environmental law but also about its relationship to other issues and to the larger structure of global political power. The implications that this has for understanding the prospects for future international collaboration are clear. In particular, our findings will be of interest to the advocates of "functionalism" who have long argued that technical (as opposed to political) and functionally specific problems are more susceptible to international management, and that technical experts (as opposed to diplomatic officials) are more amenable to international cooperation. The functionalists have also been wedded to the notion that a "common good" exists with respect to international problems and that the building of a peaceful and politically unified world order can emerge from cooperation within a variety of specific functional issue-areas. These views have been criticized on a number of grounds,[21] yet they obviously raise significant questions that must be addressed in any study of the politics of international environmental control.

The need for environmental protection is growing, and it demands a response from our international political system. To do so it is necessary to recognize explicitly the goal of a "preferred world order,"[22] a recognition necessary for both international lawyers and political scientists alike:

---

technological development—the economic process model; (2) state power in the general environment still viewed in traditional, military power terms—the overall structure model; (3) state power in the "specific environment"—the issue structure model; and (4) leverage derived from participation in international organizations—the international organization model.

21. For very good elucidations and critiques of these views, see Ernst B. Haas, *Beyond the Nation-State: Functionalism and International Organization* (Stanford, Calif.: Stanford University Press, 1964); James P. Sewell, *Functionalism and World Politics: A Study Based on United Nations Programs Financing Economic Development* (Princeton, N.J.: Princeton University Press, 1966); and Inis L. Claude, *Swords into Plowshares: The Problems and Progress of International Organization* (New York: Random House, 1964), chap. 17.

22. The Institute for World Order has sponsored a number of studies under its World Order Models Project. See *inter alia* Richard Falk, *A Study of Future Worlds* (New York: Free

> Our retreat from normative issues, and our almost total preoccupation with the analysis of things-as-they-are, seem occasionally to have erased from our minds the question of how international organizations could serve contemporary human purposes. . . . How can arrangements be developed that both meet criteria of political feasibility and promise to contribute to an eventual transformation of the organization system and therefore of characteristic outcomes?[23]

Global interdependence, though real, is today still at an early stage of evolution. Compared to the necessities of the 1980s and 1990s, those of the 1970s will seem to have been slight. Economic, political, and environmental interdependence is growing daily. Whether as a consequence of will or destiny, the political and environmental realms cannot and will not remain forever separated. To understand the trends of the present and to identify the needs of the future is the immediate task.

Press, 1975) and the journal *Alternatives*. See also Louis Rene Beres and Harry R. Targ, eds., *Planning Alternative World Futures* (New York: Praeger, 1975).

23. Robert O. Keohane, "International Organization and the Crisis of Interdependence," *International Organization* 29 (1975), pp. 360–361.

Chapter II

The International Problem of Oil Pollution

"Indecently symptomatic of the consumer societies," the oil tanker has been condemned as it fuels our affluence "regardless of the ultimate cost."[1] But what is "the ultimate cost"? Some would argue that the health and life of the oceans—and of man himself—are at stake. The issue is not a simple one, and it requires extensive information: how much oil is being shipped by sea, and what discharges result from it; where do these discharges occur; and, most importantly, what are their polluting effects on the oceans and man?

THE MARITIME TRANSPORT OF OIL

Ever-increasing quantities of oil are being transported by sea. Today, oil is the major source of fuel for ocean-going vessels,[2] but it is the vast expansion in world energy consumption—and in oil as the primary source of that energy—that has led to the tremendous increase in the transport of oil by sea. This increase has been marked by a dramatic, worldwide rise in maritime pollution.

The rapid increase in the volume of world oil consumption and exports is revealed by Table 1. For the years 1953–1973, world oil consumption went up from 649 to 2,765 million tons (an increase of 7.5 percent per year) and exports of oil rose from 236 to 1,695 million tons (an increase of 10 percent per year). The difference in the rate of increase of oil consumption as compared with oil exports has meant that whereas in 1953 only 36 percent of all oil consumed came from foreign sources, in 1973 it was 61 percent. Some of these exports were not moved by sea (particularly from Canada to

1. Noel Mostert, *Supership* (New York: Knopf, 1974), p. 65.

2. In 1914 only 3 percent of such vessels used oil as their basic fuel. By 1953 the figure was 83 percent, and today it is nearer 100 percent.

TABLE 1
World Oil Consumption and Exports, 1938–1973*

	Exports	*Consumption*	*% of Oil Exported*
1938	89	265	34
1953	236	649	36
1963	615	1,320	47
1973	1,695	2,765	61

*Figures in millions of tons
SOURCE: *B.P. Statistical Review of the World Oil Industry, 1963* and *1973* (London: British Petroleum).

the United States and from the Soviet Union to Eastern Europe), but the great majority of them were. In addition, in many regions such as Western Europe, oil is also moved by sea between the ports of individual oil-importing countries. Clearly, there is today an enormous quantity of oil being carried on the world's seas.

TYPES AND VOLUME OF OIL DISCHARGES IN THE MARINE ENVIRONMENT

With such a large ocean oil trade, pollution is inevitable. But how much? Since the late 1960s a number of studies have attempted to estimate the flow of petroleum hydrocarbons into the marine environment. All, however, differ in the sources of discharges which they include and often in their calculations of the volume of discharges from the various sources. Perhaps the most authoritative estimates were produced by a workshop sponsored by the Ocean Affairs Board of the U.S. National Academy of Sciences in 1973 (hereinafter referred to as the *NAS Report*).[3] Composed of scientists from Canada and Western Europe as well as the United States, the workshop considered many previous studies on the subject.[4] Their conclusion was staggering: in one year, 1973, 6,713,000 metric tons of oil entered the world's oceans.

Oil from ships is but a portion of this total, and any serious long-term control strategy obviously must include not only vessels but shore industries and offshore production as well. Indeed, as Table 2 indicates, 54 percent of all oil discharged into the oceans comes from "land sources" and, of this,

3. *Petroleum in the Marine Environment* (Washington, D.C.: National Academy of Sciences, 1975), p. 6. (Hereinafter cited as *NAS Report.*)
4. Two other very good studies with such estimates are K. G. Brummage, "What is Marine Pollution?" *Symposium on Marine Pollution* (London: Royal Institute of Naval Architects, 1973), pp. 1–9; and J. D. Porirelli, V. F. Keith, and R. L. Storch, *Tankers and the Ecology* (New York: Society of Naval Architects and Marine Engineers, 1971). There is an extensive bibliography of other studies in *NAS Report*, pp. 17–18.

approximately one-half originates from discharges of oily wastes into rivers. Other land-based sources include direct discharges of smaller amounts of oil from coastal refineries and nonrefining industries. These might be among the most harmful of all the sources, as they are discharged directly into the shallow and ecologically sensitive coastal areas in very concentrated form. Moreover, although the *NAS Report* focuses on the control of vessel-source oil pollution, ships are not even the only source of nonland oil pollution. Other such sources, it is estimated, account for about 11 percent of the total volume of oil discharged into the oceans, with about one-eighth of this amount coming from offshore oil production and seven-eighths from natural seepages from the ocean floor.

The sources of oil pollution on which this study will focus are those listed in Table 2 under "Marine Transportation." These account for 35 percent of all petroleum hydrocarbons entering the marine environment. These discharges, as indicated in Table 3 come from both tankers and nontankers. Some are intentionally caused (operational), and some are accidental. The intentional or operational discharges are largely a direct product of the routine operation of a ship's crew in cleaning and ballasting cargo tanks and cleaning the bilges. As with land-based pollution, it is man's intentional actions that are, by far, the most serious problem.

Operational Discharges from Tankers

Sixty-two percent of all ship-generated oil discharges are estimated by the *NAS Report* to result from routine tanker operations, which is in fact probably an underestimation. Two operations are performed by a tanker's crew during the vessel's return voyage from its unloading port that account for most of this outflow. First, after the tanker has discharged its cargo, the crew fills about one-third of the tanker's cargo tanks with seawater ("ballast") in order to maintain sufficient propeller immersion and stability. This "departure ballast" mixes with whatever oil is remaining in the cargo tanks (usually about 0.35 percent of the original cargo), and prior to arrival at the loading port, it must be removed from the tanks. On this same return voyage the crew also washes another one-third of the cargo tanks so as to clean the tanks for "arrival ballast" and to prevent the buildup of sludge. The resultant oily water residues must also be removed from the cargo tanks before the loading of a new cargo.[5]

Until the mid-1960s tankers discharged their oily water ballast and oily tank cleanings directly into the oceans. Some of these discharges occurred

5. Prior to the adoption of the LOT system (which is described further on), most tankers discharged the oily ballast during the middle of their voyage, taking on new ballast into the tanks cleaned during the first part of their voyage. They were thus able to avoid discharging the oily ballast into the harbor area at the loading port.

TABLE 2
Sources and Volume of Petroleum Hydrocarbons Entering the Oceans, 1973

Source	*Million Metric Tons*	*Percentage*
Marine Transportation		
LOT tankers	0.31	5.07
Non-LOT tankers	0.77	12.59
Drydocking	0.25	4.09
Terminal operations	0.003	0.05
Bilges bunkering	0.5	8.18
Tanker accidents	0.2	3.27
Nontanker accidents	0.1	1.64
Total	2.13	35.23
Other Marine Sources		
Natural seeps	0.6	9.82
Offshore production	0.08	1.31
Total	0.68	11.13
Land Sources		
Coastal refineries	0.2	3.27
Atmosphere	0.6	9.82
Coastal municipal wastes	0.3	4.91
Coastal, nonrefining, industrial wastes	0.3	4.91
Urban runoff	0.3	4.91
River runoff	1.6	26.17
Total	3.3	53.99
TOTAL	6.113	100.35

SOURCE: *Petroleum in the Marine Environment* (Washington, D.C.: National Academy of Sciences, 1975), p. 6.

outside fifty-mile coastal zones, as required by the 1954 Convention for the Prevention of Pollution of the Sea by Oil. Since the mid-1960s most crude-oil-carrying tankers allegedly have adopted what is known as the "load-on-top" (LOT) system.[6] By placing all oily ballast water and tank-cleaning

6. For detailed descriptions of the LOT system, see G. Victory, "The Load-on-Top System," *Symposium on Marine Pollution* (London: Royal Institute of Naval Architects, 1973), pp. 10–20; and Porricelli, Keith, and Storch, *Tankers*, pp. 6–10.

TABLE 3
Sources and Volume of Ship-Generated Oil Discharges, 1973

	Million Metric Tons	*Percentage*
Operational Discharges from Tankers		
LOT tankers	0.31	14.5
Non-LOT tankers	0.77	36.2
Drydocking	0.25	11.7
Accidental Discharges from Tankers		
Terminal operations	0.003	0.1
Tanker accidents	0.2	9.4
Operational Discharges from Nontankers		
Bilges, deballasting fuel tanks	0.5	23.5
Accidental Discharges from Nontankers		
Nontanker accidents	0.1	4.7
Total	2.13	100.1

SOURCE: *Petroleum in the Marine Environment* (Washington, D.C.: National Academy of Sciences, 1975), p. 6.

residues in a "slop tank," allowing the oil to float to the top of the oily water mixture and then decanting the water from the bottom, the LOT system permits the retention of most of the oil. When a new cargo is taken on board, it is either "loaded on top" of the oil residues in one of the cargo tanks, or the residues are retained separately in the slop tank. The system can be used only for crude and other heavy oils and cannot be employed for refined oils, which constitute approximately 20 percent of all petroleum products shipped by sea.[7] Light refined products such as gasoline do not float to the top of water as quickly as do "black" oils, and the residues of these cargoes cannot be mixed with new cargoes. To eliminate the discharge of refined oil residues, tankers would require "oily-water separators" capable of effective separation of white oils emulsified in water, storage tanks on board, and oil "reception facilities" in ports.

The *NAS Report* estimated that 80 percent of all tankers carrying crude oil used the LOT system and that it was 90 percent effective (that is, 90 per-

7. *B.P. Statistical Review of the World Oil Industry, 1973*, p. 10. The amount of refined oil remaining in cargo tanks after the cargo's discharge at an unloading port is considerably less than the amount of crude oil remaining after a comparable operation. Whereas the average for crude oils is around 0.35–0.40 percent, it is less than half of this for refined oils.

cent of the oil in the oily ballast water and oily tank cleanings was retained on board).[8] But on the basis of the authors' interviews with government and industry officials, it is clear that this estimate was greatly exaggerated in 1973 and is probably still largely inaccurate. Instead of its being 90 percent effective, secret oil industry studies in 1971 and 1972 revealed that LOT was probably only about 50 percent effective. This would have meant a discharge in 1973 of at least 1,500,000 tons as compared to the 310,000 tons estimated by the *NAS Report*. Even if the compliance of tankers has improved in recent years (as many argue), there is no authoritative way to estimate how great a change has really occurred. Moreover, considering past patterns of opportunistic behavior among shipping crews and an inadequate enforcement system, it would be naive to expect a high level of compliance.[9]

A third source of operational discharges occurs at drydocking. As part of its routine maintenance, a tanker must put into drydock for repairs about once every eighteen months, and at such a time the cargo tanks must be clean. According to the *NAS Report*, only half of these tankers cleaned their tanks and retained the residues on board or had their tanks cleaned while in drydock.[10] In the case of the others, the oily mixtures were discharged into the sea. Even here, the estimate that 50 percent retained their residues was seen by some experts as exaggerated, although it was certainly above that figure in 1978.[11]

Neither as dramatic nor devastating in the short term as accidental oil spillages, the impact of operational oil discharges often has been given inadequate treatment. It is important, however, that any future consideration of this issue start from accurate estimates as to quantity of discharge.

8. *NAS Report*, p. 9. LOT is not used (or is used very inefficiently) for vessels on short voyages (less than fifty hours) since much of the oil will not float to the top of the water in such a short period. There is a small number of tankers which for a variety of reasons have publicly refused to use LOT, and, as is pointed out further on, there is a larger number which claim they are using it but do not.

9. In 1974 Mostert in *Supership* (p. 55) made an evaluation similar to our own. Continuing evidence that ship crews will avoid the LOT requirements when they believe they can get away with it diminish claims of a high level of effectiveness. Recently, despite the widespread publicity given to the devastating effects of the *Amoco Cadiz* disaster, the French government caught four tankers (of Russian, Greek, Norwegian, and Liberian nationalities) discharging oily ballast in the vicinity of the spill. With such evidence and the weaknesses in the enforcement system, it is impossible to believe that widespread noncompliance does not occur in the open oceans—often under cover of darkness. "Tanker Captains Ignore New French Controls," *Manchester Guardian Weekly* (April 9, 1978), p. 6. The entry-into-force of the 1969 Amendments to the 1954 Convention in January 1978 is likely to have only a marginal positive effect. See Chapters 6 and 8. Data collected by the American Petroleum Institute in 1978 corroborates the preceding judgments. See footnote 100 in Chapter 4.

10. *NAS Report*, p. 9.

11. This was the judgment of government and industry officials to whom the authors spoke in 1974–1975, but there was and is no precise consensus on what percentage of tankers discharge their residues before arriving at repair yards.

On the basis of our inquiries, the amount of ocean oil discharges resulting from routine tanker operations in 1973 should have been placed closer to 3,000,000 metric tons than the previously cited estimate of about 1,300,000 metric tons. The difference is substantial.

Accidental Discharges from Tankers

Accidental discharges from tankers are of two types. One is the "terminal spill," which occurs during the loading or unloading of a tanker. These spills constitute only a very small percentage of the total pollution picture, although they can be quite serious as the oil is undiluted and is introduced directly into a harbor or port area, where it quickly fouls shore and coastal amenities. Much more important is the second type, tanker accidents at sea. But even these, using the figures of the *NAS Report*, account for only 9.4 percent of all ship-generated oil discharges and 3.3 percent of all discharges. Of course, these figures do not accurately reflect the impact of these spills. These accidents often have dramatic and devastating consequences, and they are certainly spectacular "media events." The image of a *Torrey Canyon* splitting apart under pounding seas off the southern coast of England or of an *Argo Merchant* breaking up off the northeastern coast of the United States is not soon forgotten. To the expert, they are but a part of the problem. To the public, they *are* oil pollution and they should not happen.

Over three-quarters of the oil lost by accidents is discharged as a result of structural failures, groundings, and collisions—and other less significant causes such as explosions, breakdowns, fires, and rammings. A number of studies have been undertaken with respect to the character of ship accidents, and their findings are most relevant to the control of oil pollution. First, structural failures have caused only 20 percent of all polluting incidents but have accounted for 30 percent of the total accidental oil outflow. Such structural failures are closely linked to the age of the vessels and are, therefore, likely to increase as the present fleet of supertankers approach middle and old age. Second, collisions and groundings together have constituted about 56 percent of tanker accidents and have been responsible for 47 percent of the accidental outflow. These accidents (for example, those of the *Torrey Canyon* and *Argo Merchant*) are caused largely by navigational errors.[12] While the total outflow of oil from these latter incidents is

12. These figures for the years 1969–1974 were provided to the authors by Virgil Keith. Data for the years 1969–1970 are analyzed in J. D. Porricelli and V. F. Keith, "Tankers and the U. S. Energy Situation: An Economic and Environmental Analysis," *Marine Technology*, October 1974, pp. 353–355.

not as great as that from accidents caused by structural failures, they generally occur much closer to shore (85 percent of them within about five miles)[13] and therefore cause much more damage to coastal amenities than do the structural failures, which most often occur considerable distances from shore. Interestingly, 30 percent of *all* collisions occur in the very congested English Channel and another 40 percent in the North Sea, Baltic Sea, and other waters surrounding Western Europe.[14]

Smaller and older vessels and vessels registered in "flag-of-convenience" states have by far the worst accident record. The greater frequency of accidental losses and the greater amount of oil spilled by smaller vessels are due significantly to the fact that these vessels are generally involved in coastal trading where traffic congestion is greatest.[15] The higher rate of accidents for flag-of-convenience vessels is a result of numerous factors including their greater average age and generally lower standards. The records of some of these convenience states have improved in recent years, but comparable ships registered in the developed maritime states do have significantly better records.[16] Attempts to control accidental pollution must, therefore, cover a wide range of problems: among others, structural and equipment standards, the management of ship traffic, and the technical competence of crews. Special attention must be paid to flag-of-convenience and, more specifically, to substandard vessels. Assessments of many of these strategies are not included in this study, as they have been tied almost completely to the promotion of marine safety for all ships and not to pollution prevention.

13. Porricelli, Keith, and Storch, *Tankers*, p. 17; and C. Grimes, "A Survey of Marine Accidents with Particular Reference to Tankers," *Journal of Navigation* 25 (March 1972), p. 501.

14. Robert P. Thompson, "Establishing Global Traffic Flows," *Journal of Navigation* 25 (March 1972), p. 488.

15. Data on accident and loss records by size and age are provided in Porricelli and Keith, "Tankers and Energy," pp. 353-355; *Maritime Transport, 1974* (Paris: Organization for Economic Cooperation and Development, 1975), pp. 90-91 and 97-99; and MSC/MEPC/1, pp. 5 and 9. The last is a 1977 British submission to IMCO, and the importance of size is less salient in it.

16. Data on accident and loss records by flag are provided in an OECD study (*Maritime Transport, 1974*, pp. 88-103) which covers the years 1964-1973 and a British study submitted to IMCO in 1977 (MSC/MEPC/9, pp. 8-9) which covers the years 1968-1975. The OECD study includes seven flag-of-convenience states—Liberia, Panama, Cyprus, Somalia, Singapore, Lebanon, and Honduras—as well as OECD, East Europe, and "Rest of the World" categories. The records of the flag-of-convenience states did show some improvement over the ten years. Also, the records of Liberia, Singapore, and Honduras are considerably better than those of the other four. The British study analyzes the records of the nine states with the largest tonnage (including Liberia and Panama) and groups all other countries into an "Other Flags" category. The state with the worst record by far is Greece, but Panama and "Other Flags" also have poor records. Liberia's is only a little worse than the international average but is still considerably below those of Japan, France, the U.K., and the U.S.

Operational Discharges from Nontankers

While tankers have been the main source of ship-generated oil pollution, nontankers also have made their contribution. According to the *NAS Report*, 3.5 percent of all ship-generated oil discharges comes from the discharge of oily bilge water and the deballasting of fuel tanks by nontankers. In a nontanker or dry cargo vessel, lubricating, hydraulic, and fuel oils leak from the machinery and pipes and flow into the bilge in the bottom of the ship. Water also finds its way into the bilge through leaks in the ship's hull and underwater valves, and this mixes with the oils. This oily water must, of course, be pumped periodically from the bilge in order to prevent flooding and eliminate fire hazards. The oil in the bilge water can be extracted and stored for disposal in port, but this requires the vessel to have an oily-water separator and holding tank on board. Even vessels which have such separators and holding tanks often do not use them where there are no disposal facilities in ports. Disposal of oil from bilges is not really a major source of oil pollution, but it can have deleterious effects on the coastal environment as it often takes place in or close to ports.

Another source of operational pollution from some nontankers is the deballasting of fuel tanks. After a nontanker has used the fuel in particular tanks, it sometimes fills the empty tanks with water ballast in order to maintain the ship's stability. When this water is discharged or deballasted near the end of a vessel's voyage, some of the oil that remained in the tanks is discharged with the water.[17] Dry cargo vessels are now built so that they will not have to use their fuel tanks for ballast, but there are still some older vessels which do employ them for this purpose.

Accidental Discharges from Nontankers

Nearly 5 percent of all ship-generated oil pollution comes from nontanker accidents as a result of the breaching of the fuel tanks in the bottom of a ship. The volume of all such spills is only one-half of the volume discharged as a result of tanker accidents, and they are never as large and dramatic as those from tanker accidents but, again, they are a major source of accidental discharges in coastal waters. This is due significantly to the much larger number of nontankers plying the world's oceans, many of which are involved solely in coastal trade.[18]

17. Nontankers normally must discharge the oily water ballast very near their port of destination, since they require its stabilizing weight until they arrive in the relatively calm waters of a harbor area. Unlike tankers they do not have the option of discharging their oily ballast at sea and taking on new ballast into cleaned tanks. At the same time, nontankers may not have to ballast their fuel tanks if they are fully loaded and the seas are not exceptionally rough.

18. As with tankers, nontankers which are smaller and older and are registered in flag-of-

The oil pollution problem is, therefore, a large and complex one. And the contribution made to it by ocean-going vessels is but a portion of the total picture. However, it is vessel discharges that have, almost exclusively, been the topic of international negotiations on the issue. Within these negotiations, it is operational pollution that has been the major source of debate. It accounts for by far the greatest quantity of oil entering the oceans from marine sources and, unlike accidents, it can be dealt with as an issue distinct from more general concerns on navigation, safety, ship design, and crew training and qualifications.

LOCATIONS OF OIL IN THE MARINE ENVIRONMENT

As with our knowledge of the total volume of oil discharges, our understanding of the specific concentrations of oil in different areas of the oceans is quite imperfect. Yet it is possible to establish some understanding of this matter by turning to a variety of sources. The first is anecdotal.

Observations of oil pollution are, to say the least, prolific and disturbing. Oil has been reported on the open seas, in confined water, and on shores the world over. The semi-enclosed seas around Europe, especially the Mediterranean and Red Sea, are, according to a FAO report, seriously polluted.[19] Indeed, of the high level of oil pollution in the Mediterranean, Sir Francis Chichester remarked that "time after time we sailed through patches or slicks of oil film on the surface" so that seas washing over the yacht "left clots of black oil on the deck and stained the sails."[20]

There are also reports of serious concentrations of oil in areas of the open ocean and along many of the world's beaches. During his *Ra* expeditions Thor Heyerdahl found that there was "at least a stretch of 1,400 miles of the open Atlantic . . . polluted by floating clumps of solidified asphalt-like oil."[21] In the Sargasso Sea (east of the southeastern U.S.), scientists collecting samples of marine life found that their nets became "fouled with oil and tar-thick sticky globs up to three inches in diameter" and finally were forced to abandon their project "because they were picking up three times as much oil as seaweed."[22] With respect to coastal areas the *NAS Report* noted that "anectodal evidence" indicated that most of the coast of Africa is "heavily polluted with tar,"[23] and an official of the International

---

convenience states are more likely to be lost and hence cause oil pollution as a result of accidents. *Maritime Transport, 1974*, pp. 89–103.

19. *The State of Marine Pollution in the Mediterranean and Legislative Controls* (Rome: Food and Agriculture Organization of the United Nations, 1972), pp. 25 and 55.

20. Mostert, *Supership*, p. 47.

21. U.N. doc. E/5003, May 7, 1971, *The Sea. Prevention and Control of Marine Pollution, Report of the Secretary General, Annex III: Atlantic Ocean Pollution Observed by the Ra Expedition.* See also Mostert, *Supership*, pp. 47 and 216.

22. *Wall Street Journal*, November 26, 1969.

23. *NAS Report*, p. 96.

Tanker Owners Pollution Federation Limited has admitted that "oil in the form of 'tarry lumps' can be found on most of the sandy beaches of the world."[24] Such accounts as these abound, but their importance cannot be evaluated uncritically. They are, after all, chance observations not subject to the rigors of scientific investigation, and they refer only to the visually apparent forms of oil pollution.

Unfortunately, while there have been a number of scientific studies on the concentration of hydrocarbons on the surface of the oceans, in the water column, and in sediments on the sea bottom, data are incomplete and not incontrovertible. For example, there have been no studies on pelagic tar lumps in about two-thirds of ocean space,[25] and there are serious differences over proper research techniques.[26] Difficulties persist in distinguishing petroleum hydrocarbons from biogenic ones (those which occur naturally in a specific environment), particularly at lower levels of hydrocarbon concentration and in some types of sediments.[27]

Some data are quite informative, however. The *NAS Report* has noted that there are great differences in the concentration of hydrocarbons in sediments on the ocean floor. In certain highly polluted coastal areas, concentrations range from between 100 and 1,200 parts per million (ppm), whereas in the deep marginal seas they are much less than 100 ppm. In the deep sea, they are found in densities of between only 1 and 4 ppm.[28] Studies of tar density on the ocean's surface have recorded a very high concentration of 20 milligrams per square meter (mg/m^2) in the Mediterranean, 0.98 mg/m^2 in the Norwegian coastal current, and 0.0003 mg/m^2 in the South Pacific.[29] These figures confirm the pattern made explicit in a pioneering study done on petroleum-derived hydrocarbon levels in the water columns of the Atlantic, Mediterranean, Gulf of Mexico, and Persian Gulf. Researchers there found higher concentrations along the tanker

24. J. Wardley Smith, "Oil Spills from Tankers" (London: Institute of Petroleum, 1975), p. 4. (Mimeo.)

25. *NAS Report*, p. 53. Very comprehensive tables are given in this publication detailing the available data (up to 1973) on tar densities in the world's oceans and hydrocarbons in the water column and sediments (pp. 52–57).

26. Ibid., p. 5.

27. A good understanding of this problem can be gained by consulting the following: *NAS Report*, pp. 20–21, 55, and 105; P. H. Monaghan, D. E. Brandon, R. A. Brown, T. D. Searl, and J. Eliott, *Measurement and Interpretation of Nonvolatile Hydrocarbons in the Ocean, Part I: Measurements in Atlantic, Mediterranean, Gulf of Mexico, and Persian Gulf* (Washington: U.S. Department of Commerce Maritime Administration, 1974), pp. 18–21.

28. *NAS Report*, p. 55.

29. *Ibid.*, p. 53; G. Smith, "Pelagic Tar in the Norwegian Coastal Current," *Marine Pollution Bulletin* 7 (April 1976), pp. 70–72; W. E. McGowan, W. A. Saner, and G. L. Hufford, "Tar Ball Sampling in the Western North Atlantic," in *Marine Pollution Monitoring (Petroleum)* (Washington, D.C.: U.S. Department of Commerce, December 1974), pp. 83–84; C. S. Wong, D. R. Green, and W. J. Cretney, "Distribution and Source of Tar on the Pacific Ocean," *Maritime Pollution Bulletin* 7 (June 1976), pp. 102–105.

routes in the Caribbean and Mediterranean and from the Persian Gulf to Europe. They concluded that petroleum hydrocarbons "are present in greater amounts in those geographic areas where tanker discharges are likely" and that "the estimated petroleum-derived hydrocarbon content appears to reflect the amount of petroleum transported through the area."[30]

It is possible, therefore, to determine the locations of serious oil pollution by looking at where the major international trading routes are. Tanker routes for 1976 are portrayed in Map 1. By far the greatest volumes of oil are transported from the Persian/Arabian Gulf area either west around Africa to Western Europe or east to Japan and from the Arab states bordering the Mediterranean to Western Europe. A fourth important route is from Venezuela to North America, and another of rapidly growing importance in the 1970s is from the Middle East to the United States. Clearly then, the heaviest concentrations of oil are to be found in the waters surrounding Europe, around Africa, along the southern and eastern coastal areas of Asia, and between Venezuela and the eastern U.S. coast. The predominating currents in these areas disseminate the pollution even farther,[31] but (as indicated by Map 2) there is a strong relationship between major tanker routes and the concentration of oil in the water column. Given the fact that operational discharges from tankers constitute at least 62 percent (and possibly as high as 80 percent) of all vessel-source discharges, this finding is to be expected.

The other sources of ship-generated oil pollution reinforce this pattern. The trading routes of dry cargo vessels (whose intentional discharges constitute about 5 to 10 percent of all vessel-source oil pollution) are more diversified than those of the tankers, but they involve the developed countries much more than the developing. Approximately two-thirds of dry cargo tonnage is loaded and unloaded at ports in Europe and North America,[32] and, as one would expect, a much greater volume of oil from these ships enters the coastal waters around the sea lanes between these countries. With respect to discharges resulting from accidents for both tankers and nontankers (about 10 to 15 percent of total discharges), they

30. Monaghan et al., *Nonvolatile Hydrocarbons*, p. 5. The *NAS Report* concluded that the "occurrence of pelagic tar correlates with intensity of tanker traffic in different regions of the world" (pp. 54-55); and the Group of Experts on Scientific Aspects of Marine Pollution (a body composed of scientists appointed by six U.N. specialized agencies) found that "though found widely over the oceans it is most frequent around the main routes by which oil is carried from producing areas to the refineries and along trade routes and near the population centres they connect." GESAMP doc. T5/2.01, February 5, 1974, p. 40. (distributed by IMCO).

31. As, for example, in the cases of the Aguenas Current (Mostert, *Supership*, pp. 210-211 and 213-216) or the Kuroshio Current (Wong et al., "Tar," p. 102).

32. *The Importance of Shipping* (London: International Chamber of Shipping, 1975), pp. 5-8.

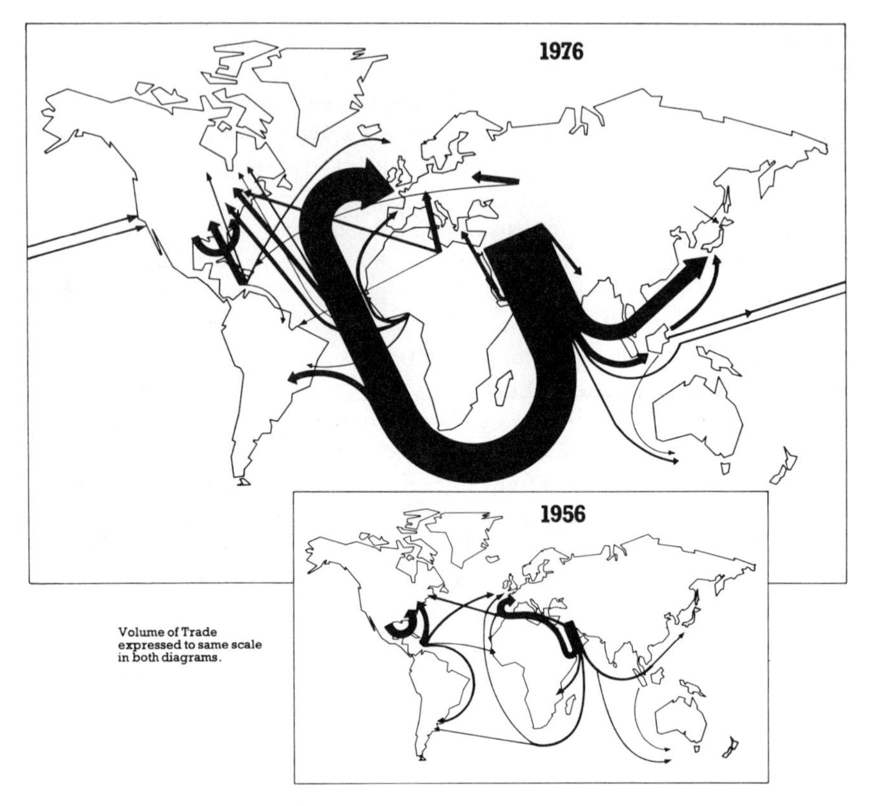

MAP 1
Main Oil Movements by Sea, 1956 and 1976
(Volume of trade expressed to same scale in both diagrams.)

SOURCE: *BP Statistical Review of the World Oil Industry, 1976*

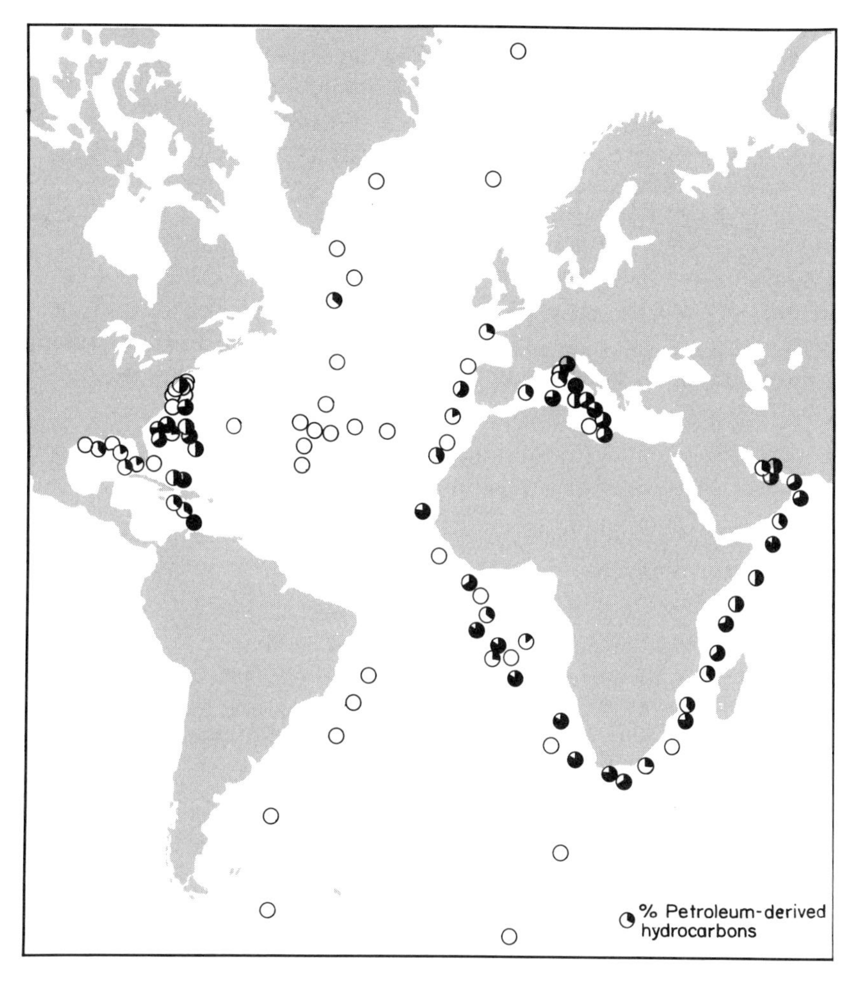

MAP 2
Estimated Proportion of Petroleum-Derived Hydrocarbons
in Total Hydrocarbons of Surface Water
(assumes natural hydrocarbons equals 20% extractable organics)

SOURCE: *Measurement and Interpretation of Nonvolatile Hydrocarbons in the Ocean, Part I: Measurements in Atlantic, Mediterranean, Gulf of Mexico and Persian Gulf* (Prepared by Exxon Research and Engineering Co. and Exxon Production Research Co. for U.S. Dept. of Commerce Maritime Administration, July 1974), p. 48.

too are concentrated along the major routes. But, while structural failures occur randomly along these paths, 70 percent of the open sea collisions have taken place in the waters surrounding Europe.[33]

Despite the lack of systematic studies on the concentration of oil in many areas of the world's oceans, the existing scientific data in combination with anecdotal evidence and a knowledge of vessel traffic routes focus attention on some obvious areas. The Mediterranean is certainly a "sick sea," as are the Persian Gulf and the other semi-enclosed seas around Europe and the Middle East. Other areas follow suit: the Atlantic coasts of Europe and North America, the coast of Africa, and the areas along the tanker route between the Persian/Arabian Gulf and Japan. There are, in addition, many port areas throughout the world which suffer from oil pollution from both vessel and land sources.

THE EFFECTS OF OIL IN THE MARINE ENVIRONMENT

The impact of oil in the marine environment is a subject of complex and controversial dimensions. Clearly, oil discharges have caused "pollution," as defined in the Stockholm Declaration:

> . . . the introduction by man, directly or indirectly, of substances or energy into the marine environment (including estuaries) resulting in such deleterious effects as harm to living resources, hazards to human health, hindrance to marine activities including fishing, impairment of quality for use as seawater and reduction of amenities.[34]

However, there is considerable disagreement over, and ignorance of, the precise nature and severity of the effects of oil. The scientific literature on the subject is extensive, but past studies have had many shortcomings;[35] and there are many areas which deserve more extensive and long-term study. The following discussion will briefly highlight the major effects of oil which have been identified by the existing literature.[36]

33. R. Thompson, "Establishing Global Traffic Flows," *Journal of Navigation* 25 (March 1972), p. 488.

34. U.N. doc. A/CONF.48/8, pp. 78–79.

35. These shortcomings are reviewed in D. F. Boesch, C. H. Hershner, and J. H. Milgram, *Oil Spills and the Marine Environment* (Cambridge, Mass.: Ballinger, 1974), pp. 4–5 and 38–40; S. F. Moore and R. L. Dwyer, "Effects of Oil on Marine Organisms: A Critical Assessment of Published Data," *Water Research* 8 (October 1974), p. 823; A. Nelson-Smith, *Oil Pollution and Marine Ecology* (London: Paul Elek, 1972), p. 99; and *NAS Report*, pp. 37, 76–82, and 88. For a recent debate on this, see P. H. Abelson, "Oil Spills," *Science* 195 (January 14, 1977); and letter in *Science* 195 (March 11, 1977), p. 932.

36. Good reviews of the effects of oil are found in *NAS Report*, pp. 73–103; Nelson-Smith, *Oil Pollution*, pp. 99–172; Moore and Dwyer, "Effects of Oil," pp. 822–826; Boesch et al., *Oil Spills*, pp. 11–46; IMCO doc. PCMP/2, January 30, 1973, *United Kingdom Study: The Environmental and Financial Consequences of Oil Pollution from Ships*, pp. 30–70; *Impact of Oil on the*

Although they account for only a small production of the oil entering the oceans, tanker accidents can cause massive localized pollution. Here, British beaches in East Anglia are covered with oil following the collision on May 6, 1978 of the Greek tanker *Eleni V* with a French vessel. Drifting slicks of operational pollution can also cause similar beach pollution stretching many miles. (Photo: AP.)

The largest of all tanker accidents has, of course, been the *Amoco Cadiz* off Brittany in March 1978 (see cover). Here are pictured a few of the 6,000 soldiers engaged in the cleanup operations of what will undoubtedly be the costliest oil pollution incident to date. (Photo: Sipa Press from Black Star.)

Oil pollution need not occur in large quantities to have serious environmental consequences. Many hundreds of thousands of seabirds die each year as a result of chronic oil pollution. Pictured above are victims of an oil slick that hit Massachusett's Martha's Vineyard in 1970. Such ecological damage is hard to compensate (Photo: UPI).

Prior to discussing the specific consequences of oil discharges, a number of variables which determine exactly how oil will affect the environment should be mentioned. First, the characteristics of the spilt oil strongly influence the damage that results. Oils are complex collections of compounds (the major classes of hydrocarbons being alkanes, alkenes, and aromatics) and the possible combinations are innumerable.[37] Some compounds are toxic and kill organisms (at least for a short while after entering seawater), while others alter the biological characteristics and behavior of organisms in a variety of ways. Still others are suspected of causing cancers. There are, however, four general types of oil most likely to be discharged from vessels: crude oil, Bunker C fuel oil, diesel fuel oil, and light petroleum products

---

Marine Environment (GESAMP Reports and Studies #6) (Rome: Food and Agriculture Organization, 1977), pp. 48–128 (hereinafter cited as *GESAMP Report*). GESAMP is the acronym for the Group of Experts on the Scientific Aspects of Marine Pollution. It includes representatives of six U.N. specialized agencies. For a description of it, see Chapter 3.

37. For discussions of the composition of oil and its varying impact, see Boesch et al., *Oil Spills*, pp. 7–8; *NAS Report*, pp. 42–43; Moore and Dwyer, "Effects of Oil," p. 820; IMCO doc. MP/CONF.8/18, August 3, 1973, *Report on the Categorization of Oils Having Regard to Their Polluting Chracteristics*; IMCO doc. MP/XIII/2(2)/5, June 1972, *United States Report on White Oils*.

Estuarial and coastal areas have a rich variety of sea life and they are the other great victim of both operational and accidental pollution. This conger eel was caught when its shore habitat was swept by oil. (Credit: B. Gysenbergh, Camera Press.)

(gasoline, kerosene). Of these, diesel fuel oil has been described as the worst in terms of toxicity,[38] although light petroleum products also are considered very toxic. Unlike the crude oils and Bunker C, these two refined oils certainly do not persist visually in the marine environment, and some doubt has even been expressed as to whether the toxic components ever reach dangerous levels for marine organisms.[39] Some have even contended that these oils are relatively harmless, but actual experience—with spills such as that at West Falmouth—challenges such a conclusion. At West Falmouth, a spill of diesel fuel oil devastated the entire contaminated area.[40] In contrast, a large spill of the much less toxic Bunker C from the tanker *Arrow* off Nova Scotia had only limited effects on the organisms in the area despite its visible persistence. At the same time, Bunker C and viscous crude oils are noted for their smothering effects on organisms in intertidal areas, and they can poison organisms at sea that come in contact with the oil soon after the spill. As for the carcinogenic elements of oil, there are disputes in the scientific community over which oils have the greatest concentration of carcinogenic fractions.[41] But while these differences do exist, it is certainly agreed that "ecological effects depend greatly on the nature of the oil spilled."[42]

Other factors also greatly influence the effects of discharges. Some of these are the biota of the area, local conditions such as temperature, currents, winds, and the extent of existing pollution, the volume and concentration of the oil discharged, and whether the location of the spill is in the open ocean or more enclosed bodies of water.[43] These factors strongly affect the "weathering" processes which gradually remove the oil spill from the environment.[44] For instance, high temperatures mean that evaporation of

38. IMCO doc. MP/CONF.8/23, October 23, 1973, p. 3; and *NAS Report*, p. 99.

39. Nelson-Smith, *Oil Pollution*, pp. 107 and 116; and IMCO doc. MP/CONF.8/18, August 3, 1973, *Report on the Categorization of Oils*, pp. 2-3.

40. M. Blumer, H. L. Sanders, J. F. Grassle, and G. R. Hampson, "A Small Oil Spill," *Environment* 13 (March 1971), pp. 2-13.

41. IMCO doc. OP/X/6/2, September 8, 1971; *NAS Report*, pp. 96-98; Blumer et al., "Small Oil Spill," pp. 2-12; J. B. Sullivan, "Marine Pollution by Carcenogenic Hydrocarbons," *Marine Pollution Monitoring (Petroleum)*, pp. 261-263; *GESAMP Report*, p. 91-92.

42. Boesch et al., *Oil Spills*, pp. 37-38. This volume provides a summary of the debate between Max Blumer, claiming oil effects are similar for all varieties of the liquid (since the same hydrocarbons are found in refined oil products as are found in crude oil) and those like Dale Straughan who contend that different oils have different effects due to their varying compositions.

43. The following sources contain good discussions of these factors: Boesch et al., *Oil Spills*, pp. 36-38; D. Straughan, "Factors Causing Environmental Changes after an Oil Spill," *Hearings Before the Committee on Interior and Insular Affairs, U.S. Senate, March 23, 24 and April 11 and 18, 1972* (Washington, D.C.: Government Printing Office, 1972), part 3, app., pp. 1775-1779; and *NAS Report*, pp. 83-84.

44. For good accounts of weathering processes, see *NAS Report*, pp. 42-71, including an extensive bibliography; and Nelson-Smith, *Oil Pollution*, pp. 71-84.

certain fractions of the oil (a process which removed a quarter of the *Torrey Canyon*'s cargo) will be fast.[45] In contrast, spillages in Arctic areas could last for decades. Biological or microbial degradation of the oil is also affected by temperature and other marine conditions.[46] One might think that as a result of such weathering processes the seas would eventually be able to cleanse themselves of oil. Yet there is a great deal of concern (and considerable differences among scientists) both as to the ability of the oceans to assimilate the increasing volume of discharges and the persistence of elements with particularly deleterious effects.[47] As a recent GESAMP report has warned, the fate of oil is "not adequately understood."[48] Moreover, many other pollutants being discharged have effects which are difficult to distinguish from those of oil.[49]

Of all the adverse effects of oil pollution, the most pathetic is the destruction of wildlife and their habitats: "in the vicinity of a spill or an area of chronic pollution . . . the effects on *all forms of marine life* may be disastrous."[50] Rich estuarine or other breeding areas are extremely vulnerable, as has been depressingly demonstrated by the *Amoco Cadiz* disaster. Unofficial reports of that spill indicate that it may have eradicated one of France's most highly prized sea bird sanctuaries in the Sept Isles archipelago at a time when its bird population (particularly the puffins) was just beginning to show signs of recovery from the *Torrey Canyon* spill a decade earlier.[51] Such damage is not restricted to calamitous spills. It has been estimated that chronic oil pollution in the North Sea and North Atlantic alone kills a staggering total of between 150,000 and 450,000 birds every year. Such losses may not easily be replenished. As the authoritative GESAMP report concluded,

45. U.N. E/5003, May 7, 1971, *Prevention and Control of Marine Pollution*, p. 28. Also see D. Pimlott et al., *Oil under the Ice* (Ottawa: Canadian Arctic Resources Committee, 1976), p. 94. At the same time, weathering might actually *increase* the menace of oil. Sunlight, for example, may intensify the toxic effects of oil (*New York Times*, June 8, 1977, p. 40). Moreover, weathered oil may, when oxidized, produce products of potential carcinogenity. See *GESAMP Report*, p. 91.

46. *NAS Report*, pp. 58–59.

47. Mostert, *Supership*, p. 186; M. Blumer, "Oil Contamination and the Living Resources of the Sea," *Hearings Before the Committee on Interior and Insular Affairs, U.S. Senate, March 23, 24 and April 11 and 12, 1972* (Washington, D.C., Government Printing Office, 1972), pp. 39–40; D. K. Button, "Petroleum Biological Effects in the Marine Environment," in D. W. Hood, ed., *Impingement of Man on the Oceans* (New York: John Wiley, 1971), p. 422; and Nelson-Smith, *Oil Pollution*, p. 204.

48. *GESAMP Report*, p. 22. See pp. 22–41 for a discussion of this point.

49. Nelson-Smith, *Oil Pollution*, pp. 167–168. The *NAS Report* has noted that it is "impossible" to determine which effects are due to oil pollution and which to other pollutants or to fishing (p. 89).

50. *GESAMP Report*, p. 76. Fur-bearing sea mammals can be severely affected as can all varieties of waterfowl, which lose the water repellancy and insulation of their feathers.

51. "The Amoco Cadiz Disaster," *Manchester Guardian Weekly*, April 2, 1978, p. 11. For a firsthand account of the effects of the disaster, see *Oceans* 11 (July–August 1978), pp. 56–61.

> [Some] sea bird populations have suffered from oil pollution to the extent that certain species or subspecies are *threatened with extinction*. For example, there is little doubt that the more southern colonies of puffins, razorbills and guillemots *are declining rapidly* on both sides of the Atlantic. Those that nest on the islands to the north of Scotland have already suffered seriously and are considered to be in a *particularly precarious position*. It is almost certain that the primary cause of these declines is oil pollution. The only satisfactory approach to the conservation of the sea birds is to reduce the number of spills and the amount of oil on the surface of the oceans.[52] [Emphasis added.]

Many specific instances of high waterfowl casualties have been reported as well.[53]

Many individuals also are concerned with the effects of oil discharges on fisheries, although research has not yet proven significant damaging effects on fish populations.[54] Fish can be killed by newly spilt oil, but the oil's tendency to float and the mobility and sensitivity of the fish usually prevent extensive losses in deeper water. Fish are, however, far more sensitive to the light refined oils (as the devastating West Falmouth spill illustrates)[55] than crude and heavy fuel oils. Even after spills of the former, *past* experience has shown that the stocks eventually recovered.[56] But the effects are by no means benign. The recovery process is slow—especially for intertidal species which are drenched with oil and can be wiped out for a number of years. Even then, full recovery cannot be assumed. One spill which caused serious damage to a large shellfish industry was that of the *Urquiola* in northern Spain. In addition, it is quite possible that there may be important, if subtle, effects on fish populations, especially in their juvenile forms.[57] Some of these effects might be identifiable only over the long term but are potentially serious threats. Finally, chronic discharges in areas very close to shore have caused a tainting of some species, making them difficult to market (for example, along the Texas coast and in the Mediterranean).

52. *GESAMP Report*, pp. 75-76. The *Torrey Canyon* spill killed between 40,000 and 100,000 waterfowl (Nelson-Smith, *Oil Pollution*, p. 161).

53. Along the tip of South Africa (where 650 tankers pass each month), chronic oil pollution threatens the jackass penguin with extinction. "Oil and Penguins Don't Mix," *National Geographic* 143 (March 1973), pp. 384-397. The damage that small spills can do is almost incredible. A small (750,000 liters) spill of light oil in Puget Sound, Washington, killed 30,000 brant geese, while it was estimated 86,000 birds were killed by a light diesel oil slick off Alaska. Both spills were in 1971 (*GESAMP Report*, p. 57).

54. *NAS Report*, pp. 89-90.

55. This spill involved 160,000 gallons of refined oil, and an oceanographer's survey within three days found 95 percent of its catch dead. Mostert, *Supership*, p. 50.

56. Boesch et al., *Oil Spills*, p. 18.

57. *GESAMP Report*, p. 62. In addition, some studies have indicated a variety of sublethal effects of oil on marine life, and their implications require extensive and long-term investigation. *NAS Report*, pp. 76-82 and 85-95; Nelson-Smith, *Oil Pollution*; Moore and Dwyer, "Effects of Oil," pp. 823-824; Blumer, "Oil Contamination," p. 44.

Examples of this do not appear to be commonplace[58] but cannot be ignored. Moreover, *major* oil discharges certainly do affect the economic interests of fishermen. This is, of course, the case for fishermen in the immediate vicinity of a large oil spill, but the implications are more widespread. Fishermen may hesitate to go to sea for fear of fouling their nets, and they may experience losses in their sales as a result of the general fear of contamination. Indeed, after the *Torrey Canyon* disaster, sales of fish were reported to have fallen markedly in both the U.K. and France.[59]

Another interest affected by oil pollution is tourism. Beaches affected by oil pollution may not wash clear for many years, and it is hardly surprising that vacationers would object to the presence of oil clots on the beaches which stain their clothing and decrease their esthetic enjoyment of the oceans. It is, however, very difficult to document economic losses to the tourist industry, apart from some direct and temporary ones following major spills. For example, there was a 20 percent drop in tourism in Santa Barbara following leakage from one of the offshore wells, but the drop was a temporary one. Likewise, there was a decrease in tourism along the British south coast after the *Torrey Canyon* spill, but the industry was back to normal after a year.[60] With regard to areas afflicted by chronic pollution, such as the Mediterranean, extensive research on the discernable effects have yet to be conducted.

An additional economic cost of oil pollution is that borne by government authorities in removing it. The intentional discharges of tankers and cargo vessels rarely cause slicks that can be removed, since the oil is usually a part of an effluent and is discharged over long distances. The massive and concentrated discharges resulting from accidents, however, must be removed in order to prevent serious damage to coastal resources. The costs of such operations are often considerable. A recent coastal spill of about 292,000 barrels of Bunker C oil at Mizushima, Japan, cost an estimated $200 million to clean up, but estimates of average costs vary widely, especially as the circumstances of each spill can diverge so much.[61] Moreover, some damage is caused by the chemical dispersants and sinking agents used in

58. *NAS Report*, pp. 89–90; *The State of Marine Pollution in the Mediterranean and Legislative Controls* (Rome: Food and Agriculture Organization, 1972), pp. 25–26; and Mostert, *Supership*, pp. 46–47. See also *GESAMP Report*, pp. 118–119.

59. IMCO doc. PCMP/2, February 1973, p. 68.

60. Nelson-Smith, *Oil Pollution*, p. 170.

61. C. W. Nicol, "The Mizushima Oil Spill—A Tragedy for Japan and a Lesson for Canada," Environmental Protection Service, West Vancouver, Canada, May 1976, pp. 20–21. The Maritine Safety Agency was awarded Can. $165 million for its expenses, and an additional Can. $43 million was given to fishermen and other people for their cleanup work. Also, Can. $57 million was given to fishermen, fish markets, hotels, and so on as compensation for damage.

the cleanup operations.[62] Some international agreements have been concluded to compensate parties for the cleanup costs and some of the damages resulting from these accidental spills, but nothing has been done to compensate the victims of chronic oil pollution.

The impact of oil pollution on marine organisms as well as its economic and esthetic costs for man are important and disputed issues, but the greatest controversy regarding the effects of oil concerns its impact on human health.[63] Oil does contain some carcinogens, and it has been argued that these can be ingested by marine organisms, thus becoming concentrated in the food chain. Through man's consumption of fish, these carcinogenic agents then become a hazard to man himself.

There are a number of problems with this charge. Although there is no proof that the contrary is true, it has yet to be established that food web concentration occurs with these carcinogens the way it does with chemicals like DDT. As the *NAS Report* points out, "There is almost no reliable information concerning the possible concentrating or diluting mechanisms for these compounds in the food chain of marine or terrestrial organisms."[64] Moreover, many would consider oil pollution a relatively minor source of carcinogens in the oceans.[65] In addition, other foods also contain carcinogens, and the *NAS Report* notes that it "makes no sense to stop eating fish for fear of their possible carcinogenic content and replace the fish by another food source that poses an equal or even greater danger."[66] While at the moment there is no firm basis for judging that the discharge of oil into the marine environment poses a threat to man's health, scientific knowledge of this area is scanty and scientists themselves believe that all should be "seriously concerned."[67] After all, for many substances—such as DDT, mercury, PCB, Freon, and others—their cumulative effects were quite remote in time and often in location from their immediate impact.

There is clearly much still to be learned about the effects of oil in the

62. Boesch et al., *Oil Spills*, pp. 27–28; *NAS Report*, p. 84; Nelson-Smith, *Oil Pollution*, pp. 173–213; R. A. A. Blackman, "Toxicity of Oil-Sinking Agents," *Marine Pollution Bulletin* 5 (October 1974), pp. 116–118; *GESAMP Report*, pp. 26–27.

63. See the following for an overview of this issue: *NAS Report*, pp. 96–99; M. Blumer, "Scientific Aspects of the Oil Spill Problem," *Environmental Affairs* 1, pp. 57–58; Nelson-Smith, *Oil Pollution*, pp. 95–98; Boesch et al., *Oil Spills*, pp. 31–34; C. E. Zo Bell, "Sources and Biodegradation of Carcinogenic Hydrocarbons," *Proceedings of Joint Conference on Prevention and Control of Oil Spills, June 15–17, 1971* (Washington, D.C.: American Petroleum Institute, 1971), pp. 441-451; *GESAMP Report*, pp. 91–116.

64. *NAS Report*, p. 97. See also *GESAMP Report*, p. 100.

65. IMCO doc. PCMP/2, p. 18; Zo Bell, "Carcinogenic Hydrocarbons," pp. 441 and 451; *NAS Report*, pp. 96–97; *GESAMP Report*, pp. 92–93. Biosynthesis and land drainage are seen as the major contributors.

66. *NAS Report*, pp. 98–99.

67. Ibid., p. 98.

marine environment. However, from what is already known of its effects on sea life and waterfowl and its economic costs to man, the wisest policy for the present is that which was expressed by a delegate to a recent international conference:

> . . . the matter is certainly one on which reasonable men can differ. My delegation is not reluctant to state its view is based on the fundamental presumption that doubts should be resolved in favour of protecting the environment. To paraphrase a comment of the distinguished delegate of the U.K. . . . pollution should be adjudged guilty until proven innocent.[68]

These sentiments were echoed by the scientists convened by the National Academy of Sciences: ". . . it seems wisest to continue our efforts in the international control of inputs and to push forward research to reduce our current level of uncertainty."[69] Without such an approach, the problem can only get worse.[70]

68. Transcript of Committee II Proceedings, 1973 IMCO Conference, October 16, 1973, p. 4.

69. *NAS Report*, p. 107. A brief outline of needed research on chronic oil pollution is contained in P. H. Monaghan and C. B. Koons, "Research Needed to Determine Chronic Effects of Oil on the Marine Environment," *Marine Pollution Bulletin* 6 (October 1975), pp. 157–159. A serious problem in on-going research may be that a great deal of it is sponsored by the oil industry. More funding should be made available by public authorities. Boesch et al., *Oil Spills*, pp. 41–43.

70. It should perhaps be noted at this point that many economists would argue that the best way to resolve the various problems discussed in this chapter would be by means of "cost-benefit analysis." This study does not employ such an economic evaluation. Despite great concern over the costs of alternative solutions, a more than impressionistic application of cost-benefit techniques was rare in the negotiations and always reflected the political preferences of the analyst. In any event, the usefulness of the technique is questionable given its very basic assumptions. On this latter point, see R. C. d'Arge and E. K. Hunt, "Environmental Pollution, Externalities, and Conventional Economic Wisdom: A Critique," *Environmental Affairs*, Vol. 1 (1972), p. 266.

Chapter III

IMCO and Its Members

Most international regulations for the protection of the marine environment center on the single problem of vessel-source oil pollution. These regulations have emanated from the Intergovernmental Maritime Consultative Organization, the only international organization with a long and continuous history of environmental regulation. Today, more global bodies are involved with environmental issues than ever before, although their orientation is largely the collection and exchange of information and technical assistance. No other global agency compares to IMCO in the formulation of international environmental conventions.

THE HISTORY AND FUNCTIONS OF IMCO

Prior to the Second World War, international maritime problems were dealt with by a large number of disparate bodies, most of which were non-governmental. However, during the war an authority was needed to coordinate all aspects of Allied shipping, and this took the form, first, of the combined Shipping Adjustment Board and then, in 1944, of the United Maritime Authority. The United Maritime Authority met for the last time in February 1948 and was succeeded by the United Maritime Consultative Council (UMCC), a temporary body set up to deal with the problems that arose in reestablishing normal peacetime activities.

At this time, the desirability of a permanent intergovernmental organization in the shipping field was apparent, and the member states of the UMCC recommended convening an international conference to set up such an organization. Governments were invited to attend the United Nations Maritime Conference in Geneva in February 1948 to consider the creation of such a body within the U.N. system, and after seventeen days of debate, the Convention of the Intergovernmental Maritime Consultative Organization was ready for signature. The organization itself came into being a decade later, in 1958, after the requisite twenty-one nations had ratified the convention. The convention has subsequently been amended in

1974, 1975, and 1977, although only the first set of amendments has entered into force (on April 1, 1978).[1]

The basic functions conferred on IMCO in 1948 were not very different from those delegated to other specialized agencies in their particular fields—namely, to pass recommendations, convene conferences, draw up conventions, and facilitate consultations among member states (article 3). At the same time there was more resistance to the organization's becoming an effective regulatory body than was the case with other agencies. A reflection of this sentiment was the interjection of the term *consultative* in the title of the organization and the explicit requirement in the convention that the organization abstain from dealing with "those matters which appear to the Organization capable of settlement through the normal process of international shipping business" (article 4).

The resistance to IMCO on the part of many shipping states was not surprising. For over fifty years such a body had been opposed on the grounds that "the creation of a central body might introduce political distractions into a sphere which is essentially technical."[2] This is a continuing argument of the maritime states, although the increasing political significance of shipping is indisputable. The problem was aggravated by the fact that not only were "technical matters . . . particularly with regard to maritime safety and navigation" included within the jurisdiction of the new maritime agency but so too were commercial matters such as "the removal of discriminatory action and unnecessary trade restrictions" and "unfair restrictive practices by shipping firms" (article 2). Powerful shipping industries in many maritime states viewed interference by intergovernmental agencies in the competitive relations and pricing system within the industry as anathema. Governments were, therefore, wary of (indeed, opposed to) this facet of IMCO's jurisdiction. But the body that was directly responsible for the IMCO Convention, the United Maritime Consultative Committee, was under the control of the more broadly based United Nations Economic and Social Council (ECOSOC). Consequently, the commercial lobby did not succeed initially in excluding the broader economic authority from the IMCO constitution of 1948.

1. The original convention is contained in the IMCO publication *Basic Documents I* (March 1968). On IMCO's creation, see D. J. Padua, "The Curriculum of IMCO," *International Organization* 14 (autumn 1968), pp. 524–547; and Harvey B. Silverstein, *Superships and Nation-States: The Transnational Politics of the Intergovernmental Maritime Consultative Organization* (Boulder, Colo.: Westview, 1978), pp. 10–23. The 1974 amendments are in Res. A.315 (ES.V); the 1975 amendments, Res. A.358 (IX); and the 1977 amendments, Res. A.400 (X).

2. Sir Osborne Mance and J. E. Wheeler, *International Sea Transport* (London: Oxford University Press, 1945), p. 7, cited in R. McLaren, "An Evaluation of a Secretariat's Political Role within the Policy Process of an International Organization—The Case of IMCO" (Ph.D. dissertation, University of Pittsburgh, 1972), p. 90.

But, with their more direct influence on the implementation of the convention, industry opposition was triumphant in subsequent years. To begin with, it delayed the organization's actual coming into being for ten years, until 1958. The organization would be effective only if most shipping states belonged to it so the convention required ratification by twenty-one states, including seven with registered shipping in excess of one million tons each. Thus, despite the ratifications of the U.S., U.K., France, and the Netherlands in the several years following the 1948 conference, it was not until a further group of shipowning states including Greece, Norway, Panama, Japan, Italy, and the USSR signed in 1957 and 1958 that the organization came into being. (Denmark, Sweden, Finland, Liberia, and Germany followed suit as soon as IMCO was formally in existence. This meant the inclusion of all major maritime states.)

Second, by the time IMCO did actually start to function, almost all the prospective members of the organization had been convinced of the futility of IMCO's entering the field of commercial regulation. The split between shipowning and shipusing interests has been an omnipresent source of irritation in commercial maritime affairs. With the shipuser's great dependence on shipowners, the latter have usually had their way. Thus, those provisions conferring authority over commercial matters have been a dead letter, and to date there has been no serious attempt within IMCO to resurrect the organization's explicit jurisdiction in the field. But the problem did not go away and, with the growth of Third World influence, commercial shipping issues such as discriminatory trade practices became more pressing than ever. As a result, this field was taken over in 1965 by the U.N. Conference on Trade and Development (UNCTAD), a forum dominated by developing states and one infinitely more hostile to traditional maritime interests than IMCO.

Despite the convention's silence on the pollution-control function (a matter which will be altered with the entry-into-force of the 1975 Amendments), the scope of IMCO's jurisdiction over technical matters has been extended to include this as well as marine safety.[3] In fact, IMCO's role in regulating ship-generated marine pollution was accepted even prior to its formal creation in 1959. At the 1954 Oil Pollution Conference, it was agreed that the "bureau powers" for the newly created convention should be transferred from the United Kingdom to IMCO when it came into existence. IMCO sponsored a conference in 1962 to amend the 1954 Convention and set up a Subcommittee on Oil Pollution under the Maritime Safety

3. Some commentators have argued that this lack of explicit recognition of the pollution-control function has hindered IMCO's development of it. See Eldon V. C. Greenberg, "IMCO: An Environmentalist's Perspective," *Case Western Reserve Journal of International Law* 2 (winter 1976), p. 135.

Committee in 1965; but pollution control was still a relatively minor concern of the agency. With the *Torrey Canyon* disaster in 1967 and the increased concern for environmental protection in the developed countries following this and other developments in the late 1960s, the situation changed rapidly. In 1969, IMCO sponsored a conference on coastal state intervention and compensation in oil pollution incidents; in 1969, it further amended the 1954 Convention; and it organized the major Conference on Marine Pollution in 1973 and the Tanker Safety and Pollution Prevention Conference in 1978.

IMCO's structure has undergone substantial alterations to accommodate its new orientation. It created an ad hoc Legal Committee to examine the legal aspects of the *Torrey Canyon* problem in 1967, and this has since become a permanent organ of the agency. In addition, the Subcommittee on Oil Pollution was renamed the Subcommittee on *Marine* Pollution (SCMP) in 1969 so as to broaden its scope. As the importance of pollution control continued to grow, this committee was itself replaced in 1973 with a new Marine Environment Protection Committee (MEPC), a permanent subsidiary body of the Assembly (unlike its predecessors, which answered to the Maritime Safety Committee, MSC). Subsequently amendments to the IMCO Convention have given the MEPC formal institutional equality with the MSC. With these developments pollution prevention and the promotion of maritime safety now have equal formal status within the organization.

Former IMCO Secretary-General Sir Colin Goad observed in an interview that the *Torrey Canyon* disaster and the resulting increase in the organization's involvement in the marine pollution area was a "godsend" for IMCO. During the mid-1960s there had been considerable concern on the part of a number of Secretariat members and national representatives that IMCO might not be able to survive as a specialized agency within the U.N. system given the narrow scope of its activities. In particular they feared being brought directly under one of the bodies set up and controlled by the General Assembly. The new involvement of the organization in the pollution field now gave it "a second arm" which is continually increasing IMCO's status and salience within the U.N. system.

IMCO's activities in the field of marine safety do, however, still constitute the largest portion of its work. But, as was mentioned in Chapter 1, this study will not be concerned with the negotiations in this area since pollution prevention was a subsidiary concern to that of improving safety and reducing accidents. Furthermore, the number of agreements on such matters is so numerous and their provisions so complex that an adequate treatment of them would require a separate study. (The first conference significantly concerned with the prevention of polluting accidents was held in 1978, and this is treated in Chapter 4.)

At the same time, the major contributions of IMCO in the field of marine safety can be identified by looking at a number of the conventions which have developed out of its meetings and conferences.[4] Probably the most important is the International Convention for the Safety of Life at Sea, 1960 (revised in 1974), which deals with such things as structural requirements for ships and life-saving equipment. Former Secretary-General Colin Goad noted that this convention, including its application and amendment, was almost the *raison d'être* for IMCO until the mid-1960s. More recently, important products of IMCO's work have included the International Convention on Load Lines, 1966 (which sets the maximum draft that ships may have when loaded) and the Convention on the International Regulations for Preventing Collisions at Sea, 1972.[5] These latter regulations (which came into force in July 1977) will incorporate over sixty vessel traffic-routing schemes. IMCO organs also have passed numerous recommendations which have often been incorporated into the national legislation of states.

Other essential elements in marine safety are the standards for the training and certification of crews.[6] Past attempts at regulation of this area always have encountered opposition from both shipowners (who are concerned with wage levels) and labor unions (which are eager to protect jobs). Consequently, until 1978 the only international conventions in the area were the somewhat forgotten ILO Officers Competency Convention of 1936 and the Convention Concerning the Certification of Able Seamen of 1946. What crew standards existed were, for the most part, controlled by national legislation. Of late, however, IMCO has become increasingly involved in the area through its Subcommittee on Standards of Training and Watchkeeping and a conference in June 1978 to draft a Convention on Training and Certification of Seafarers. However, the resultant convention only stipulates the *types* of knowledge required for officers and crew and will not accept procedures to assure particular *levels* of knowledge. Without a global exam-setting and -marking system or rigorous IMCO monitoring of national systems for judging the qualifications of ship personnel, it is doubtful

4. A list of all these conventions can be found in *The Activities of the Inter-Governmental Maritime Consultative Organization in Relation to Shipping and Related Maritime Matters* (London: Inter-Governmental Maritime Consultative Organization, 1974), pp. 37–39.

5. For a stinging criticism of IMCO conventions affecting the design, construction, and operation of tankers, see Mostert, *Supership*, pp. 156 and 198 ff. By rule 10 of the Collision Regulations, all traffic separation schemes adopted by IMCO will be made mandatory. For a history of the regulations, see *IMCO News*, 1977 (no. 2), pp. 4–8.

6. A good study on this topic was submitted to IMCO by France. See MP/XIII/2(9)/9.

whether such a convention will have a major effect. IMCO is, through its technical assistance to developing countries and through the ILO-IMCO Joint Training Committee, also trying to assist countries in their individual efforts to improve national standards. However, given the magnitude of the problem, the contribution through this channel will be slight. In general, therefore, it will still be very much up to individual governments and industry to set their own standards. In some countries concern with this issue has significantly increased, but in others the will to embark on a strong course of action does not presently exist.

THE STRUCTURE OF IMCO

All of IMCO's organs are, in varying degrees, concerned with marine pollution. The three traditional, deliberative bodies (the Assembly, Council, and Maritime Safety Committee) as well as the more recently institutionalized Legal Committee, Marine Environment Protection Committee, and Technical Cooperation Committee deal with the issue. There also have been, as mentioned, a number of subsidiary bodies created to deal particularly with pollution questions and to recommend courses of action to the higher organs or directly to member states. Chart 1 indicates how the identity, composition, and structure of these bodies have altered over time.

The Assembly is supreme governing body of IMCO. It consists of the representatives of all member governments and meets once every two years, although it can be called into "extraordinary session" by at least one-third of its members or by the Council. The matters with which it deals are both substantive and organizational. It alone can pass resolutions to be forwarded to member governments as IMCO "recommendations." A simple majority is required to pass these resolutions. This is virtually the full extent of the Assembly's substantive activities, as it usually just works as a rubber stamp for proposals referred to it by other bodies in the organization. The Assembly recommendations are nonbinding, but they often are incorporated into national legislation or international conventions at a later date. The Assembly is also empowered to adopt amendments to the conventions, and these are also forwarded to treaty adherents for ratification. The Assembly's organizational responsibilities include such tasks as electing the Council (and previously the Maritime Safety Committee), establishing subsidiary bodies, approving the appointment of the secretary-general, and determining budgets. Contributions to the budget are assessed on the basis of the percentage of total tonnage registered in each state.

CHART 1
IMCO Structures Concerned with Marine Pollution

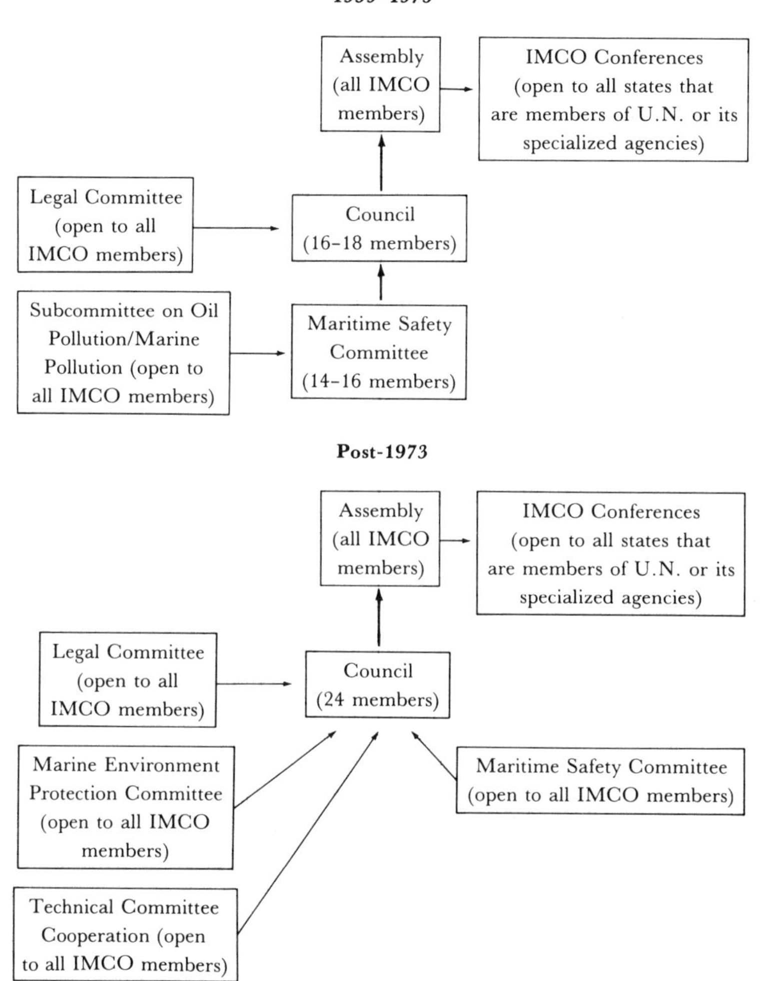

NOTE: The 1974 Amendments to the IMCO Convention, which increase the size of the Council to twenty-four entered into force on April 1, 1978. The 1975 Amendments, which bestow institutional equality with the MSC on the MEPC and Legal Committee, and the 1977 Amendments, which bestow such equality on the Technical Cooperation Committee, had not entered into force as of 1978. These bodies report to the Assembly through the Council. However, the Council can only comment on their technical reports and cannot alter them. All IMCO bodies (except conferences) can pass recommendations by the affirmative votes of a majority of the states present and voting.

The Council is IMCO's governing body between meetings of the Assembly. It generally meets twice a year, but it also can be called into "extraordinary session," as it was following the 1967 *Torrey Canyon* disaster. It is responsible for receiving and commenting on reports of the MSC, MEPC, Legal Committee, and Technical Cooperation Committee for transmission to the Assembly; reviewing reports of its committees (for example, the Facilitation Committee); proposing substantive recommendations and the organization's budget to the Assembly; and administering IMCO personnel, including the appointment of the secretary-general with the approval of the Assembly. The Council is IMCO's central policy organ, since it establishes the work program for other bodies and approves both administrative and substantive recommendations to be considered by the Assembly. Representatives to its meetings tend, therefore, to be drawn from national foreign ministries and from among high officials of transportation departments.

The nature of the Council membership has changed over the history of the organization. When the IMCO Convention was drafted at the United Nations Maritime Conference in 1948, there was a problem of achieving a balance within the organization between nations which were the suppliers of shipping services and those which were the consumers. The outcome was articles 17 and 18 of the Convention providing for a Council of sixteen members—six states with the largest interest in providing shipping services, six with the largest interest in international seagoing trade, and four from the general membership. In view of the continuing expansion of IMCO's membership, these articles were amended in 1964 to provide for a somewhat more equitable geographical distribution. The Council was enlarged to eighteen members, with six to come from the general membership "which have special interests in maritime transport or navigation, and whose election to the Council will ensure the representation of all major geographical areas of the world." In 1974, the articles were again amended so as to increase the number of states on the Council to twenty-four and the number of members from the general membership to twelve.[7] With the entry-into-force of these amendments on April 1, 1978, the executive body of the organization has escaped the formal dominance exercised by the developed maritime states.

The Maritime Safety Committee (MSC) until 1978 consisted of fourteen to sixteen states, of whom eight were to be the largest shipowning countries. It is now open to all members (article 28). It normally meets twice a year. Delegates to it are usually technical experts from national transport ministries. This is the main technical organ of the agency, and its work "covers

7. Res. A.315 (ES.V).

various fields such as aids to navigation, construction and equipment of ships and off-shore drilling units, rules for preventing collisions at sea, dangerous cargoes in packages and in bulk, fire protection, maritime safety procedures and requirements, marine casualty studies, search and rescue, and any other matters connected with IMCO.''[8] Although not formally equal to the Council in the IMCO Convention, the MSC is able to ''provide machinery for performing any duties assigned to it'' (article 29[B]). Its more detailed work is carried out through numerous subcommittees such as the now-defunct Subcommittee on Marine Pollution (SCMP).[9] Like the present MSC, these subcommittees have always been open to any IMCO member wishing to participate. (This was potentially an important provision since the subcommittees do most of the drafting of IMCO regulations and conventions, but few nonmembers attended.)

The Legal Committee was set up under the auspices of the Council in May 1967 to examine possible changes in maritime law after the *Torrey Canyon* disaster.[10] The committee has considered the legal aspects of issues such as liability and compensation for pollution damage from oil and other substances, wreck removal and salvage, ships in foreign ports, passengers and baggage, the new Ocean Data Acquisition System (ODAS), general limitations on maritime liability, and the enforcement of IMCO conventions. With respect to pollution, it often has had to provide legal advice to the SCMP and later the MEPC, since both bodies have been concerned with the same enforcement and jurisdictional issues. The committee also maintains liaisons on work in international maritime law with other United Nations bodies and agencies, especially the United Nations Conference on the Law of the Sea, UNCTAD, and UNCITRAL, and with nongovernmental organizations such as the Comité Maritime International.[11]

While the Legal Committee has been largely responsible for IMCO regulations on liability and compensation for marine pollution damage, other aspects of the pollution issue have generally fallen to the *Subcommittee on Oil Pollution* (1965) and its successors, the *Subcommittee on Marine Pollution* (1969) and the *Marine Environment Protection Committee* (1973). First meeting in 1965, the SCOP was created to oversee the application and revision of the 1954 Oil Pollution Convention (and its 1962 and 1969 Amendments).

8. *IMCO and Its Activities* (London: IMCO, 1974), p. 7.

9. Other subcommittees are more directly oriented to problems of navigation: Safety of Navigation; Radio Communications; Life-Saving Appliances; Training and Watchkeeping; Ship Design and Equipment; Fire Protection; Subdivision, Stability, and Loadlines; Containers and Cargoes; and Carriage of Dangerous Goods. Many of the subcommittees have obvious implications for environmental protection.

10. It was originally an ad hoc committee. With the entry-into-force of the 1975 Amendments to the IMCO Convention, it will have equal status with MSC and MEPC.

11. See *IMCO and Its Activities* (London: IMCO, 1974), p. 30.

Its immediate successor, the SCMP, was responsible for the formulation of the 1973 Convention for the Prevention of Pollution from Ships. As subcommittees of the MSC, both the SCOP and SCMP were open to all members of IMCO—unlike the MSC itself.

The Marine Environmental Protection Committee was created by the Assembly in November 1973.[12] It was meant to replace the SCMP, but it was made directly responsible to the Assembly. It will be given equal status with the MSC within IMCO with the entry-into-force of the 1975 Amendments to the IMCO Convention. Like the SCMP it is open to all IMCO members, but more states should participate in it since it has greater status and is not a subcommittee of the MSC.

The proposal for the creation of the MEPC came from the United States in June 1973. With an eye to the upcoming Law of the Sea Conference, the U.S. government felt that upgrading and institutionalizing IMCO's pollution-control function would forestall potential demands for either an extension of coastal state jurisdiction to control pollution or for the creation of a new pollution prevention agency. In particular, the Americans hoped that the new body would increase IMCO's attractiveness as an environmental organization for the developing countries.[13]

As Sir Colin Goad, the former secretary-general, has commented, the MEPC has immensely enhanced IMCO's status by supplementing the MSC and putting pollution control on at least a formally equal basis with marine safety. The MEPC is concerned with the environmental aspects of shipping and with all facets of the implementation of the 1973 Convention for the Prevention of Pollution from Ships, including undertaking technical studies (on such subjects as the "tagging" of oil cargoes for identification and setting specifications for discharge and monitoring equipment). Most importantly, the MEPC can act in a "quasi-legislative" capacity with respect to the amendment of the 1973 Convention. That convention permits amendments to technical provisions adopted by the appropriate body (such as the MEPC) to enter into force unless objected to by a certain percentage of the parties to the convention. This "tacit acceptance" approach to amending international treaties is a major innovation since the traditional "explicit acceptance" approach has resulted in delays in bringing amendments into force, impeding the updating of technical provisions. Finally, the MEPC coordinates the pollution-control activities of IMCO with other organs and acts as a clearing house for information sent to it by governments and other international organizations.

A final standing IMCO body concerned with pollution control is the

12. Res. A.297 (VIII).

13. Interviews with American and other government officials. For detailed outline of the scope of the MEPC's work, see MEPC I/10, March 8, 1974. For a critique of its performance, see Greenberg, "IMCO: Environmentalist's Perspective," pp. 148–149.

Technical Cooperation Committee (TCC). Originally created by the Council in 1969 to respond to the growing desire of the developing countries for technical assistance for their nascent shipping industries, it was in 1977 made a "principal organ" like the MSC, MEPC, and Legal Committee, reporting to the Assembly through the Council.[14] The funding which it oversees comes from the U.N. Development Program, the U.N. Environment Program, and "funds in trust" provided by individual IMCO members. The increasing importance of the TCC is due, in part, to the interest of Secretary-General Srivastava and has significantly increased the interest of the developing countries in the organization.

The ultimate IMCO deliberative "organ" is the IMCO-sponsored conference called to conclude a treaty or revise an existing one. Such conferences are convened by the secretary-general with the approval of the Assembly, though the impetus for them comes from the deliberations of the MSC, MEPC, or Legal Committee. This conference process is the basis of IMCO's law-creating process. Unlike some other agencies (such as the International Civil Aviation Organization), IMCO's authority is not of a "legislative" or even presently a "quasi-legislative" nature but is of a voluntary, treaty-making character.[15] IMCO members and members of other U.N. agencies are invited to an IMCO conference, where the approval of two-thirds of the participants is required for each of the proposed articles of a draft convention. Even then, under the terms of the treaty, the final product of the conference becomes international law only when an agreed number and/or group of states ratify the convention. Often this has occurred only after a delay of many years.

In addition to the deliberative organs of IMCO, the other important organ of IMCO is its *Secretariat*. As with the agency generally, the development of the Secretariat was hindered by the conviction of most early parties to the IMCO Convention that the agency should not be given an active or influential role. Indeed, the United Kingdom actually proposed that there should be no independent secretariat at all and that the agency should be run by individuals supplied by the member governments on a voluntary basis. Four years after the organization began its operation, its Secretariat included only twenty professional members, and today, with about eighty professional staff (and about twice as many service staff), it is still one of the smallest specialized agencies in the United Nations system.

The official functions of the Secretariat are to provide administrative

14. Res. A. 358 (IX).

15. For a thorough discussion of the distinctions among these powers, see Edward Yemin, *Legislative Powers in the United Nations and Specialized Agencies* (Leyden: A. W. Sijthoff, 1969); and C. H. Alexandrowicz, *The Law-Making Functions of the Specialized Agencies of the United Nations* (Sydney: Angus and Robertson, 1973), especially pp. 4, 66–69, and 152–156. With the entry-into-force of the 1973 Convention, its "tacit acceptance" procedure for amending the Convention will bestow "quasi-legislative" powers on the Organization.

services for all meetings of IMCO deliberative bodies, to submit reports on matters under discussion, and to assist the deliberative bodies in coordinating their activities with those of member states and other U.N. agencies. As a rule, policy initiatives have not come from the Secretariat, as the member states have discouraged this, but suggestions of Secretariat members have on occasion facilitated agreements.

Since 1958, four men have served as secretary-general of IMCO: Ove Nielsen (Denmark), 1958–1962; Jean Roullier (France), 1962–1967; Colin Goad (United Kingdom), 1967–1973; and since 1974, C. P. Srivastava (India). Mr. Goad (now Sir Colin Goad), who was also deputy secretary-general under Mr. Roullier, has complained that the bias against an activist Secretariat was also applied to the secretary-general, who was expected to concern himself solely with administrative and personnel matters. His "political" activities were confined basically to explaining the utility of the organization to governments and negotiating with other U.N. agencies over jurisdictional matters. During Mr. Goad's tenure, the scope of the secretary-general's activities was not significantly altered except in representing the organization abroad. Concerned for the very survival of IMCO after UNCTAD's Committee on Shipping had been created in 1965, Mr. Goad attempted to increase IMCO's political presence by visiting those states whose support was crucial to its existence and by intensifying his participation in United Nations bodies. This more active role on the part of the secretary-general was facilitated by the new demands placed on the whole organization as a result of the 1967 *Torrey Canyon* disaster.

In more recent times, the roles of the Secretariat and the secretary-general have continued to expand (if slowly) in both marine safety and marine pollution. After the Law of the Sea Conference, this trend will likely accelerate, especially as that conference explicitly confers standard-setting jurisdiction over shipping on "the competent international organization" (IMCO) for most ocean space (see Chapter 6). Moreover, as the general level of interest in shipping rises among the developing states, there is likely to be more interest in IMCO. Incumbent Secretary-General Srivastava of India is well qualified to respond to this new interest. As a national of a developing maritime state, Mr. Srivastava is concerned to bring developing states increasingly into IMCO's activities. Moreover, having previously been president of India's national shipping company and active in conferences sponsored by UNCTAD's Committee on Shipping, the secretary-general is well versed in many other organizational aspects of the problem.

The changes in IMCO's structure which have been wrought by the 1974, 1975, and 1976 Amendments (including the change in name to "International Maritime Organization") have given those bodies concerned with marine environmental protection greater status and even some greater

IMCO's Secretary-General C. P. Srivastava meets United Nations Secretary-General Dr. Kurt Waldheim outside IMCO's present headquarters on Piccadilly in London. (Photo: UPI.)

powers (''tacit acceptance'' of amendments). But the changes are neither so sweeping nor important as really to herald a new beginning for the organization as an independent environmental legislative authority instead of one dependent on governmental consent. Ultilmately IMCO is, as ever, dependent on changes in values and policies among the member governments.

THE IMCO MEMBERSHIP

IMCO's membership has greatly expanded over the past two decades. From its original composition of only 27 states in 1958, it expanded to 102 in 1977. In addition, in the early stages, the membership was composed predominantly of developed, European states with large maritime interests, but this was soon changed with the influx of nonmaritime developed states and, increasingly in recent times, of developing states as well. Chart 2 indicates the changes in the organization's composition. While states have been grouped into three broad categories, it would be quite misleading to see

With growing demands being made on IMCO in recent years, IMCO has been pressured to expand. Pictured here is a model of the new headquarters to be built on the south bank of the Thames across from the Houses of Parliament. This facility should be in use by the mid-1980s by which time the agency will also have a new title, the International Maritime Organization (IMO). (Photo: Rose and Dyble.)

these groups in IMCO as discrete, cohesive blocs. Shared interests frequently transcend these broad groupings, and divergent interests coexist within them. As with all international organizations, IMCO has its own "particular environment"

Some Attributes of IMCO Members

Certain characteristics of a state help to explain its interests and role in IMCO. Two attributes in particular are recognized by the IMCO Convention as entitling member states to important roles in the decision-making structure: shipownership and volume of sea trade. Shipownership involves both tankers and nontankers (that is, dry cargo ships). Needless to say, domestic tanker tonnage is of special significance in affecting a state's attitude toward proposals for oil pollution prevention. Table 4 lists the thirty states with the greatest tanker tonnage between the years 1955–1975, a period in which total tanker tonnage increased sevenfold, from 26,454,641 grt (gross registered tons) to 150,057,269.

A number of interesting trends are evident in this table. Throughout the entire twenty-year period, the number of large tankerowning states has remained virtually unchanged. The number of states with at least a 1 percent share of total world tonnage only increased from thirteen to fifteen, and there were only minor changes in the composition of this group. Their share of total tonnage also fell by only 2 percent, from 92 percent to 90 percent, in that time. In 1975, only two developing states, Liberia and Panama, had more than 1 percent of the total world tanker tonnage, and their vessels were largely controlled by American and some Greek companies.[16]

Control of oil shipments has evidently remained the province of but a handful of states, mostly developed. However, within this small group, some dramatic shifts have occurred. Many states have retained a fairly stable share of tanker tonnage, but some—notably, the United States and Panama—have declined significantly while others—Liberia, Japan, and the USSR—have grown tremendously. Liberia is now in a class by itself and, with Japan, the U.K., and Norway, constitutes the central core of large maritime (that is, shipping) states.

Particularly in the earlier years, a substantial volume of polluting oil also emanated from nontankers. The figures for the world nontanker fleet are contained in Table A in Appendix 1, and they reveal a slightly greater dispersion than is the case with tankers. In general, the same developed and flag-of-convenience states dominate the scene, but some other states—India, Singapore, and Canada—also have shares greater than 1 percent of the total tonnage.

In looking at a state's trading interests, one must look at nontanker shipments as well as oil exports and imports. Tables C and D in Appendix 1 show that, as expected, the developed states of Western Europe, North America, Japan, and Australia again dominate both in the trade in dry cargo goods and in the importation of oil. The pattern of such trade has been fairly consistent over the years, but recently the United States and Japan have topped the list of importers. Here again, there are only a very few developing countries with shares greater than 1 percent of the total.

16. Analyses of the development and policies of the flag-of-convenience fleet can be found in B. A. Boczek, *Flags of Convenience: An International Legal Study* (Cambridge, Mass.: Harvard University Press, 1962); Erling D. Naess, *The Great PanLibHon Controversy: The Fight over Flags of Shipping* (Epping, Essex: Gower Press, 1972); Samuel S. Lawrence, *U.S. Merchant Shipping Policies and Politics* (Washington, D.C.: Brookings Institution, 1966); the report by the OECD Ad Hoc Group on Flags of Convenience, March 14, 1972 (mimeo.); and *Maritime Transport 1974* (Paris: OECD, 1975), pp. 88–105.

Chart 2

IMCO Membership[1]

Group[2]	*Original Members*		*1959–1965*				*1966–1976*			
Western	Canada	(1948)	Denmark	(1959)			Malta	(1966)		
Europe	Netherlands	(1949)	Finland	(1959)			Hong Kong	(1967)		
and	United Kingdom	(1949)	F.R. Germany	(1959)			Cyprus	(1973)		
Others	United States	(1950)	Sweden	(1959)			Austria	(1975)		
(WEO)	Belgium	(1951)	Iceland	(1960)			Portugal	(1976)		
	Ireland	(1951)	New Zealand	(1960)						
	France	(1952)	Spain	(1962)						
	Australia	(1952)								
	Israel	(1952)								
	Switzerland	(1955)								
	Italy	(1957)								
	Greece	(1958)								
	Norway	(1958)								
	Japan	(1958)								
	Turkey	(1958)								
Soviet	USSR	(1958)	Bulgaria	(1960)			Hungary	(1970)		
Bloc			Poland	(1960)			D.R. Germany	(1973)		
(SB)			Czechoslovakia	(1963)						
			Romania	(1965)						
Developing	Burma	(1951)	Ghana	(1959)	Cameroon	(1961)	Cuba	(1966)	Thailand	(1973)
(Dev.)	Argentina	(1953)	India	(1959)	Morocco	(1962)	Lebanon	(1966)	Zaire	(1973)
	Dominican Rep.	(1953)	Liberia	(1959)	Nigeria	(1962)	Singapore	(1966)	Oman	(1974)
	Haiti	(1953)	Yugoslavia	(1960)	Korea	(1962)	Maldives	(1967)	Sudan	(1974)

Honduras	(1954)	Ivory Coast	(1960)	Algeria	(1963)	Peru	(1968)	Tanzania	(1974)
Mexico	(1954)	Kuwait	(1960)	Brazil	(1963)	Uruguay	(1968)	Colombia	(1974)
Ecuador	(1956)	Senegal	(1960)	Syria	(1963)	Saudi Arabia	(1969)	Congo	(1975)
Egypt	(1958)	Indonesia	(1961)	Tunisia	(1963)	Barbados	(1970)	Ethiopia	(1975)
Iran	(1958)	Cambodia	(1961)	Philippines	(1964)	Libya	(1970)	Guinea	(1975)
Pakistan	(1958)	Madagascar	(1961)	Trinidad-Tobago	(1965)	Malaysia	(1971)	Venezuela	(1975)
Panama	(1958)	Mauritania	(1961)			Chile	(1972)	Bangladesh	(1976)
						Eq. Guinea	(1972)	Gabon	(1976)
						Sri Lanka	(1972)	Jamaica	(1976)
						China	(1973)	Papua New Guinea	(1976)
						Iraq	(1973)	Bahrain	(1976)
						Jordan	(1973)	Cape Verde	(1976)
						Kenya	(1973)	Surinam	(1976)
						Sierra Leone	(1973)	Bahamas	(1976)

Change in IMCO Membership over Time

	Original Members	*1965*	*1976*
WEO	55.6%	37.3%	26.5%
SB	3.7%	8.5%	6.8%
Dev.	40.7%	54.2%	66.7%

1. The year in which each IMCO member accepted the IMCO Convention is in parentheses. As of March 1977 there were one hundred and one IMCO members and one associate member (Hong Kong).

2. The countries have been grouped using the formal UNCTAD categories. Also, a few countries have been transferred from one grouping to another (for example, Israel from developing to WEO).

TABLE 4
World Tanker Registry, 1955–1975*

Rank	*1955*	%	*1960*	%	*1965*	%	*1970*	%	*1975*	%
1	United Kingdom	19.89	LIBERIA	17.32	LIBERIA	19.30	LIBERIA	22.44	LIBERIA	27.70
2	United States	16.34	United Kingdom	17.00	Norway	15.16	United Kingdom	13.97	Japan	11.68
3	Norway	15.77	Norway	15.37	United Kingdom	14.42	Japan	10.71	United Kingdom	10.73
4	LIBERIA	8.90	United States	11.18	United States	8.20	Norway	10.28	Norway	8.92
5	PANAMA	8.07	PANAMA	6.03	Japan	6.90	United States	5.44	Greece	5.53
6	Italy	4.57	France	4.48	PANAMA	4.44	Greece	4.50	France	4.62
7	France	4.36	Italy	4.16	France	4.15	France	4.04	PANAMA	3.69
8	Sweden	3.41	Japan	3.78	*Soviet Union*	3.85	*Soviet Union*	4.02	United States	3.44
9	Netherlands	3.38	Netherlands	3.27	Italy	3.61	PANAMA	3.82	Italy	2.71
10	Japan	2.81	Sweden	3.18	Greece	3.10	Italy	3.19	*Soviet Union*	2.47
11	Denmark	1.91	Greece	2.21	Netherlands	2.78	Netherlands	2.30	Sweden	2.02
12	ARGENTINA	1.28	Denmark	2.10	Sweden	2.54	German F.R.	1.91	German F.R.	1.82
13	German F.R.	1.09	*Soviet Union*	1.67	Denmark	1.77	Sweden	1.87	Netherlands	1.76
14	Canada	0.91	German F.R.	1.35	German F.R.	1.60	Spain	1.65	Spain	1.70
15	*Soviet Union*	0.80	Spain	1.08	Spain	1.13	Denmark	1.56	Denmark	1.44
16	Spain	0.80	ARGENTINA	0.83	ARGENTINA	0.99	Finland	0.78	SINGAPORE	0.96
17	Greece	0.68	BRAZIL	0.72	BRAZIL	0.75	BRAZIL	0.66	Finland	0.76
18	BRAZIL	0.64	VENEZUELA	0.62	Finland	0.58	ARGENTINA	0.58	BRAZIL	0.69
19	HONDURAS	0.58	Canada	0.53	Belgium	0.50	Bermuda	0.54	Bermuda	0.68
20	VENEZUELA	0.54	Finland	0.41	VENEZUELA	0.36	KUWAIT	0.49	INDIA	0.44
21	MEXICO	0.49	Belgium	0.38	MEXICO	0.32	Belgium	0.35	KOREA (Rep. of)	0.43
22	Belgium	0.37	MEXICO	0.30	Canada	0.30	INDIA	0.33	CHINA	0.41
23			Portugal	0.30	Portugal	0.27	KOREA (Rep. of)	0.33	KUWAIT	0.41
24			Turkey	0.16	INDIA	0.25	YUGOSLAVIA	0.30	ARGENTINA	0.36

25			YUGOSLAVIA	0.13	Australia	0.25	Canada	0.29	Cyprus	0.35
26			*Poland*	0.12	Turkey	0.21	VENEZUELA	0.29	Portugal	0.34
27			TAIWAN	0.11	KUWAIT	0.18	Portugal	0.29	Belgium	0.25
28			Israel	0.10	*German D.R.*	0.17	MEXICO	0.28	TAIWAN	0.22
29			EGYPT	0.09	BAHAMAS	0.15	TAIWAN	0.27	Turkey	0.22
30			CHILE	0.08	EGYPT	0.14	Australia	0.22	VENEZUELA	0.21
Others		2.40		0.90		1.60		2.30		3.00
World Total GRT		25,454,641		41,465,102		55,046,070		86,139,853		150,057,269

NOTE: In this chart, developing states are capitalized, Soviet bloc states are in italics, and WEO are in standard print. GRT = gross registered tons.

*States are ranked according to their percentage of gross registered tanker tonnage. *Lloyd's Register of Shipping Statistical Tables*, which was used for all years, included data on only twenty-two states in its 1955 edition.

The major maritime *exporters*, however, constitute a very different group (Table B, Appendix 1). Ten are Middle Eastern developing states, and the rest are largely from the developing world as well. In general, the flow is markedly from the developing to the developed states.

The nationality of oil companies and their ownership of tankers in their own and foreign states reinforce these patterns in ship registry and trade. In his study of the international oil industry, Neil Jacoby has analyzed the roles of the major companies in the various sectors of the industry.[17] The production and refining sectors (where effective control lies over the export and import of oil) reveal the overriding importance of American and, to a lesser extent, British firms. In the production sector in 1972, ten of the largest twenty firms were American, and they controlled 50 percent of all oil produced. British Petroleum and Shell (jointly owned by British and Dutch interests) controlled an additional 26 percent. However, the post-1973 nationalizations of many production facilities have led to a major transformation of this situation.[18] In the refining sector there has been less industry concentration, although the United States and the United Kingdom still control a large share. The five largest American companies were among the twenty largest in the world, controlling 29 percent of the refining. British Petroleum and Shell controlled an additional 20 percent, with the remainder being divided among a large number of firms, most of which were still in the developed nations.

Another sector where the oil industry has a large direct interest is in the ownership of tankers. Thirty-one percent of all tanker tonnage is *directly* owned by oil companies, and a further large segment is owned by them through "dummy corporations." The "seven major" oil companies account for a large share (71 percent) of this ownership.[19] As the interests of the oil company shipowners often differ from those of the independent shipowners, it is significant that of all the large registry states it is only in the United Kingdom that the oil companies have substantial direct ownership of tankers (see Table E in Appendix 1). The United Kingdom, with its large B.P. and Shell tanker companies, registers 35 percent of the tankers of the "seven majors." Interestingly, the flag-of-convenience state Liberia

17. Neil H. Jacoby, *Multinational Oil: A Study in Industrial Dynamics* (New York: Macmillan, 1974). The leading firms in the production and refining sectors are listed on pp. 192–193 and 198–199.

18. While nationalizations and the purchase of large shares of oil company holdings began after the formation of OPEC in 1968 (the most notable being the nationalization of B.P. holdings by Libya in 1970), it was not until after the 1973 Arab-Israeli war that a major revolution in this area occurred. Jacoby, *Multinational Oil*, pp. 112–115 and 255–306; Christopher Tugendhat and Adrian Hamilton, *Oil: The Biggest Business* (London: Eyre Methuen, 1975), part 2; and articles in *Petroleum Economist*, 1974–1976.

19. M. A. Adelman, *The World Petroleum Market* (Baltimore: Johns Hopkins University Press, 1972), p. 105; Jacoby, *Multinational Oil*, p. 177.

registers 17 percent, including large segments of Exxon, Mobil, Gulf, and Standard Oil (California) fleets (Table F, Appendix 1). Were the clandestine registries also known, these figures would likely be much higher. It is evident, however, that these states and those others in Table 3 with a high percentage of such registries are particularly susceptible to oil industry pressures on tanker pollution issues. The independent tankerowners, on the other hand, exercise their greatest influence in the Scandinavian states and Greece.

In review, it is clear that a dozen or so developed states have a disproportionately great interest in the shipping and oil industries and in the consumption of the products they bring. At the same time, their interests are not identical, and their policies frequently diverge. The developing states are, in contrast, virtually shut out from these industries. Liberia and Panama are important states of registry, but their fleets are foreign-owned. Apart from these flags of convenience, a number of developing countries—India, Brazil, Argentina, Venezuela—have begun to develop fleets, although not significantly in the tanker field. But, as is evident from their demands in UNCTAD's Committee on Shipping, their aspirations for these industries are by no means satisfied. It is really in the area of the oil *export* trade that the "developing" (in particular the Middle Eastern) states have a large interest.

Participation in IMCO Bodies

Participation in IMCO organs is extremely uneven. This has particularly been the case in the Council and Maritime Safety Committee by virtue of their formal restrictions on membership. The various pollution committees, the Legal Committee, and the IMCO conferences have all had an open membership policy, but the actual record of participation there is highly uneven as well.

Past criteria for membership in both the Council and Maritime Safety Committee have ensured the domination of the organization by a limited number of states with large maritime interests. Chart 3 illustrates this point dramatically. All the major shipowning nations (except the flag-of-convenience state Liberia) and the largest trading countries (including Belgium, Canada, and Australia as well as the major European states, which are also large shipowners) have been represented almost continuously on the Council and, to a slightly lesser extent, on the MSC.[20] In contrast, the developing states were drastically underrepresented until 1978, especially as they have constituted the numerical majority of the

20. For a detailed breakdown of national shipowning interests from 1950 to the present, see the tables in this chapter and in Appendix 1.

Chart 3
Membership in IMCO's Elective Bodies

Western Europe and Others					*Council*								*MSC*	
	1959	*1961*	*1963*	*1965*	*1967*	*1969*	*1971*	*1973*	*1975*	*1977*	*1961*	*1965*	*1969*	*1973*
Canada	X	X	X	X	X	X	X	X	X	X	X	X	X	X
France	X	X	X	X	X	X	X	X	X	X	X	X	X	X
Germany (FR)	X	X	X	X	X	X	X	X	X	X	X	X	X	X
Japan	X	X	X	X	X	X	X	X	X	X	X	X	X	X
Norway	X	X	X	X	X	X	X	X	X	X	X	X	X	X
U.K.	X	X	X	X	X	X	X	X	X	X	X	X	X	X
U.S.	X	X	X	X	X	X	X	X	X	X	X	X	X	X
Greece	X	X	X	X	X	X	X	X	X		X	X	X	X
Italy	X	X	X	X	X	X	X		X		X	X	X	X
Netherlands	X	X	X	X	X	X	X				X	X	X	
Australia	X	X	X	X	X	X	X							
Belgium	X	X	X	X		X	X							
Sweden	X	X	X	X	X								X	
Spain													X	X
Malta										X				
% of membership	86.7	81.3	81.3	81.3	66.7	66.7	66.7	61.1	50	33.3	71.5	71.5	75	62.5
Soviet Bloc														
USSR	X	X	X	X	X	X	X	X	X	X	X	X	X	X
Poland					X	X	X	X	X					
Romania										X				
% of membership	6.7	6.3	6.3	6.3	11.1	11.1	11.1	11.1	11.1	8.3	7.1	7.1	6.3	6.3

Developing														
India		X	X	X	X	X	X	X	X	X				
Argentina	X	X							X	X	X	X	X	X
Brazil					X	X	X	X	X	X				
Pakistan											X	X	X	
Algeria							X	X	X	X				
Egypt (UAR)					X					X			X	X
Liberia										X	X	X		X
Ghana						X	X							
Indonesia								X	X	X				
Nigeria								X	X	X				
China									X	X				
Yugoslavia														X
Kenya										X				
Kuwait										X				
Malagasy Rep.										X				
Mexico										X				
Peru										X				
% of membership	6.7	12.5	12.5	12.5	22.2	22.2	22.2	27.7	38.8	58.3	21.4	21.4	18.8	31.3

NOTE: The elections for the Council are for two years and those for the MSC were for four years. The last election for the MSC was in 1973 since on April 1, 1978 it became open to all members.

IMCO membership since 1963. Until 1975 they constituted less than 28 percent of the Council membership, and during the period 1973–1977, the last years when the MSC was a limited membership body, the figure there was only 31 percent. The entry-into-force of the 1974 Amendments in 1978 has significantly altered this situation. The MSC is now an open membership body, and 58 percent of the Council members elected in 1977 were developing countries. Their potential for affecting the future of the organization has greatly increased.

An issue of particular interest in the past "membership politics" of the bodies was the exclusion of the Liberian delegation from the Council and the attempt to exclude it from the MSC as well. In 1959, two flag-of-convenience states, Liberia and Panama, had a registered tonnage which placed them among the top eight members of IMCO. Yet a number of European countries challenged their right to be considered among the eight largest "shipowning" countries because, of the vessels registered in these two states, few were actually owned by their nationals. The U.K., in particular, argued that "shipowning" meant that the real ownership of the ships must rest with companies or nationals of the state seeking representation. The United States (a large portion of whose shipping interests had sought refuge under these flags to avoid American taxation and union requirements) supported the Liberian claim in the face of vehement European opposition.

An advisory opinion was sought from the International Court of Justice to resolve the dispute. The court looked to the long tradition of determining shipownership in light of the place of *registry* and ruled in favor of Liberia and Panama. The 9–5 decision was certainly correct in light of traditional maritime practice, but it was a severe shock to the European states. Skeptical of IMCO in the first place, their suspicions about the new agency were more than confirmed.[21]

By 1961, Panamanian registry had slipped below eighth place so that country never took its seat on the MSC. Liberia has been represented on the MSC but, until November 1977, never on the Council. The criterion for membership there was much more vague—the convention refers to six states with the "largest interest in providing shipping services" and not to the largest "shipowners."

Clearly then the formal structure of IMCO's policy-making organs previously ensured continuing control by the developed states as long as they remained a cohesive bloc. Some developing states with maritime interests

21. See *IMCO Case* (Advisory Opinion by the International Court of Justice on the *Constitution of the Maritime Safety Committee of the Intergovernmental Maritime Consultative Organization*, International Court of Justice Reports, 1960, p. 150). See also Boczek, *Flags of Convenience*, pp. 125–155.

TABLE 5

Participation in IMCO's Open Committees and Conferences

	SCOP/SCMP and Legal Committee, 1965–1973	*MEPC and Legal Committee, 1974–1976*	*Oil Pollution Conferences*	
			1954, 1962	*1969, 1971, 1973*
Western Europe and others	68% (3.12)	70.2% (3.36)	73.6% (7.6)	77.7% (10.2)
Soviet bloc	36.5% (2.71)	41.3% (2.77)	53.9% (3.9)	63% (4.9)
Developing countries	9.48% (1.86)	19.6% (1.86)	25.5% (2.56)	41.5% (3.74)

NOTE: The percentage figure indicates the percentage of committee meetings or conferences which the members of each group attended. The figure was obtained by dividing the actual number of attendances by the number of possible attendances.

The figure in parentheses is the average size of delegation. It was obtained by dividing the total number of delegates sent by members of each group by the total number of meetings or conferences they attended.

The data was obtained from IMCO documents.

—such as Argentina, Brazil, and India—participated fairly regularly in the Council and MSC but not in numbers that posed any threat. It is only the recent fragmentation of the developed group and the altered membership criteria of the elective committees that have changed this picture.

Given the record of past participation in the open committees, however, there is little immediate chance of significant change. As Table 5 reveals, even in these bodies only a fraction of the IMCO membership designates representatives to the meetings. The developed states heavily predominate. Furthermore, the delegations they send are substantially larger so that they cover a very broad spectrum of expertise. Many of the smaller states simply designate local embassy officials as their representatives to the meetings, and even they often do not actually attend.[22]

The explanation for this inactivity is easy to discern. Most developing states are very far from London, so the costs of participating in regular meetings of the organization are inhibiting. Moreover, it is the developed states and their industries that possess the necessary expertise. Not surprisingly, therefore, one finds that the developing states have participated more actively in the political/legal bodies—the Council and Legal Committee—than in the technical/scientific bodies—the MSC and MEPC.[23] In recent years, however, developing state participation does seem to be increasing generally and particularly in the MEPC.

Nongovernmental Participants

It is through IMCO's committees and conferences that transnational, nongovernmental organizations have access to the agency. The work of IMCO obviously has great significance for many industries throughout the world, and IMCO itself is heavily dependent in its deliberations on the expertise of the industries it regulates. As a result, "consultative" status is extended to approximately forty intergovernmental (IGO) and nongovernmental (NGO) organizations representing a variety of industrial, commer-

22. There also have been significant differences in the levels of participation among states in the three groupings. The average size of U.K. delegations to all IMCO bodies has been generally almost double that of any other state. The delegations of the U.S., France, Norway, Sweden, Japan, the Netherlands, Italy, and Germany (FR) have been decidedly larger than those of other developed countries. Among the developing countries, Liberia's delegations have been decidedly the largest, although even they have only been equal to an average developed country. Next to those of Liberia have been the delegations of Brazil, Argentina, Egypt, Indonesia, India, Mexico, and Kuwait. On the nature of participation, see also Harvey B. Silverstein, "Technological Politics and Maritime Affairs—Comparative Participation in the Intergovernmental Maritime Consultative Organization," *Journal of Maritime Law and Commerce* 7 (January 1976), pp. 367–407.

23. See Chart 3 and Table 5. In comparison, the developed states showed a preference for the scientific/technical bodies.

cial, and environmental interests.[24] This observer status, while it does not confer voting rights, does allow an organization to attend and participate in all formal meetings.

Despite the large numbers with "consultative" status, it is only a handful of organizations representing the shipping, oil, and marine insurance industries that have had a significant impact on pollution regulations. By far the two most important bodies have been the International Chamber of Shipping and the Oil Companies International Marine Forum.

One of the oldest and most important NGOs is the International Chamber of Shipping (ICS), an organization composed of thirty national shipowner associations. Its policies are strongly influenced by the Western European associations, particularly by the General Council of British Shipping in whose building its headquarters are located.[25] The participation of the associations varies a great deal, with the most active being those from the U.K., Norway, Sweden, Netherlands, Denmark, U.S., France, and Germany. Of particular importance to the ICS are the interests of the non-oil-company ("independent") tankerowners, for they are more numerous and powerful in most national shipowner associations. The ICS pools the resources of its constituent associations in the formulation, coordination, and dissemination of positions with respect to national and international shipping regulations, including those concerned with pollution control, and it has been active in the work of the Legal Committee, SCMP, MEPC, and IMCO conferences. Its greatest influence is still, however, exerted outside the IMCO forum through the coordination of its constituent organizations in lobbying national governments.

In the late 1960s the oil industry increasingly found that it was unable to secure representation of its interests as *both* a shipowner and a cargo-owner through the ICS. As a result, in 1970 it formed its own independent organization, the Oil Companies International Marine Forum (OCIMF).[26] The

24. The basic purposes for providing NGOs with consultative status as well as the rules governing their participation are set forth in *Basic Documents, II* (London: IMCO, 1969), pp. 66–69.

25. Seventeen of the thirty member associations are Western European (with two from both Italy and Finland) and seven are from developing states. The secretary-general and chairman have always been British nationals. The authors obtained most of the information about the ICS from government officials familiar with its operations. Some information was secured from ICS staff.

26. American and British companies have unquestionably exerted the greatest influence within OCIMF, with American companies predominating. Observers have also noted that as between individual companies represented, there is a correlation between size and willingness to pay for pollution control.

The chairmanship of the General Committee has alternated between American and British nationals—the incumbent in 1978 being Capt. A. Dickson of Shell Marine International. The three vice-chairmanships are always divided among representatives from the western hemisphere (generally the U.S.), the Far East (generally Japan), and Europe. In 1978, the twelve

OCIMF membership in 1978 included forty-five groups of companies representing approximately 80 percent of the total volume of oil shipped by sea. All of the "big seven" oil companies (five of which are American) are members. The European and Japanese firms outnumber the American ones, but the latter—and especially the five largest—are generally more influential in OCIMF councils. OCIMF produces technical reports, sets forth policies favored by the industry, and lobbies both national governments and delegates at IMCO meetings to secure acceptance of these policies. It has been an omnipresent participant in the Legal Committee, SCMP, MEPC, and MSC and has been instrumental in shaping recent conventions on pollution prevention standards and pollution liability.

Probably the most conservative force in international shipping is the marine insurance industry. Its most powerful members are the British and Scandinavian Protection and Indemnity Associations (P & I Clubs) and the London-based Lloyd's insurance syndicates, who virtually control the setting of ship insurance rates. At IMCO they are represented respectively by the Baltic International Maritime Conference (BIMC) and the International Union of Marine Insurers (IUMI). In fact, the IUMI is usually represented by Lloyd's officials. The two organizations, however, do not play an activist role in IMCO meetings. Their representatives attend mainly to keep track of any proposals which might affect their interest in liability for pollution damage. If they want to influence the negotiations, they and their members generally lobby governments directly (particularly the British and Norwegian).

The Comité Maritime International (CMI) is unique among the "consultative" organizations at IMCO. Not only does it represent the entire range of interests involved in shipping, but it has also formulated many of its own conventions. Indeed, this is an organization with a great history of its own dating back to 1898. Conceived as a forum in which to harmonize shipowner and shipuser interests, the CMI has for over eighty years drafted many "private law" conventions on the responsibilities and rights of nongovernmental parties involved in maritime trade.[27] These conventions, upon completion, have been submitted to governments for their approval.

---

elected members came from the U.S. (six), Europe (four), Japan (one), and Brazil (one). On substantive issues each member has one vote, but on constitutional issues voting is weighted according to volume of oil transported by sea. The financial headquarters of OCIMF is in Bermuda. However, its operational office is in London since its representatives must constantly attend IMCO meetings. The executive secretary in 1978 was C. A. Walder, who is a British national and was between 1963 and 1970 the Marine Manager of Esso (U.K.).

27. It has dealt with such matters as the rules on collisions, salvage, limitations of liability, bills of lading, maritime liens, arrest procedures, and the liability of operators of nuclear ships. For a brief history of the CMI, see Albert Lilar and Carlo Van den Bosch, *Le Comité Maritime International, 1897–1972* (Belgium: Comité Maritime International, 1972). Lilar was the president of the CMI for much of its recent history.

Such widely accepted maritime laws as the "Hague-Visby Rules" have emanated from the CMI. As IMCO's work in pollution control began, in 1967, to stray into the field of "private" law, the relationship between the two bodies became a competitive one. This was so in the negotiations on pollution liability that followed the *Torrey Canyon* incident. At that time many maritime states wanted the CMI—and not IMCO—to deal with all legal matters touching on private (that is, nonstate) interests, although the debate was soon resolved in IMCO's favor.

Another organization concerned with IMCO's work on environmental protection is the International Association of Classification Societies (IACS). It is composed of the nine largest classification societies (the largest being Lloyd's Register of Shipping and the American Bureau of Shipping), whose tasks are to assure that vessels are built to particular specifications and to certify that they meet stipulated construction and equipment standards throughout their years at sea.[28] In carrying out periodic inspections and issuing certificates, they operate as agents of both shipowners and flag states. They are run by committees which are largely composed of shipowners (other representatives being insurance officials and naval architects), but the interests of their permanent staffs do not always coincide with those of shipowners. The major function of IACS (formed in 1968) has been to unify rules so as to assure vessel safety and to prevent competition based on lower or laxer requirements. Its representatives have attended most meetings of the MSC and MEPC, and it was very active in the 1978 Tanker Safety and Pollution Prevention Conference—especially in the negotiations on the inspection and certification of tankers. The classification societies also exert considerable influence through the participation of many of their officials on national delegations.

One transnational environmental group, the Friends of the Earth (FOE), has had consultative status at IMCO since 1973. Its influence in the Legal Committee, MEPC, and conferences has been negligible as a result of infrequent attendance and inadequate expertise. At the same time, national environmental groups do exercise a tremendously important influence on IMCO's work through their impact on the formulation and domestic acceptance of certain government policies, particularly those of Great Britain in the 1950s and early 1960s and the United States in the 1970s. They have not, however, directly participated in the organization in a "consultative" role, so the representation of environmental interests has been left to governments.

28. The other seven members are Bureau Veritas (France), Det Norske Veritas (Norway), Germanischer Lloyd, Nippon Kaiji Kyokai, Polish Register of Shipping, Register of Shipping USSR, and Registro Italiano Navale. For additional information, see "IACS," *Surveyor* (American Bureau of Shipping), November 1977.

OTHER ORGANIZATIONS AND THEIR RELATIONS TO IMCO

A vast number of United Nations agencies and interagency and special project bodies are concerned with marine pollution. Each deals with a different aspect of the problem or, at least, has a different approach to it. Indeed, there seems to be an almost meaningless scramble of organizations, leading one author to comment,

> . . . the proliferation of organizations, proposals, and their attendant maze of alphabet soup acronyms . . . is a symptom of a lack of political will and the inability of the sovereign nation-state system to adapt to today's needs. If governments were serious about international environmental reform, and generally recognized the long-term hazards of marine pollution they would choose an organization and use it. . . . The international organizations cannot be blamed for the confusion; they have only done what has been expected of them.[29]

It is not the purpose of this section to explain this proliferation of activity. Rather, it will review the activities of a number of bodies concerned with the marine pollution problem in order to place IMCO's activities in an organizational context.

The United Nations Conference on the Law of the Sea (UNCLOS)

IMCO's work is intricately interwoven with the broader concerns of the United Nations Conferences on the Law of the Sea. From the first two conferences in 1958 and 1960 to the most recent one (which held its seventh session in the spring, 1978), the work of UNCLOS has coincided with the nearly twenty-year life span of IMCO. The work of the two bodies is distinct, however.

The law of the sea conferences are temporary U.N. bodies convened under the authority of the General Assembly to redraft as necessary the basic "constitutional" law of the oceans. As such, the work of UNCLOS is most concerned with allocating jurisdictional rights and duties among states and not with particular technical questions of implementing these rights. In the marine pollution area, UNCLOS III has been almost solely concerned with shipping and with the need to divide jurisdiction over the setting and

29. Robert A. Schinn, *The International Politics of Marine Pollution Control* (New York: Praeger, 1974), p. 125. For an analysis of U.N. agencies in this field and a prescription regarding their organization, see Nancy D. and Christopher C. Joyner, "Prescriptive Administrative Proposal: An International Machinery for Control of the High Seas," *International Lawyer* 8 (January 1974), pp. 57–73. For a description of these bodies, see "Annotated Directory of Intergovernmental Organizations Concerned with Ocean Affairs," U.N. doc. A/CONF.62/L.14, August 10, 1976.

enforcement of antipollution standards among flag, coastal, and "port" states.

In drafting specific technical rules for flag and coastal states alike, IMCO therefore operates within the overall legal constraints set by the law of the sea. UNCLOS decisions do not dictate the details of IMCO's regulatory work, but they do help to shape them. At the same time, IMCO is itself frequently drawn into the larger jurisdictional debates that are supposedly the sole prerogative of the UNCLOS forum. Between sessions of UNCLOS, states often seek favorable interpretations or applications of existing law of the sea by IMCO. Given the powerful precedent any IMCO actions would set, a complex interrelationship exists between the two bodies. (See Chapter 6 for a fuller discussion.)

The United Nations Environment Program (UNEP)

The United Nations Conference on the Human Environment was held in Stockholm in June 1972 after four years of preparation. Although the conference covered all forms of pollution, marine pollution was given special treatment. Clearly an international issue, it was becoming an increasingly grave problem in which there was a definite "absence of effective coordinating mechanisms at both the national and international levels."[30] In addition, therefore, to producing a multiplicity of principles, declarations, and recommendations, the conference highlighted the need for a new intergovernmental body to coordinate the actions of the existing U.N. agencies.[31]

The United Nations Environment Program (UNEP) was created for this purpose. Acting not as a new specialized agency but as a "switchboard" with a "coordinating and nonoperational" function, the new organization is composed of a secretariat, various committees, and a fifty-eight member governing council through which direct access to government members is maintained. Through its executive director, the organization has responsibility for coordinating environmental programs in the United Nations system under the guidance of the governing council. The executive director also administers the multipurpose Environment Fund.[32]

Like the United Nations itself, UNEP lacks the authority to order the existing U.N. specialized agencies to follow particular courses of action or

30. L. G. Engfeldt, "The United Nations and the Human Environment—Some Experiences," *International Organization* 27 (summer 1973), p. 408.

31. L. B. Sohn, "The Stockholm Declaration on the Human Environment," *Harvard International Law Journal* 14 (summer 1973), pp. 423–515.

32. Maurice Strong, "Concept of UNEP as Leader and Catalyst," *U.N. Chronicles* 12 (May 1975), p. 34; Engfeldt, "United Nations," p. 408.

to collaborate with one another in prescribed ways. It has therefore sought to find and fill gaps outside the programs of present agencies, to encourage interagency cooperation on existing or new programs, and to encourage the implementation of regulations approved by the agencies.[33] One area where UNEP has furthered coordination of existing programs and added some modest elements of its own is environmental monitoring. As part of its Earthwatch program, it has moved quickly to establish both the Global Environmental Monitoring System (GEMS) and the International Referral System (IRS). As well, UNEP encourages the conclusion of international and regional agreements for the world's oceans and gives financial assistance to a variety of programs through its Environment Fund, which is generally aimed at getting agencies to initiate new programs or alter environmental priorities.[34]

During its first few years of operation, UNEP has had some impact on IMCO. A jurisdictional dispute arose as a result of UNEP's promotion of regional pollution accords, especially in the Mediterranean area. This has largely been resolved by a recognition that the regional accords would be carried out in consonance with IMCO's global regulations and responsibilities in the area of ship-source pollution. A formal understanding to this effect was reached between the two agencies in 1976, so that UNEP's role is to act as "catalyst and coordinator" and IMCO as the "substantive agency for programme implementation."[35] UNEP is a source of funds for IMCO's environmental technical assistance activities, although the two organizations did conflict over their rights to send advisers to assist countries with vessel-source pollution (a responsibility now recognized as IMCO's). Furthermore, UNEP has actively urged member states to adopt IMCO regulations incorporating higher environmental standards and has backed certain proposals under consideration by IMCO bodies. For example, not only did UNEP's then executive director, Maurice Strong, address the

33. David S. Zalob, "The U.N. Environment Programme: Four Years after Stockholm," *Environmental Policy and Law* 2 (June 1976), p. 50.

34. See, for example, "The Environment Program—Report by the Executive Director," U.N. doc. UNEP/GC/14/Add.2, December 1973, p. 102. See also IMCO doc. MEPC I/Inf.B(6), March 1974. Interviews.

35. "Memorandum of Understanding Concerning Cooperation between the Inter-Governmental Maritime Consultative Organization and the United Nations Environment Programme," November 9, 1976 (issued by IMCO). Seven areas of "mutual interest" are outlined including international conventions, regional action plans, national legislation, and technical assistance. Also see the 1977 documents A. X/22/1 and MEPC VI/5. Information on the Mediterranean dispute and its resolution was obtained through interviews with governmental and international officials. On the Mediterranean programs, see U.N. docs. UNEP/WG2/5 (1975), UNEP/1G.11/3 (1977), and UNEP/1G.11/4 (1978); and Peter H. Sand, "Protection of the Marine Environment," *Environmental Policy and Law* 1 (February 1976), p. 154. See also MEPC III/4 on the cooperation for the Mediterranean agreement. IMCO contributed a draft protocol on the prevention of oil pollution.

delegates at the beginning of IMCO's 1973 Conference, but the organization also submitted a memorandum commenting on specific proposals at the conference.[36] To date it has not had noticeable influence on states' policies toward IMCO conventions, but it has been given little leverage to do so.

United Nations Development Program (UNDP)

UNDP was formed in 1965 when the General Assembly decided to merge the Expanded Program of Technical Assistance (EPTA) with the United Nations Special Fund. It primarily provides funds for technical assistance and preinvestment surveys, and in 1973 its $313 million budget constituted 80 percent of the assistance which the U.N. agencies provided.[37] It has administrative responsibility for the overall supervision of the money it spends, but it relies on the specialized agencies for the "execution" of particular projects. The agencies are represented on an advisory body, the Interagency Consultative Board.

IMCO is an executing agency of the UNDP responsible for technical assistance in the field of shipping, and it receives most of its technical assistance funds from that source. In this reliance on the UNDP, IMCO is fairly typical of United Nations agencies which, with the exception of the World Health Organization, do not allot funds from their regular budgets for technical assistance. It is particularly in the field of navigational expertise and training that UNDP concentrates its funds, with funding for pollution-control projects coming from UNEP's Environment Fund. UNDP-funded projects do not, therefore, have as a *primary* purpose the control of pollution, but insofar as they foster expertise on shipping matters within recipient states, they can have important environmental consequences.

United Nations Conference on Trade and Development (UNCTAD)

The first United Nations Conference on Trade and Development was held in Geneva from March to June 1964, and it led to the establishment by the General Assembly of a permanent body with the same name. Its principal objective is the international redistribution of income through changes in the international trading system.[38] Since its inception, it has been the

36. MP/CONF/INF.16.

37. *Yearbook of the United Nations 1973* (New York: United Nations, 1976), p. 327. IMCO received only $1.7 million. The only U.N. specialized agency receiving less was the Universal Postal Union, which got $1.6 million.

38. R. N. Gardner, "The United Nations Conference on Trade and Development," in R. N. Gardner and M. F. Millikan, eds., *The Global Partnership: International Agencies and*

most important international agency through which the developing countries have tried to alter international economic relations in their favor.

It has been as a result of the creation of the Committee on Shipping in 1965 that UNCTAD has had an influence on IMCO. Within the committee the main objectives of the developing countries have been to reduce transportation costs and freight rates and to "increase their participation in the shipping industry by acquiring new ships and gaining admission to liner conferences."[39] In order to realize these goals the developing states have sought a voice in framing new international legislation on shipping as well as increased technical assistance to develop their own merchant fleets. UNCTAD's tasks on these issues clearly overlap with functions which had originally been delegated to IMCO in its constitution but which, as a result of the opposition of the major maritime states, had been excluded from the agenda of the organization. In fact, the creation of the Committee on Shipping can in large measure be attributed to the hiatus existing in the U.N. system as a result of the failure of IMCO to concern itself with the "commercial" aspects of shipping.[40]

IMCO's Secretariat did not attempt to protect its legal jurisdiction in this field—it would have been a bitter and futile task—but sought instead to guarantee the reservation to IMCO of any activities with respect to maritime safety and navigation. This was not an area of particular interest to the leaders of UNCTAD, and it was possible for the secretaries-general of IMCO and UNCTAD to agree informally, in autumn 1965, to a functional division of powers between the two organizations.[41] This accommodation has worked reasonably well, although disputes (for example, rules concerning containerization) still do occur. The demands of the developing states in UNCTAD for assistance in building their shipping industries have encouraged IMCO to develop its own technical assistance program. To this end, IMCO has an agreement with UNCTAD on the types of assistance which each organization will provide.[42] Recently, both organizations have

---

Economic Development (New York: Praeger, 1968), p. 123. For a recent study comparing IMCO and UNCTAD (and UNCITRAL), see Evan Luard, *International Agencies: The Emerging Framework of Interdependence* (London: Macmillan, 1977), pp. 44–62.

39. "Shipping," *International Conciliation* 579 (September 1960), p. 164.

40. R. L. Friedheim, "International Organizations and the Uses of the Oceans," in R. S. Jordan, ed., *Multinational Cooperation* (London: Oxford University Press, 1972), p. 229; Iqbal Haji, "UNCTAD and Shipping," *Journal of World Trade Law* 6 (January-February 1972), pp. 58–118.

41. While UNCTAD would be concerned with economic/commercial matters, IMCO's area of responsibility would be technical/navigational concerns. There have been some conflicts over this division of responsibilities. See Silverstein, *Superships and Nation-States*, pp. 131–133. For a recent accord, see the 1977 document A X/22, annex 7.

42. *Technical Assistance in Maritime Transport* (London: IMCO/UNCTAD Publication, 1974). The 1965 accord was an adoption of a previous U.N.-IMCO agreement.

participated in a UNDP-funded program to develop national shipping fleets and crews in the Caribbean area.[43]

United Nations Education and Scientific Organization (UNESCO)

UNESCO's activities in the marine pollution field are carried out by the Intergovernmental Oceanographic Commission (IOC), a semi-autonomous subsidiary body of UNESCO. Basically a scientific research organization, it promotes and coordinates scientific investigation and monitoring of the oceans, the international exchange of oceanographic data, and education in ocean research. It has contributed to the drafting of a convention creating a large monitoring system, Ocean Data Acquisition System (ODAS), and has sponsored the International Decade of Ocean Exploration as the "accelerated phase" of the Long-Term Expanded Program of Ocean Research (LEPOR).[44]

In order to facilitate the participation of other U.N. specialized agencies in research projects which are a part of LEPOR, an Inter-Secretariat Committee on Scientific Programs Relating to Oceanography (ICSPRO) was created. IMCO has been a member of this committee and had maintained one professional officer fulltime in the Secretariat. An IMCO-IOC committee also is studying the legal implications of the Oceans Data Acquisition System, and IMCO is participating in two of the IOC's major undertakings—the Global Investigation of Pollution in the Marine Environment (GIPME), which is a framework for the conduct of baseline studies by governments on the extent of pollution and its effects, and the Integrated Global Ocean Station System (IGOSS), which includes a pilot oil-pollution-monitoring project.[45]

Food and Agriculture Organization (FAO)

The interest of the FAO in marine pollution is rooted in its concern for ocean fisheries. To date, this has primarily manifested itself in research into

43. An interesting side effect of the advent of the Committee on Shipping was that many officials from developed maritime states, having observed the increasing stridency of the developing states on shipping questions, hesitated to restrict IMCO in other areas. This was particularly evident in the late 1960s when the Comité Maritime International was hoping to exclude IMCO from considering some areas of private law having an impact on environmental issues under IMCO's auspices. See A. Mendelsohn, "The Public Interest and Private International Maritime Law," *William and Mary Law Review* 10 (summer 1969), p. 811.

44. U.N. doc. UNEP/GC/14/Add.2, December 1973, p. 100. See also UNESCO *Courier*, January 1977.

45. See MEPC I/8/Add. 1 (February 25, 1974), MEPC III/4/2 (May 16, 1975), and MEPC III/4/3 (June 11, 1975).

the effects of various pollutants on fish, leaving the issue of the control of the pollutants themselves to governments and to international agencies such as IMCO. In 1970, it held a Technical Conference on Marine Pollution and Its Effects on Living Resources and Fishing in Rome, at which some four hundred delegates from all over the world reviewed many diverse aspects of the pollution/fisheries problem. The FAO also has an Advisory Committee on Marine Resources Research (ACMRR), which has worked in conjunction with the IOC and UNEP in making a detailed survey of living marine resources threatened with depletion. The results of such research can be very influential in IMCO's pollution-control policies, although FAO has little direct interaction with IMCO. FAO has, however, sent observers to IMCO meetings dealing with pollution more frequently than any other specialized agency.

World Meteorological Organization (WMO)

The WMO is a U.N. specialized agency which was established to expand and improve meteorological services and to encourage research and training in meteorology. With regard to the environment, it is involved in monitoring the atmosphere, assessing the level and effects of atmospheric pollution (on the oceans as well as land), and examining the effects of weather on the oceans. Its importance to IMCO lies in assessing the nature and behavior of marine pollutants from varous sources. The WMO is cooperating through its Executive Committee Panel on Meteorological Aspects of Ocean Affairs in UNEP's Earthwatch program. This includes monitoring "marine pollution, particularly oil pollution from ships. . . . The relevance of WMO's activities in this connection arises from the relationship between movement and dispersion of oil slicks and current weather conditions."[46] Although WMO's work clearly has much relevance to IMCO, there is little direct contact with IMCO except through joint participation in ICSPRO and GESAMP and in UNESCO's projects GIPME and IGOSS. At the same time its work on the monitoring of pollutants does affect perceptions of the character of the pollution problem and the requirements for enforcing pollution regulations.

International Labor Organization (ILO)

The ILO's maritime program is concerned primarily with the conditions of work of ships' officers and crews and with raising crew standards.[47] It is

46. D. A. Davies, "The Role of the WMO in Environmental Issues," *International Organization* 26 (spring 1972), pp. 327–336. See also U.N. doc. UNEP/GC/14/Add.2, December 1973, p. 104, and IMCO docs. MEPC I/8/1 (March 6, 1974) and MEPC III/4/3 (June 11, 1975) outlining WMO's activities.

47. *Winds of Change* (Geneva: International Labour Organization, 1971); *Report of the*

this latter matter, in particular, that is relevant to vessel safety and pollution prevention and, since 1964, there has been a joint committee of the ILO and IMCO which has adopted guidelines on crew training and standards. Although only guidelines, they have been accepted as basic standards in many countries. The committee also has been empowered since 1974 to try to resolve differences between the two organizations. Through its participation on the joint committee, the ILO was actively involved in IMCO's development of the 1978 Convention on Training and Certification of Seafarers, the first comprehensive convention of its kind. It also maintains its own continuing program on the question of substandard crews and ships.[48]

Joint Group of Experts on the Scientific Aspects of Marine Pollution (GESAMP)

GESAMP was set up in 1968/69 as a U.N. interagency advisory body (without independent authority) to undertake studies on marine pollutants and to recommend approaches to their control. Its initial members were IMCO, FAO, UNESCO, and WMO, and they were later joined by IAEA, WHO, UNEP, and the U.N. itself. Its Administrative Secretariat is at IMCO in London. An administrative secretary at IMCO coordinates its operation.

With the creation of GESAMP, it was recommended that participating agencies disband their own scientific advisory bodies and transfer their functions to GESAMP in order to avoid duplication of efforts. As a result, scientists appointed by the various agencies now meet within GESAMP in their individual capacities and work on specific problems presented by the sponsoring agencies. With a budget of approximately $50,000 the scientists meet only about five days annually, but intersessional working groups do communicate a great deal by written correspondence. Ostensibly, at least, GESAMP is the major scientific advisory body on marine pollution for all the participating agencies, and it therefore has an important coordinating function. One commentator has advised that "any movement to increase GESAMP's scope and authority (especially to link it permanently with decision-making mechanisms) increases rational policy and decreases political irresponsibility. Environmental policy making is only as good as its primary scientific information system."[49]

GESAMP has undertaken work on the impact of oil in the marine environment (at the request of the FAO), coastal water quality criteria, the

---

Director General: Report I, International Labour Conference, 62nd (Maritime) Session, 1976, (Geneva: International Labour Office, 1976), pp. 67–69.

48. See MEPC VI/10/1 (October 5, 1976).

49. Schinn, *International Politics*, p. 124.

disposal of wastes by dumping, and pollution arising from exploration and exploitation of the seabed.[50] It also undertakes a critical review of methods used in its field and maintains an up-to-date list of harmful chemical substances for incorporation into the 1973 Convention for the Prevention of Pollution from Ships. IMCO has relied on this work for the identification of those noxious and hazardous cargoes which are potential pollutants. GESAMP remains the central scientific body for IMCO's work in the pollution field.

These international bodies are not the only ones impinging on the pollution control activities of IMCO, but they are certainly the most important global ones.[51] Of the nine, the activities of four (UNESCO, FAO, WMO, and GESAMP) are relevant almost solely to the acquisition of knowledge on the marine pollution problem. A fifth (UNDP) is important for the technical assistance function of IMCO, but, as noted, it is less important than UNEP in providing assistance specifically for environmental protection.

Three of the four remaining organizations impinge mainly on IMCO's rule-making function. UNCTAD's Committee on Shipping could stray into the setting of norms on technical shipping matters which impinge on pollution control, and the ILO's concern for the training and competence of seamen could lead to some jurisdictional conflict with IMCO. Both are matters requiring continuing coordination between the agencies, but the basic framework of this coordination does seem to have been agreed on. The single organization whose functions overlap most importantly on a continuing basis with those of IMCO is UNEP. Its monitoring activities, technical assistance program, support for comprehensive regional pollution arrangements, and coordinating role all provide the basis for a rich continuing relationship with IMCO.

In recent years there has been some overlap between the activities of IMCO and the U.N. Law of the Sea Conference with respect to environmental rule-making—but largely in the areas of standard-setting and

50. See IMCO doc. MEPC V/4/1 (May 6, 1976).

51. Two organizations of developed Western states have had some influence. They are NATO's Committee on the Challenges of Modern Society (CCMS) and the Organization for Economic Cooperation and Development (OECD). The CCMS has discussed different regulatory approaches to marine pollution and has sponsored technical studies. The OECD has promoted research and the formulation of general principles on transfrontier pollution (which have some limited applicability to ocean pollution). It also has completed a significant study on flags of convenience. M. Sudarakis, "NATO and the Environment: A Challenge for the Challenger," *Environmental Law and Policy* 2 (June 1976), pp. 69 ff; R. Train, "A New Approach to International Cooperation: The NATO CCMS," *University of Kansas Law Review* 22 (winter 1974), pp. 167–191; Patrick Kyba, "CCMS: The Environmental Connection," *International Journal* 29 (spring 1974), pp. 256–267; Gunnar Randers, "NATO's International Governmental Cooperation in Environmental Management," Allen V. Kneese et al., eds., *Managing the Environment: International Economic Cooperation for Pollution Control* (New York: Praeger, 1971), p. 343; and S. McCaffrey, "The OECD Principles Concerning Transfrontier Pollution: A Commentary," *Environmental Law and Policy* 1 (June 1975), pp. 2–7.

enforcement jurisdictions and not on technical shipping regulations. While jurisdictional matters have, as a rule, been thought by governments not to be within IMCO's purview, there is always the tendency to seek short-term gains on specific issues. Indeed it is almost inevitable that the organization will become involved in such matters, for the law of the sea is always developing. However, wherever the need for broad law-of-the-sea revisions is perceived and work undertaken in a U.N. conference, IMCO's more limited authority circumscribes its activities (see Chapter 6). But when UNCLOS III is terminated, demands will arise again for IMCO to act in the face of ambiguities or omissions in the law of the sea.

PART TWO

The Formulation of the Environmental Regime

Chapter IV

Discharge Standards and Pollution-Control Technologies

> Two slicks of "dirty brown" oil fouled the waters off an 85-mile stretch of the Florida keys only a few miles offshore from the nation's only living coral reefs. The oil—probably pumped overboard during the night when a passing tanker flushed out its hold—stretched from Key Largo to Marathon and from Big Pine to Key West.
>
> No ships responsible for the spill had been located, although Coast Guard planes were scouting the Gulf of Mexico and the Atlantic Ocean as far north as Cape Canaveral for suspect vessels. . . . It was "unlikely" that the ship or ships responsible for the spills would ever be found.
>
> ("Finding Polluter of Keys Unlikely, Coast Guard Says," *Miami Herald*, April 30, 1977, pp. 1 and 144)

From the Persian Gulf to the coast of Florida, such spills are familiar flotsam trailing in the wake of the tanker trade. Unlike the more dramatic shipping accidents, this pollution is "operational," just another part of the business. It is also, as we have seen, the single most significant source of ship-generated oil pollution and the main focus of IMCO's long history of environmental regulation. In contrast, accidental pollution has been a topic of more recent vintage and has been treated largely as a by-product of regulations on navigation, safety of life at sea, crew training, and traffic control. The control of operational pollution will be the central concern of this chapter, although some recent attempts to stem accidental spills also will be considered.

Attempts to conclude an international agreement to control ocean oil pollution extend back over fifty years. In a pattern that was to become remarkably consistent over time, the first international conference occurred in 1926 in response to one state's concern with its own coastal problem—in this case, the American government's opposition to operational discharges by nontankers off its east coast.[1] Such was the chronic state of this pollution

1. The following information on the negotiations during the 1920s and 1930s is largely taken from Sonia Zaide Pritchard, "The International Politics of Oil Pollution Control: 1920 to the Load-on-Top System" (Ph.D. dissertation, Department of International Relations, London School of Economics, 1976), parts 1–3. The texts of the conventions analyzed in this

that after World War I, demands for national action escalated sharply and forced the passage by the United States Congress of the Oil Pollution Act of 1924. Realizing the international ramifications of the issue, President Harding also authorized his secretary of state to convene the first international conference on the problem. It was held in Washington in June 1926, and its purpose was not to produce an international treaty but to formulate a draft convention for submission to a subsequent international meeting.

The American proposal was to limit the rate of operational discharges throughout the oceans, and it did have some support. Both Canada and the United Kingdom recognized the legitimacy of the American environmental concerns, and they joined with it in advocating the installation of the necessary technological equipment (oily-water separators) on existing ships. However, no other states suffered from similar pollution (or if they did, were not under any compelling pressure to do something about it), and they opposed the substantial economic costs that the proposal would have entailed. As a result, the conference failed to back the U.S. position, although a compromise was accepted which prohibited polluting discharges with an oil concentration above 500 ppm within zones extending fifty miles from coastlines. It was also agreed at the meeting that the U.S. would call a second conference to incorporate the compromise into an international treaty.

Prospects for the acceptance of such fifty-mile zones in a treaty seemed quite good at the end of the 1926 meeting, but support for it soon began to dissipate. A zonal approach would still have required extra on-board equipment, and the industry was not enthusiastic about that.[2] Even the support

---

chapter can be found in several sources. The 1954 International Convention on the Prevention of Pollution of the Sea by Oil is in 327 *United Nations Treaty Series* (UNTS) 3; the 1962 amendments to the 1954 convention—600 UNTS 332; the 1969 amendments to the 1954 convention—9 *International Legal Material* (ILM) 1; the 1971 amendments to the 1954 convention—9 ILM 25; the 1973 International Convention for the Prevention of Pollution from Ships—12 ILM 1319; and the 1978 Protocols to the 1973 convention and to the 1974 International Convention on Safety of Life at Sea—17 ILM 546.

2. The technological requirements for zonal and complete "prohibitions" for tankers and nontankers did not change from the 1920s through the early 1960s. A knowledge of these will facilitate an understanding of the negotiations analyzed in this chapter. Where there was a requirement for controlled discharge by *nontankers* (for example, a specified parts per million of oil in an effluent) throughout the oceans (commonly but inaccurately referred to as *complete prohibition*), oil residues would have to be retained on board. This would require that the vessels have oily-water separators and holding tanks on board and that ports have oil "reception facilities." The same requirements existed in the case of *wide prohibition zones* if, as was sometimes the case, nontankers could not discharge oily ballast water in their fuel tanks until they reached port. This would also hold for coastal trades where the vessels never left the zones. Since the 1960s most nontankers have been built so as not to have to use fuel tanks for ballast, so this problem is disappearing. However, nontankers are still faced with the problem of oily bilge water for which they still generally require a small separator and holding tank.

In the case of *tankers*, compliance with a *complete prohibition* would have required (at least

of the American government began to wane as a result of some positive effects from its own 1924 legislation and from the "voluntary cooperation" of its national shipowners. Moreover, experiments conducted at the time by the American Bureau of Standards seemed to indicate that previous concerns about the persistence of oil in the sea were exaggerated. The Americans never called the subsequent meeting to approve a treaty, and the compromise agreement did not become international law.

The 1926 exercise did produce some beneficial results, however. Upon a request by the British government to the International Shipping Conference (an organization of national shipowner associations), the shipowners of seven major maritime countries *voluntarily* accepted a fifty-mile discharge prohibition zone. It was not an ambitious start to international collaboration, but it was a first step—and the only one for the next twenty-eight years.

One additional faltering attempt was made in the interwar period, and this time as a result of a British initiative. Like the United States, Britain soon found its coastline becoming so polluted that public outcry demanded reform. In 1934 it called on the ill-fated League of Nations to promote an international accord on oil-pollution control, and it created an ad hoc committee of experts to suggest possible courses of action. Again, Britain backed the old—and expensive—proposal for controlled discharges throughout the oceans. This required the installation of oily-water separators on ships and waste "reception facilities" in ports, and again the majority of participants sought the less expensive solution (at least for certain vessels) of controlled discharge in coastal zones. A draft convention was circulated to governments in November 1935, but larger events intervened: Germany and Italy refused to attend any future pollution conference. These refusals doomed any prospect for international agreement, as no other maritime state (including Britain) would sign an accord to which all major maritime powers did not adhere.

Intergovernmental discussions began again only after the Second World War when, in 1949, the Transport and Communications Commission of the U.N. Economic and Social Council reviewed the activities of the League in order to determine the scope and ramifications of the oil pollution problem.[3] Even at that time, the United Nations secretary-general

until the late 1960s) extensive reception facilities at loading ports where they could have discharged their residues. Strict adherence would also have demanded oily-water separators, although this depended on the allowable concentration of oil in a discharge. However, compliance with *wide coastal prohibition zones* did not generally demand onboard separators and reception facilities since the tankers could clean some tanks at sea and fill these tanks with ballast before entering the coastal zones. One major exception was tankers involved in coastal trades which stayed within zones.

3. The information on the United Nations' activities and the events leading up to the convening of the 1954 Conference are taken from "Introductory Memorandum Circulated by

pressed the case for international action, but the Commission concluded that it would be premature to attempt an international conference. Instead governments should form their own national committees to examine the problem, while the Commission would set up an international committee of experts to evaluate alternative courses of action.

Throughout these rather removed bureaucratic discussions, British concern was mounting at a much more immediate level. Governmental officials were painfully aware of increased fouling of local beaches from passing ships and of the killing of thousands of seabirds in the coastal environment. Public outspokenness was growing rapidly and, in 1952, a powerful environmental lobby, the Coordinating Advisory Committee on Oil Pollution of the Sea (CACOPS), was created to pursue the case. Labor M. P. James Callaghan was elected its president, and the organization began actively to lobby government officials. The new organization also sponsored its own conference in October 1953, a gathering which was attended by a large number of British naturalists and environmentalists.[4] By this time, the British government was itself concerned that similar pressures elsewhere could result in uncoordinated regulations by other governments which could hinder free international shipping. And, of course, it felt that any regulations which Britain imposed on its own fleet should also be borne by the fleets of other countries so as not to undermine the competitive position of the British industry.

With these considerations in mind, the British government again seized the initiative for international action, and the Ministry of Transport created a Committee on the Prevention of Pollution of the Sea by Oil under the direction of Percy Faulkner. The *Faulkner Report*, issued in July 1953, presented a thorough analysis of the problem and made many specific recommendations that it argued should be incorporated into an international convention.[5] Coincidentally, it was at the CACOPS conference in October 1953 that the Ministry of Transport chose to announce formally its decision to convene an international conference on the problem. All parties to the 1948 Safety of Life at Sea Convention or state attendants at the safety conference were to be invited, as was every maritime state with a gross registered tonnage exceeding 100,000 tons. Real international action was finally to be taken. Or so it seemed.

the Government of the United Kingdom," 1954 Conference Documents, February 1954, pp. 1-4; and Pritchard, "International Politics," part 4.

4. *Report of Proceedings, International Conference on Oil Pollution of the Sea, 27 October 1953* (London: Coordinating Advisory Committee on Oil Pollution of the Sea, 1954).

5. Ministry of Transport, *Report of the Committee on the Prevention of Pollution of the Sea by Oil* (London: Her Majesty's Stationery Office, 1953) (hereinafter cited as the *Faulkner Report*).

THE 1954 CONVENTION ON THE PREVENTION OF POLLUTION OF THE SEA BY OIL

The three-week conference was convened on April 25, 1954 with thirty-two states, representing 95 percent of world shipping tonnage, in attendance.[6] Again, the basic British proposal was for a ''general prohibition'' (controlled discharges throughout the oceans) on all operational discharges by both tankers and nontankers. Only discharges with an oil concentration of less than 100 parts per million (ppm) were exempted. This position had grown directly out of the recommendations of the *Faulkner Report*, a summary of which the British delegation circulated to all participants at the conference.[7] Indeed, prior to the conference, Mr. Faulkner and other officials of the Marine Division of the Ministry of Transport had undertaken extensive negotiations to persuade other delegations of the wisdom of their broad approach.[8] Domestically, the extensive British proposal did have at least superficial industry support,[9] despite the projected cost for the oil industry of building reception facilities at loading terminals and the obvious delays that would be encountered by tankers and nontankers at reception facilities in ports.[10] At the conference, however, the British delegation was

6. Of the thirty-two states in attendance, eighteen were developed states from Western Europe, North America, and Australasia; three were from Eastern Europe; four from Asia; six from Latin America; and one from Africa.

7. ''Introductory Memorandum Circulated by the Government of the United Kingdom,'' 1954 Conference Documents, February 1954. Unlike present-day conferences, there was cnly a single document circulated to participants prior to the 1954 Conference, and there were no formal preparatory meetings. (All 1954 Conference documents are available in typescript in the IMCO library.)

.8. These negotiations included a visit to Washington by Mr. Faulkner and his aides. The American attitude was not positive, given the local success of their voluntary zonal prohibition, and their lack of support was fatal to the British proposal. Interview, Sir Gilmore Jenkins, head of U.K. delegation at 1954 Conference.

9. Oil industry support for the building of tanker reception facilities (see *Faulkner Report*, p. 28) did not indicate a firm commitment to the British proposal given the fact that the British oil companies (B.P. and Shell) did not fully own the loading terminals but only shared them with non-British companies whose lack of support for the position rendered its acceptance unlikely. See A. Logan, ''The Working of the International Convention from the Point of View of British Tanker and Oil Companies,'' *Report of the Proceedings, International Conference on Oil Pollution of the Seas, 3-4 July 1959, Copenhagen* (London: Advisory Committee on Oil Pollution of the Seas), pp. 34-35. On the other hand, nontanker owners genuinely supported the British conference proposal for the installation of oily-water separators given the imminence of legislation to require them on all *British* ships—the Oil in Navigable Waters Act of 1955. International acceptance of these separators would protect the British ships from a competitive disadvantage. D. C. Haselgrove, ''The Oil in Navigable Waters Act of 1955,'' *Report of the Proceedings, International Conference on Oil Pollution of the Seas, 3-4 July 1959, Copenhagen* (London: Advisory Committee on Oil Pollution of the Seas).

10. The delays in port associated with the use of reception facilities have been a major deterrent to the use of this procedure. A ship's master is responsible for getting the cargoes loaded and delivered on time and is loath to be shunted from dock to dock in order to discharge the oily residues. As a result, there is a tendency for these to be flushed out with the ballast outside the port.

predictably unsuccessful, having to fall back yet another time. The proposed compromise called for an overall fifty-mile "prohibition zone" and some larger regional prohibition zones[11] and a general ban on discharges by tankers which were going to ports with reception facilities. (Both the terms *general prohibition* and *prohibition zones* are misleading in that the proposals permitted discharges under 100 ppm in either the entire ocean or the designated coastal areas.)

The debate on this new package focused on a number of contentious issues: the seriousness of the problem, past experience with zones, the technical feasibility of suggested preventive measures, and cost.[12] With regard to the problem itself, it is interesting that experience with substantial pollution did not seem to have been a determinative influence on the positions of participants. Indeed, of the eleven states which later supported a general prohibition, only five of them (Germany, the U.K., the Netherlands, Ireland, and Israel) described the pollution of their own coasts as serious, while two of the states in the opposing coalition, Denmark and Sweden, complained about extensive coastal pollution. Indeed, the Danish spokesman commented, "Oil pollution had had deplorable effects along a great deal of the Danish coast and the position was going from bad to worse. Last summer most beaches were unusable because of the nuisance and holiday resorts had suffered economic loss."[13] Conversely, many supporters of the general prohibition—New Zealand, Canada, Poland, and the USSR—reported that oil pollution was not a significant problem on their coasts.

Despite this anomaly, debates on the *nature* of the oil pollution problem in general revealed a parallel between the alternative a state favored at the meeting and the information they seemed to possess on the persistence, behavior, and effects of oil in the marine environment. For example, the British delegation stressed both the persistence of oil in the oceans and its long-term movement over vast areas, and the Soviet delegate, also a supporter of the general prohibition, commented,

> The Soviet Union has undertaken research work of different kinds and observations, which have proved that oil and oil residues have a pernicious effect on fish, and particularly on fish roe and spawn. Roe and spawn perish in oil-contaminated water. Oil pollution of breeding

11. The British wanted regional prohibition zones in the Atlantic to 40° west off the British Isles, in the North Sea, and in an area off the Australian coast adjacent to the Great Barrier Reef.

12. The debates frequently pitted Britain against the other major shipowning states—France, United States, Norway, and Denmark.

13. General Committee, April 27, 1954, p. 3. Denmark opposed not only a general prohibition but also broad zonal prohibitions and the suggested regional prohibition in the North Sea. It did, however, support an extended 100-mile zone off its own coast.

areas hampers the reproduction of fish and results in a decrease in their number. Fish change their migration routes, keeping away from oil-polluted areas.[14]

These views were challenged by those favoring a less radical solution. The Americans and particularly the French questioned the validity of British laboratory experiments and noted that no scientific evidence existed on the movement of oil in the oceans and on its harmful effects on marine life.[15] The British response to this criticism was to fly delegates to its western coastline to witness the fouling of its beaches. If domestic experience had little effect on states' policies, this predictably had even less. Only one delegation,[16] the Dutch, were converted to the British position.

In the main, however, it was differing perceptions of cost/benefit that dictated the choice between a general prohibition and a system of zonal controls. The Americans claimed that the establishment of coastal prohibition zones by voluntary cooperation between the government and industry had proven very effective in controlling the problem, so that no more expensive alternative was justified. The British and some other Western European delegates (all of whose countries bordered waters used by an increasing number of tankers carrying crude oil and by vessels of more varied nationalities) disagreed. The existing system was, for them, quite inadequate.[17] Even though industries and governments would have to bear the cost, it was necessary to do so, said the British delegation, in order to protect the marine environment. Most of the other delegates objected strongly to the imposition of these costs.[18] The Norwegian delegate, for example, believed that shipowners would unfairly have to bear the brunt of the costs. The retention of oil residues on board tankers and their discharge at loading ports would double the time of tankers in ports and force tankers on short voyages to stay at sea for a longer period in order to facilitate an oil/water separation.[19] At the same time, the delegate was so unconcerned about the *social* costs of oil transportation that he deplored the very suggestion that "the shipping industry was being asked to assume a financial

14. General Committee, April 30, 1954, p. 2.

15. Statement circulated at the request of the French delegation, General Committee, April 30, 1954; General Committee, May 5, 1954, p. 5.

16. Statement in the General Committee, May 5, 1954, p. 5.

17. See the debates in the Subcommittee on Zones and American statements in General Committee, April 30, 1954, pp. 6–8.

18. Ten states spoke out against the cost of reception facilities; seven against the cost of oily-water separators; and four against the cost of longer voyages and delays in ports. Proceedings of Subcommittees on Oily-Water Separators, Tankers, Zones, and Port Facilities.

19. Tanker Subcommittee, April 29, 1954, p. 1. Although Britain attempted to rebut these criticisms, there was no conclusive information either way.

burden in order to obviate the financial losses which it had been stated had been suffered by the tourist industry."[20]

Differences over the desirability of a general prohibition for nontankers also mirrored differing perceptions of the capabilities of existing technologies—in particular, oily-water separators. The opponents of a general prohibition argued that the separators were unreliable and, in fact, did not work at all for some oils.[21] Indeed, a demonstration of a British separator resulted in a malfunction that allowed a very visible escape of oil into the Thames—to the acute embarrassment of the British delegation.

Most conference delegates were hostile to both the general prohibition and to the zonal compromise. In the end, the basic fifty-mile prohibition zone was accepted,[22] but two of the three proposed regional prohibitions (in the North Sea and Western Atlantic) were significantly reduced,[23] and a proposed ban on discharges by tankers going to ports with reception facilities was completely rejected.[24] The conference also accepted provisions which effectively reduced both the obligation and the ability of nontankers to comply with any of the zonal prohibitions.[25] Not only were technological requirements for nontankers excluded from the provisions of the convention but, when going to ports without reception facilities, the nontankers would be allowed to discharge in the zones.[26] This was a necessary legal loophole, but it was also a fundamental recognition of the weakness of the regulatory system. When the real product of the meeting was evaluated, the

20. Subcommittee on Oily-Water Separators, April 28, 1954, p. 3.

21. Proceedings of the Subcommittee on Oily-Water Separators.

22. The definition of a prohibition zone (an area in which a vessel cannot discharge oil in concentrations greater than 100 ppm) as well as the exception for nontankers going to ports without reception facilities are contained in article III. The geographical scope of these zones is described in annex A.

23. The opponents of broad regional prohibition zones defeated the proposal for the North Sea and secured in its place a 100-mile zone off its southern perimeter. They also defeated a suggestion of a Western Atlantic zone to 40°W and substituted one that extended only to 30°W and would not necessitate costly detours by ships. The vote on the former was 9-5-9 and on the latter 7-9-13; General Committee, May 10, 1954, pp. 9–10 and 6–7. See Chart 4.

24. Although there were very few ports with reception facilities, opponents interpreted the provision as a first step toward a general prohibition and defeated the resolution in plenary. The vote was 10-14-14; General Committee, May 10, 1954, pp. 5–6. See Chart 4.

25. The weaknesses in the regulations pertaining to nontankers were manifold, and their effect was to emasculate the concept of coastal prohibition zones for these vessels. On the positive side, contracting parties were obligated to prevent oil from the machinery getting into the bilges or to install a small separator to extract the small amounts of oil from the bilge water (article VII). However, the Convention did not require vessels to install separators to handle oily ballast water and only required states to install reception facilities in "main ports," for which no definition was provided (article VIII). The British tried to have such ports defined as those handling more than 750,000 tons of cargo a year (General Committee, May 10 and 11, 1954), but this proposal and any others that would have inserted precise definitions were defeated—thus allowing future parties to the convention to follow whatever course they wanted.

26. Article II(2)(*b*).

only costs imposed by the convention were those resulting from the need for tankers to spend extra time outside the prohibition zones in order to discharge oily ballast water and tank cleanings.

The entire 1954 regulatory system was a reflection of the fact that most governments were still not willing to accept any important control costs themselves or even to impose such costs on their industries.[27] Many were not aware of (or, at least, concerned about) pollution problems, and they did not assign a high priority to the environmental benefits which an effective general or zonal prohibition would bring. But, for the first time, an international convention on oil pollution had been formulated. At long last, governments were being forced to take *some* action at the international level. The extent of their response was not impressive, but the fact that they did respond at all was an important milestone in the development of international environmental regulation.

Through their statements and recorded votes on the central issues, we can identify the major policy orientations of the participants at the conference.[28] Chart 4 summarizes these. Clearly the debate on the desirability of a general prohibition was the central issue during the conference, particularly in its early stages, and to forbid discharges by tankers going to ports with reception facilities would have been a concrete step toward implementing such a prohibition. The acceptance of regional prohibition zones off the western coast of the U.K. and in the North Sea would also have had a comparable, if more geographically limited, effect.

The supporters of the general prohibition were, interestingly enough, a motley assortment of Western European, Commonwealth, and Communist bloc states.[29] Apart, however, from the United Kingdom, Germany, and the Netherlands, the major international shipping states opposed the general prohibition and in this they were supported by the developing states.

27. A number of restrictions on the operation of the Convention deserve mention. First, the Convention is applicable only to discharges of "crude oil, fuel oil, heavy diesel oil, and lubricating oil" [Article I(1)]—thus excluding refined or "nonpersistent" oils. Second, ships being used as naval auxiliaries, ships of under 500 tons, and ships engaged in the whaling industry are not covered by the convention [article II(*i–iii*)]. Third, the Convention does not apply to a vast realm of discharges that are "for the purpose of securing the safety of the ship, preventing damage to the ship or cargo, or saving life at sea" and to "the escape of oil, or of an oily mixture from damage to the ship or unavoidable leakage" [article IV(1) and (*b*)]. Fourth, it does not apply to vessels on the Great Lakes between the United States and Canada [article II(*iv*)].

28. Four indicators have been chosen in determining the policy orientations: (1) statements for or against prohibition; (2) votes on the U.K. resolution to forbid discharges by tankers going toward ports with reception facilities; (3) votes on the U.K. resolution for an extended zone off the British Isles to 40°W; and (4) votes on the U.K. resolution for a North Sea prohibition zone.

29. Poland and the USSR. They were strongly in favor of a general prohibition and only opposed the regional zones because they viewed them as inadequate.

CHART 4

State Positions on General and Zonal Prohibitions at 1954 Conference

	Statement on desirability of a general prohibition	*Vote on U.K. res. to forbid discharges by tankers going to ports with reception facilities*	*Vote on U.K. res. for extended zone off British Isles to 40°W*	*Vote on U.K. res. for North Sea prohibition zone*
General Prohibition				
Australia	X	X	X	X
Germany	X	X	X	X
Ireland	X	X	X	X
New Zealand	X	X	X	X
U.K.	X	X	X	X
India	X	X	X	A
Netherlands	X	A	X	X
Canada	X	A	A	X
Israel	X	A	A	X
Poland	X	X	A	A
USSR	X	X	A	A
Leaning toward Zones				
Sweden	X	O	A	X
Brazil	O	X	A	A
Portugal	O	X	A	
Venezuela	A	A	A	
Mexico	A	O	A	A
Zones				
Chile	O	O	A	
Greece	O	O	A	A
Italy	O	O	A	A
Yugoslavia	O	O	A	
France	O	O	O	A
U.S.	O	O	O	A
Spain	O	O	O	
Belgium	O	O	O	O
Denmark	O	O	O	O
Finland	O	O	O	O
Japan	O	O	O	O
Norway	O	O	O	O
Outcome of Vote		10-14-14	17-9-13	9-5-4

Key

X = Statement in Favor or "Yes" vote

O = Statement against or "No" vote

A = No statement, abstention vote

Spaces between groups of states delineate those groups with overall similar policies (voting patterns).

As we have seen, British policy at the time was influenced by the relatively high level of pollution along its shores and by the pressure of domestic public interest groups. The two European maritime states that supported Britain's position were, not surprisingly, also countries that bordered the heavily used English Channel and North Sea. The fact that eight other states—most of whom did not have serious pollution problems—also supported the British initiative was due both to British diplomatic persuasion (many were Commonwealth countries) and to the fact that those countries lacked large shipping and oil industries. By the same token, the opposition of most participant states can be attributed to the unwillingness of their governments to assume the necessary costs themselves (for reception facilities for nontankers) or to impose the necessary costs on their industries. They were, as a result, quite unwilling to approve all but the most modest costs, and even then they wanted someone else to assume them.

THE 1962 AMENDMENTS

If participants at the 1954 Conference did not explicitly recognize the inadequacy of their product, they at least provided for its revision. A resolution of the meeting requested the "Bureau Power," the United Kingdom, to call an amending conference in three years' time, but it was not until the middle of 1958 that the convention entered into force. Only then had the required ten states (five of which each had to have at least 500,000 tons of ship tonnage) ratified the convention. During that same year, the IMCO Convention also obtained the necessary number of ratifications, so IMCO now assumed the role of "Bureau Power." Finally, by the fall of 1959 the IMCO Secretariat was able to recommend that the new conference be held. The time suggested was early 1962, and it was approved by the Assembly.

In the interval, two events occurred which reminded states of the oil pollution problem. In 1958 and 1960 the Conference on the Law of the Sea was held in Geneva, producing four new international conventions, among them the Convention on the High Seas. Clearly referring to the 1954 Convention, article 24 of the High Seas Convention requested that states draw up regulations to prevent pollution of the seas by oil *subject to existing conventions on the matter*. Second, in July 1959 the renamed British Advisory Committee on Oil Pollution of the Sea (ACOPS) sponsored a conference in Copenhagen to reconsider various aspects of the oil pollution problem. Attended by official and unofficial representatives from eleven countries, this new meeting passed a number of resolutions including a request urging an immediate amendment of the 1954 Convention to extend the zones.[30]

30. MSC II/11, pp. 9, 17-18. (All subsequent document citations refer to IMCO documents unless preceded by a different prefix—for example, U.N. doc.)

The conference was convened in the last week of March 1962. Thirty-eight states were now in attendance, although the regional distribution was similar to that of the 1954 meeting.[31] The Western powers and Soviet bloc constituted a numerical majority and still dominated the debate. Only sixteen of the participants were parties to the 1954 Convention, and twelve of these submitted statements on their experiences with its operation.[32]

A host of amendments was submitted. As was to be expected, the most important proposals concerned either changes in the existing zonal system and the technologies required for its implementation or a general prohibition. On the latter, the key proposal was again a British one. For all *new* vessels over 20,000 grt, it was recommended that operational discharges in excess of 100 ppm be banned everywhere. It was mainly tankers that reached this size, so the impact of the proposal on dry cargo ships (which were, in any event, increasingly less important as a source of oil pollution) was minimal.

Behind all the proposals for the strengthening of the 1954 Convention was the recognition that there had been no real improvement for any state since 1954. In fact, in most areas—particularly the Mediterranean—the problem had become much worse. The survey of conference participants documented a problem of alarming dimensions.[33] Almost all European countries (and to a lesser extent the North American ones) faced much-increased coastal oil pollution. For one thing, maritime oil movements had doubled during the previous eight years. But, for Mediterranean countries, it was actually the establishment of a pollution prohibition zone in the adjacent Atlantic coastal area that was a major source of their problem. In order to avoid longer voyages outside of the Atlantic zones during their trips from northeast Atlantic ports to the Middle East, tankers found it necessary to dump their residues in the "free zones" in the middle of the Mediterranean. As John Kirby, a former executive with Shell Marine International, commented, "Oil companies had to put the oil somewhere between leaving a U.K. or other Western European port and the Suez Canal. The Mediterranean was the logical answer."[34] Kirby estimated that of the approximately one million tons of crude oil being discharged annual-

31. There were eighteen states from Western Europe, North America, and Australasia, six from Eastern Europe, six from Asia (including Japan), four from Latin America, and four from Africa. Only half the participants spoke with any regularity. Many spoke not at all.

32. The twelve states were those that had ratified the convention prior to 1958. They included ten Western European states as well as Canada and Mexico. Four states ratified the Convention in the year preceding the 1962 Conference. They were the United States, Poland, Kuwait, and Iceland. Liberia ratified immediately before the meeting, although its instrument of ratification was not yet deposited. By article XVI of the 1954 Convention, only parties to the Convention could amend it so the legal power of amendment at the Conference was ultimately in the hands of the sixteen or seventeen states. They did, in the end, simply approve the decisions of the larger group.

33. 1962 CONF/I.

34. "Background to Progress," *The Shell Magazine* (London), January 1965, pp. 25-26.

ly into the oceans in the early 1960s, perhaps one half of it was being put into the Mediterranean. Needless to say, the policies of many coastal Mediterranean states underwent some revision after the entry-into-force of the 1954 regulations.

To strengthen the zonal system of the 1954 Convention, it was proposed that the zones be extended from 50 to 100 miles, supplemented in some select areas by zones of even greater breadth. The extensions were ostensibly to occur only for states that justified the need for them, but the conference approved such requests with little discussion. Some countries and shipping interests were opposed to the zonal extensions, though experience with the 1954 Convention should have alleviated their fears. Only a small number of prosecutions were made under the convention, and none had taken place for violations beyond the territorial seas (where the power to regulate existed for flag states).[35] Hence, the larger zones were accepted. Predictably enough they were concentrated around the developed —and polluted—countries of Western Europe and North America, but large areas in the middle of the Mediterranean were still left open to discharges.[36]

Unfortunately, the vision of progress was more apparent than real. Even at this conference, no provisions were made to force shipowners and governments to install the technologies that would enable ships—particularly nontankers—to comply with the zones. Despite numerous reports and debates, most of the conference participants gladly accepted American assertions of the continued technical shortcomings of existing oily-water separators, a claim reinforced by yet another malfunction of a British demonstration separator.[37] In the case of one technical requirement, reception facilities, the conference actually took a step backward. The old article 8 had required that governments ''ensure the provision'' of facilities in main ports, but this was now replaced with the stipulation that they only ''shall take all appropriate measures to promote'' their installation. Both the United States and Great Britain had sought this amendment for their own particular reasons,[38] and even those states which opposed it were reluctant to accept any obligation to build the facilities themselves. Indeed, as

35. 1962 CONF/II. Only in the cases of the U.K. and West Germany was there any evidence of fairly rigorous enforcement.

36. One-hundred-mile zones were granted to Iceland, Norway, the United States, Spain, Portugal, the states on the Mediterranean Sea, Adriatic, Red, Black, and Azov seas, and the Persian Gulf. Areas off India, the Malagasy Republic, Canada, and Australia also were covered, as was the Northeast Atlantic zone unsuccessfully sought by the U.K. in 1954. The entire North and Baltic seas were now made complete prohibition zones as well. Annex A.

37. A malfunction in the separator necessitated a delay of a week in the demonstration, by which time any impact it might have had was lost.

38. The U.S. opposed article VIII now because of its own lack of federal control over American port authorities. In fact, it had submitted a reservation with respect to article VIII when it ratified the Convention a year earlier. Similarly, the U.K. had experienced difficulties in securing the construction of reception facilities by port authorities except where they could ensure that they would be commercially viable through sale of the refined residues.

the French delegate commented, France's support for the old wording was

> . . . not intended to impose obligations directly on governments but rather to allocate responsibility for providing adequate facilities to the bodies in charge of port administration. Those bodies should arrange for financing the necessary installations, and private companies could also assist by providing suitable facilities. The French government would assist by facilitating loans to cover expenditures incurred by Chambers of Commerce and might provide grants for first installation.[39]

By not accepting meaningful obligations to install reception facilities, the governments at the conference were ensuring that adequate facilities would not be built and that many vessels would not be able to comply with the discharge standards. Many nontankers would still have to discharge their ballast in the usual fashion—close to shore in the prohibition zones. Moreover, without additional technologies, new tankers would be unable to implement the general prohibition that was to be applied to them upon the amendments' entry-into-force.[40]

Despite the obvious inconsistency in their policies, it was the British delegation that proposed the general prohibition on polluting discharges for *new* vessels over 20,000 tons. It was to apply only to ships ordered after the entry-into-force of the amendments, but it was a start toward a general prohibition. The U.K. delegate was frank in his assessment of the situation: "If oil pollution was to be overcome, it would necessarily cost something. A start must be made somewhere, and the large tankers were certainly responsible for most of the pollution. It was essential to take a step towards total prohibition."[41] In making the proposal the British government, however, was certainly aware that several preconditions for its successful implementation did not exist. Very large separators with a throughput capacity of about 500 tons per hour were probably required to maintain oil discharges below 100 ppm, although delegates often claimed a sufficient separation would occur in a cargo tank. Also, large reception facilities at oil-loading terminals were absolutely necessary for tankers to dispose of the residues. These costs and others were not stressed by the U.K., but they did not escape the attention of other governments.

The proposal drew sharp reactions from a number of maritime states—Japan, the Netherlands, Norway, and the U.S.—which cited its practical,[42]

39. 1962 CONF/C3/SR.6, p. 8.

40. May 1968, as it turned out.

41. 1962 CONF/C3/SR.7, p. 8.

42. Where there were no reception facilities at the loading terminal, the ship would have to retain residues on board—at great cost (1962 CONF/C3/SR.7, pp. 7–8; and C3/WP4/annexes 1 and 2). Where there were such facilities, a tankerowner (particularly an independent tankerowner) could not assure the acceptance of the residues if he did not control the facility. Furthermore, carriage of these residues would result in higher charges for passage through the Suez and Panama canals (1962 CONF/C3/WP4/annexes 1 and 2).

economic,[43] and technical[44] shortcomings. Yet it did obtain sufficient backing to be included in the amendments—if only at a price.[45] Nascent environmentalism was, as ever, tempered by economic caution. It was clear that the oil industry and not governments would be expected to foot the bill for the reception facilities. The conference participants explicitly rejected any obligation on contracting parties to build them. They even accepted a sweeping Norwegian amendment that allowed a master to exempt his vessel from compliance with the prohibition if "special circumstances made it neither reasonable nor practicable to retain the oil or oily mixtures on board."[46]

Throughout this debate, the industry was strangely silent. Indeed, although they were certain to be saddled with most costs from the new amendment, some industry officials—particularly those from Shell and B.P.—supported the U.K. position. In fact, the American companies knew little about it, and even in Shell and B.P. the issue was dealt with by low-level officials with little interest or guidance from high executives. Their support for the British position was largely a matter of goodwill toward a government under domestic pressure. Indeed, it was not until after the 1962 Conference that the oil industry, led by Shell, came to understand the economic implications of what governments apparently expected of it. Only then did it devote real attention to the problem.[47] The result was the development of an alternative and potentially very effective system of pollution prevention.

With no recorded votes at the conference, only those states participating in the debate could be identified with the various alternative proposals. As is evident from Chart 5, Britain was able to secure the support in 1962 of those Mediterranean countries (particularly France and Italy) which had

43. The Dutch delegation in particular complained about the competitive disadvantage they would be under as parties to the amendments vis-à-vis nonparties (1962 CONF/C3/SR.7, pp. 7–8). The Norwegians stressed that this would be greater for independent shipowners than for oil industry ships which had a larger and more diversified economic base (1962 CONF/C3/SR.7, p. 8).

44. It was strenuouly argued that a 100-ppm limit was not technically feasible and that there were substantial offloading problems (1962 CONF/C2/SR.6 and 7).

45. The basic discharge regulations are in article III and the geographical scope of the new zones in annex A. There were also a number of minor changes in the 1954 Convention. First, the exception which allowed tankers to discharge sediments from cargo tanks was dropped. The French proposed this alteration within the Convention, and as a result of sound scientific evidence they were able to bring the other delegations around to their point of view. Second, the size of the tankers covered by the Convention was changed from those above 500 tons to those above 150 tons [article II(1)(*a*)]. Thus, only the smallest coastal tankers were exempted from the regulations. Third, article VII was altered by the provision that "carrying water in oil fuel tanks should be avoided if possible." Most nontankers built from the 1960s on were constructed so as not to have to use their fuel tanks for ballast.

46. Article III(*c*). The discharges over 100 ppm had to be outside the coastal prohibition zones.

47. Interview, John Kirby, former Marine Coordinator, Shell Marine International.

opposed it in 1954. The Commonwealth and Communist bloc states also continued to support Britain. The latter group was easily able to do so as their limited international maritime activities ensured that it would be the Western maritime states and not themselves that would be affected by any changes. Britain's major opponents were all states with large shipping and/or oil industries who clearly recognized that it was their nationals who would have to bear the brunt of the cost.

THE 1969 AMENDMENTS

The history of the 1969 Amendments is quite different from that of the 1954 Convention and the 1962 Amendments. Private industries (especially the oil industry) were now important participants in the formal negotiations, which were themselves carried out within IMCO bodies over a five-year period rather than a month-long international conference.[48]

Prior to 1962 the oil and shipping industries had paid little attention to the creation of the international regulations. It was only the beginning of a potentially expensive general prohibition scheme and the realization that public and governmental priorities were shifting that finally forced the industry, and particularly the British-based Shell Oil, to act.[49] Their response was to seek a cheaper and more practical alternative, and it was decided that the load-on-top system (LOT) provided the answer.[50] John Kirby, marine coordinator of Shell Marine International, was charged with obtaining its international acceptance.

Shell could respond promptly to the 1962 Amendments, as it had actually developed the LOT approach as far back as 1953. At that time the system had been shelved because refineries had objected to the salt content of the

48. The final acceptance of the amendments occurred at a meeting of the IMCO Assembly rather than at an IMCO-sponsored conference. This procedure was in consonance with one of the methods of amendment set forth in article XVI of the 1962 Amendments to the effect that amendments could be presented to the contracting parties for ratification on the approval of two-thirds of the members of both the Maritime Safety Committee and the Assembly.

49. Speaking about Shell's reactions to the economic implications of building reception facilities in the Middle East, John Kirby of Shell Marine International has remarked, "It was only our close study of the solution recommended by the Conference of discharging oil ashore into slop facilities that drove us toward the load on top method." He also noted that apart from the economic costs of the facilities, Shell was also very worried about whether they could actually be built at offshore terminals and what the oil companies could do with the oily residues once they had received them at the oil loading terminals. "Background to Progress," *The Shell Magazine* (London), January 1965, p. 25.

50. The LOT system has been described in Chapter 2. The oil/water mixtures from the top of ballast water and from tank cleanings are put in a cargo tank or a special, heated slop tank. The oil is allowed to float to the top and then the water is decanted from the bottom of the tank. The concentrated oily residues with a small salt water content are generally then left in the tank and the new cargo is loaded on top of these. Sometimes the residues are kept separate from the cargo in the slop tank.

CHART 5
State Positions on British Proposal
for Vessels above 20,000 Tons, 1962

For	*Against*
Australia	Japan
Canada	Netherlands
France	Norway
Germany	U.S.
Ireland	
Italy	
Kuwait	
USSR	
U.K.	

crude oil they would have had to accept.[51] Now, with governments assigning a higher priority to pollution, Shell's refineries were directed to adapt their operations and install technologies allowing them to receive such oil. After ten years of inaction, it took only a few months until it was "discovered" in *early 1963* that both the tanker and refinery operations could be successfully adjusted to utilize the LOT system.

Having established the technological feasibility of the system, Kirby approached British Petroleum and quickly won its support. The task was now to market the system to other oil corporations and to the independent shipowners. A joint approach was immediately made to the major American oil companies. They were soon able to secure the backing of the largest American oil company, Esso (now Exxon), and on June 17, 1964 these three companies announced that they were adopting the LOT system of pollution prevention on tankers. The selling of LOT to the other American majors was carried out by Esso, and it soon recruited Standard Oil of California, Mobil, and Texaco. Kirby, meanwhile, had turned his attention to the European oil companies and rapidly converted them as well.[52]

It was, however, the cooperation of the largely European independent tankerowners that was crucial to the success of the program. These operators accounted for two-thirds of the world tanker tonnage, and it was they

51. Even before the passage of the 1962 Amendments, Shell tanker specialists had begun to turn back to LOT as possibly the best future pollution-prevention alternative. This had followed an unsuccessful four-year experiment with the installation of a small refinery on board a tanker for converting cargo residues into fuel oil. Interview with Shell official.

52. Interviews with John Kirby and Maurice Holdsworth, Shell Marine International.

who were most intransigent. Unlike the oil companies, competitive shipping was the entire business of these shipowners, so that the extra work for the crews, the extra fuel to operate the pumps, and the 5 percent Suez Canal surcharge for tankers not free of oil were substantial deterrents.[53] These direct costs were crucial for independent tankerowners. So too were the indirect costs that would result from the loss of flexibility at being cut off from those refineries still not accepting salty residues. But the major oil companies had become convinced that it was in their long-run interest to adopt the system rather than be forced to accept the very expensive alternative inherent in the 1962 Amendments. Shell, British Petroleum, and a number of the other major oil companies, therefore, agreed to reimburse the independent tankerowners for any direct expenses that could be shown to be attributable to the adoption of LOT![54]

The selling of LOT to the oil companies and the independent shipowners was so successful that, by early 1965, the Swedish and British governments were able to bring to IMCO a description of the development and wide acceptance of the system.[55] At that time, the oil companies could already claim that 60 percent of all tanker tonnage was employing the system and another 18 percent was considering it. An important point about the development of LOT is that it was done completely independently of governments and in a very short time. In fact, the oil companies had adopted a system which by their own admission violated both the 1954 Convention (then in force) and the 1962 Amendments then being ratified by many governments.[56] Thus, their actions can almost be viewed as an assumption of international legislative power—or at least forcing the hands of governments by presenting them with a *fait accompli*. It was, however, only because governmental enforcement of the existing regulations was so poor

53. Initially the Egyptians wanted to impose much higher tolls for tankers which did not have completely clean tanks and were hence not gas free, but Kirby personally was able to persuade them to settle for 5 percent. This arrangement came into effect in July 1964.

54. Kirby noted, "It is fairly obvious—particularly so in these days of weak markets—that no independent owner is going to adopt a radically new practise if it is going to cost him money." "Background to Progress," *The Shell Magazine* (London), January 1965, p. 25. The indirect costs are the more significant, so that tankerowners will simply not use LOT on those spot charters going to refineries that refuse to accept salty residues.

55. SCOP 1/21, p. 4. At this stage, it seems as if these two governments were acting simply as informational channels to IMCO for the oil companies (United Kingdom) and the independent shipowners (Sweden).

56. The oil industry admitted that the LOT system would lead to discharges in which the oil content was greater than 100 ppm—particularly when the water near the water/oil interface was being discharged. Hence, tankers on some voyages (especially short-haul) wold have violated the 100-ppm stipulation in the 1954 Convention when tankers were in the coastal prohibition zones. It also would have led to violations of the regulation in the 1962 Amendments (which came into force in 1968) that the new tankers over 20,000 grt could *never* discharge more than 100 ppm (SCOP I/21, p. 3).

(particularly outside a state's territorial waters) that the industry was able to implement its own alternative.[57]

Unlike the negotiations for the 1962 Amendments, all the formal deliberations leading to the acceptance of the industry proposals occurred within regular IMCO bodies and not at an IMCO-sponsored conference. Interestingly, LOT was discussed initially at the *first* meeting of the Subcommittee on Oil Pollution (SCOP) in February 1965. At the time, government delegates were very skeptical about its operational effectiveness, and they feared that it would violate existing regulations.[58] To resolve the uncertainty, the subcommittee set up a working group to examine the LOT system for its pollution prevention potential and for its compatibility with the 1954 Convention and the 1962 Amendments. The working group particularly focused on the problem of the system's violation of the 100-ppm standard, and the report, which was not reviewed until 1967, produced no surprises. It duly noted that the only way that the system could ensure that discharges were less than 100 ppm was by installing an oily-water separator and an oil-monitoring system.[59] An effective oil-monitoring system still had not been developed and oily-water separators were not always effective. However, the oil companies had, by now, firmly taken the initiative, and they persuaded the British government to experiment with them at the government's Warren Springs Laboratory. With the greater interest of Britain's domestic oil industry, British environmental policy at IMCO became more sensitive to its views. The Warren Springs tests focused on measuring the amount of oil discharged over a set distance and not, as in the past, on the concentration of oil in a measured discharge of seawater. They established that an instantaneous rate of discharge of 60 liters per mile would produce a sheen on the water which would break up and disperse in a period of only two to three hours.[60] As a result of these experiments

57. The attention of governments was then focused on bringing the 1962 Amendments into force, and this was assured in May 1967 after the required twenty-two states had ratified the amendments. The actual entry-into-force was one year thereafter in May 1968. The preemption by industry of government was so successful that one of the leading British governmental experts on oil pollution commented in 1975 that he did not think that there was a tanker over 20,000 dwt in the world complying with the 1962 Amendments despite the fact that they had been law for seven years. (Interview with Gordon Victory, former official, Marine Division, U.K. Department of Trade.)

58. Most of our information on the views of states in the meetings of the Subcommittee on Marine Pollution which discussed the LOT system has been taken from minutes of one national delegation. Official minutes were not recorded.

59. OP/WG.I/II; OPII/2. An oily-water separator was viewed as desirable since it would produce a cleaner separation of oil and water in the tanks—thus preventing the discharge of a considerable amount of oil with the water—and an oil monitoring system was recommended since it would indicate when the oil content was exceeding 100 ppm.

60. G. Victory, "The Load-on-Top System, Present and Future," *Proceedings of a*

the British government now became a strong advocate of LOT and of a new international regulatory scheme. Another factor encouraging British support for the new system was the realization that separators could not be built large enough for the VLCCs (over 200,000 dwt) which began to appear in the mid-1960s. Without such separators it was impossible for such tankers to comply with the 100-ppm rule in the 1962 Amendments.

In January 1968, the U.K. formally submitted a resolution to the meeting of the Subcommittee on Oil Pollution urging governments to promote the adoption of the LOT system. By a vote of 10-5-4, the resolution was adopted, although there was still substantial opposition to it.[61] Between that meeting and the next one in December 1968, governments examined this new panacea—all the while being bombarded by the intensive lobbying of the oil companies. In a paper presented to an ACOPS conference during that year, John Kirby of Shell projected that the LOT system would cope with 80 percent of potential operational pollution from tankers.[62] He also argued that to continue to aim for a 100-ppm discharge limit was, quite simply, extravagance. The cost of installing the necessary separators and monitors for the world's tankers was astronomical—he estimated 125 million. This expenditure was just not justified by the small additional reduction in oil pollution: "the price of perfection is extraordinarily high."[63]

By the December 1968 meeting of the SCOP, the opponents of the LOT system were convinced. They would now accept it provided that the 60 liters per mile requirement was supplemented by stipulations that the total volume of oil discharged would not exceed 1/15,000 of a tanker's cargo-carrying capacity and that the tankers discharge only clean ballast within a fifty-mile zone adjacent to all coasts.[64] The oil companies opposed the

Symposium on Marine Pollution, Royal Institute of Naval Architects (London), February 1973, pp. 10–11.

61. OP IV/9; and minutes of a national delegation. The two strongest supporters of the resolution were the U.K. and France. The U.S., which was then beginning to adopt a strong environmentalist stance, opposed the sacrifice of the 100-ppm standard and the broad coastal prohibition zones as a retrograde step. (In fact, the proposed narrow coastal zones had much more stringent controls in that virtually only "clean ballast"—instead of 100 ppm—could be discharged.) Opposition also came from Germany, Sweden, the USSR, and Japan. The USSR opposed it because LOT required voyages of at least two days and the voyages of most of its tankers were of a shorter duration. It preferred reliance on reception facilities. The Japanese opposition stemmed from the fact that their refineries still refused to accept crude oil with a salt content.

62. The other 20 percent was composed of tankers traveling to repair ports, tankers carrying specialized crudes and refined oils which could not be mixed with the residues of other cargoes, and tankers on voyages of less than thirty-six hours. J. H. Kirby, "The Clean Seas Code: A Practical Cure for Operational Pollution," *Proceedings of the International Conference on Oil Pollution of the Sea* (Rome), October 7–9, 1968, pp. 206–207.

63. Ibid., p. 215.

64. OP V/12, p. 5. The U.K. proposed a *complete* prohibition zone (only clean ballast) of

1/15,000 stipulation. They knew that tankers could not comply with it in circumstances such as short voyages or rough weather. More importantly, it also meant that violations could now actually be detected by port inspections of the volume of oil on board. Both the oil and shipping industries and the British government opposed the retention of broad coastal zones with regulations much more stringent than the old ones (that is, only clean ballast rather than discharges below 100 ppm). Their tests, they argued, indicated that tankers' compliance with the sixty liters per mile standard beyond fifteen miles from shore would prevent pollution. Also, they stated that fifty-mile zones would certainly cause more inconvenience for the industry: some tankers would have to make detours in order to avoid illegal discharges within them. Despite these criticisms, the additions to the proposed regulations were finally approved in order to secure acceptance of the whole LOT system. In October 1969 they were formally adopted by the IMCO Assembly.[65] A revolution in pollution regulations had been achieved.

Amendments to the discharge regulations for nontankers were also at issue during this period, and some important changes were made. Polluting discharges were prohibited throughout the oceans (and not just in the zones). Nonpolluting discharges were defined for nontankers as those occurring at a rate of less than sixty liters per mile but, unlike the new regulations for tankers, the old 100-ppm content standard was still to apply. Furthermore, the discharges at this rate and concentration were to be made "as far as practicable from land."[66] However, the conference participants again failed to require those technologies necessary for compliance (oily-water separators and reception facilities). They were ensuring that many vessels would be unable to comply.

With the unanimous acceptance of the amendments in October 1969, the oil and to a lesser extent the shipping industry had won their battle to defeat the potentially expensive 1962 tanker regulations. At the same time they were able to do so only by developing and implementing a new and potentially very effective system of pollution prevention. Its effectiveness was based not only on the reduction in oil discharges which it promised, but also on its potential enforceability. Prosecutions of violations of the 1954

---

ten miles and a maximum-volume discharge of 1/10,000 of cargo capacity but finally accepted fifty miles and 1/15,000.

65. Resolution A.176 (VI). At the Maritime Safety Committee meeting in February 1969 there was an unsuccessful attempt by the U.K., Norway, and the Netherlands to alter the width of the zones from fifty miles to twenty-five miles. The International Chamber of Shipping was then advocating fifteen-mile zones. France, Sweden, and the U.S. successfully led the opposition to the proposed change. MSC XIX/SR.4, pp. 9–11; and MSC XIX/13/5.

66. Article III(*a*). Another minor change which was accepted for both nontankers and tankers was that the exception of residues of fuel and lubricating oil be eliminated. Article IV(*c*) of the 1962 Amendments was thus eliminated.

and 1962 regulations had required establishing that vessels were discharging an effluent with an oil content greater than 100 ppm—an almost impossible task. The new tanker regulations could be enforced by inspecting the volume of residues in the tanks at oil-loading terminals. Despite this feature the major industrial interests and most governments were happy with the outcome of the negotiations. Not only did the prospects for pollution mitigation look bright as a result of the development of LOT, but the oil companies and governments were free of the economic burdens of building reception facilities. And shipping companies were to be spared the delays in ports where reception facilities existed and the reduction in cargo-carrying capacity where reception facilities did not exist. If the prospects looked too ideal, they were. There were many disguised weaknesses in the LOT system, and these were soon to be revealed.

THE 1971 AMENDMENTS

Unlike provisions of the 1954 Convention and its amendments which concerned operational pollution, the major 1971 amendment was directed at the mitigation of pollution damage resulting from tanker *accidents*.[67] This issue first emerged as an important concern of governments following the *Torrey Canyon* disaster of 1967. There was much fertile ground for improvement in areas such as crew standards, navigational aids, traffic control, and separation schemes, and the *Torrey Canyon*'s catastrophic grounding rammed home this need. Even so, these were extremely technical issues of relevance to all shipping and not just tankers, and they were dealt with at IMCO as they had traditionally been, as safety and not pollution-prevention issues. It was into the Maritime Safety Committee with its plethora of subcommittees that the organization funneled these issues. New regulations on some of these questions emerged during the next decade as amendments to safety conventions or solely as Assembly recommendations.

One proposal was an exception in that it was seen purely as a pollution-prevention issue and was integrated into the 1954 Convention—namely, setting limitations on cargo tank sizes or, in the technical terminology of the naval architects, the "hypothetical oil outflow" of cargo tanks.[68] This matter was considered in 1970 by the Subcommittee on Ship Design and

67. One other noncontentious amendment extended the tanker discharge prohibition zone off the Australian coast adjacent to the Great Barrier Reef from 50 to 150 miles. Resolution A/232 (VII).

68. *Hypothetical oil outflow* refers to the outflow which would result from damage to specific parts of the ship. With respect to ship-side collision damage, it would in general be equivalent to the capacity of two adjacent wing tanks.

Equipment of the MSC, which provided an evaluation of six alternatives which the MSC might want to adopt.[69]

The MSC's deliberations in early 1971 focused on two alternatives put forward by the subcommittee. The first was supported by the U.S., the U.K., and the majority of the committee members and limited the "hypothetical outflow" to 30,000 cubic meters regardless of the deadweight tonnage of the tanker.[70] The second alternative, supported by France, Japan, and Greece, permitted an increase in the size of the cargo tanks proportional to the increase in the deadweight tonnage of the tanker. This latter proposal was the most lenient of the ones discussed in the subcommittee, but even it demanded smaller tank sizes for vessels above 200,000 dwt than were then being built. It was, however, the first alternative which was accepted by the MSC and forwarded to the Assembly for approval.[71]

The issue was not an insignificant one, and the proponents of the minority proposal dug in their heels against the MSC recommendation. The Assembly meeting of October 1971 was the scene of a sharp confrontation. Ostensibly the debate was concerned only with the probable effects of tank-size limitation on pollution abatement and with the costs of the various proposals for shipbuilding and oil transportation. And these were certainly important considerations. With regard to pollution abatement, the French delegation pointed out that a reduction in tank size increased the surface and corner areas within the cargo tanks on a vessel. This, they argued, could actually increase operational pollution and the risk of explosion.[72] In fact, these arguments had great merit, but only if the tanks were inadequately cleaned and LOT improperly employed.

As ever, a major area of contention concerned the effects of the regulations on the cost of shipbuilding and, therefore, the price of oil for the consumer. Estimates of cost increases diverged greatly, with the U.K. anticipating an additional cost of 2 to 3 percent[73] and France projecting an increase of over 10 percent (thereby endangering the economic viability of tankers over 250,000 dwt).[74] Definitive estimates of these costs are difficult

69. Document DE VI/11 submitted to the MSC meeting in March 1971.

70. Only one of the options suggested by the subcommittee, that for a limit of 20,000 cubic meters, was more stringent. It was initially supported by Egypt, Sweden, the USSR, and the U.K.

71. The vote in the MSC was 13-2-1, with France and Greece voting against and Japan abstaining. See MSC XXIII/SR.4, p. 15.

72. A VII/9/1, p. 7.

73. A VII/SR.9, p. 11.

74. A VII/9/1, p. 4. Furthermore, differences clearly existed as to the relative value of the expenditure regardless of the cost in absolute terms. Brazil's delegate felt that even a 2 percent increase was a "very large sum for a developing country" (A VII/SR.9, p. 16), while the United States saw the cost as "negligible in comparison with the cost of the damage done by an unrestricted outflow of oil" (A VII/SR.9, p. 14).

to make, and one member of the IMCO Secretariat actually agreed with the French prognosis.[75] However, a 1971 study done for the American shipbuilding industry projected costs and effects closer to those anticipated by the backers of the MSC proposal:

> The effect on transportation costs is less than 2% for the 300,000 dwt tanker and a maximum of 17.5% for the million ton tanker. . . . Even for Europe and Japan, the additional transportation cost for the million ton tanker with enough bulkheads to minimize the loss after accidental oil spills, amounts to a maximum estimate of 2/10 of 1 cent per gallon of crude oil. The million ton tanker would still transport oil at a lower cost than a 300,000 tonner.[76]

Higher costs could, the French delegation asserted, have important environmental effects. Any economic disincentives to building larger tankers would result in an even greater number of smaller tankers, and these would be more likely to be involved in accidents. This viewpoint was by no means accepted by most of the other participants, but in any event it was not the environment that was France's first concern.[77] Other economic interests pressed more strongly on both France and Japan.

In 1971, only France and Japan could accommodate "ultra-large crude carriers" (ULCCs) in their ports,[78] and therefore it was these countries which would have been immediately penalized by the regulations.[79] While

75. Interview. The official saw the impact of the MSC proposal as making tankers over 500,000 tons uneconomical.

76. Edwin M. Hood, "Ecology, Shipbuilding Prices, and Consumer Cost," *Report on the Sixteenth Annual Tanker Conference* (Washington, D.C.: American Petroleum Institute, Division of Transportation), May 10–12, 1971, p. 203. During the October 1971 discussion, the Panamanian delegate put forward what must be one of the most unlikely suggestions ever made in an IMCO meeting! He suggested that "if the number and size of tankers were to increase, thus increasing the possibility of accidents, the effects of such accidents might be minimized if a device were to be installed on such tankers to explode, set fire or otherwise neutralize the cargo in case of emergency. Such a device would be operated by remote control, and would be so located so that it could be activated by the last person to leave the ship without endangering safety." A/VII/SR.9, p. 13.

77. France did cite a study showing that accidents in the last ten years had decreased in frequency with an increase in the size of the vessels. Others countered that the lack of maneuverability, deeper draft, and greater tank sizes could cause unprecedented catastrophes in the future unless large tankers were built with smaller tanks. See A/VII/SR.9.

78. "Ultra-large crude carriers" describes oil tankers with a cargo carrying capacity in excess of 400,000 tons. It is possible for oil discharge ports to accommodate very large tankers by "lightering" them (transferring some of their cargo to smaller tankers outside of the port so as to reduce their draft), but this practice is costly and also poses some navigational and safety problems, especially for the ULCCs.

79. Japan's attachment to the French proposal was not surprising. It was, after all, Japanese shipyards that initiated the supertanker boom. The vast majority of ULCCs have been built in Japan, and the expertise for their construction remains largely Japanese. Moreover, with the long haul from the Persian Gulf, Japan benefits more than most from the "economies of scale" represented by the large vessels.

there were no prospects of American facilities accommodating tankers of even 300,000 dwt,[80] the French were in the process of expanding their oil discharge terminals at Antifer (near Le Havre) and Fos (near Marseille) to accommodate tankers above 500,000 dwt. Indeed, this expansion program reflected a desire not only to bring about a reduction in the cost of oil for France but also to make France both the major oil entrepôt for Western Europe—in particular, for southern Belgium and the German Ruhr—and the major Western owner of ULCCs. A French government publication set forth this strategy with great candor:

> The geographical handicap that placed the Ruhr 250 km from Rotterdam and 500 km from Le Havre will thus be largely offset by economies on the cost of sea transportation. Over and above the satisfaction of national needs, this international role is a new and fundamental aspect of the Antifer project: the port of Le Havre will be in a position to participate on exceptionally competitive terms in the supply of crude oil to much of Northwestern Europe, more especially to Wallonia and the Ruhr. The extension of the port's oil hinterland beyond the French frontiers thus becomes a practical possibility.[81]

France was profoundly committed to the success of its proposal and put its full diplomatic muscle behind it. Belgium and other Francophone states were heavily lobbied, as was the Soviet bloc.[82] Early voting in the Assembly, however, resulted in the failure of both the MSC and the French alternatives. The original French proposal (now cosponsored by Brazil—a developing state with maritime ambitions) was soundly rejected,[83] and the

80. Hood, "Ecology," p. 203.

81. "Antifer—A New Oil Terminal," A/101/74 (London: French Embassy Press and Information Service, 1974), p. 8. The publication also noted (p. 2) that if in 1980 all of France's oil supplies were carried by gigantic tankers, there would be a saving of 1,000 million F per annum. French government officials indicated to the authors that it was indeed the interest in making Antifer the major oil entrepôt for Western Europe (and thus replacing Rotterdam's preeminent position at the time) which prompted its vehement opposition to the MSC proposal. Recently the shipping magazine *Fairplay* has noted that the French domestic shipping industry "has received a good deal of attention at the highest political level. President Giscard d'Estaing recently stated that the promotion of France's maritime transport facilities should be considered a matter of permanent concern." ("French Owners Count Their Blessings," *Fairplay International Shopping Weekly*, January 22, 1976, p. 5.) Furthermore, any success which the French would have had in making France the major oil entrepôt for Europe would have redounded to the benefit of its own tanker industry, since 67 percent of all oil brought into French ports must be carried in French flag vessels and they had already begun the expansion of their ULCC fleet. The world's largest tankers, *Batilus* and *Bellaya* (542,000 dwt), are part of the French fleet.

82. A Belgian official has commented that this was the only occasion that he recalled when the French Foreign Ministry made strong diplomatic representations to the Belgian government requesting that Belgium support France in a vote in IMCO. Six other Francophone states, formerly French colonial territories, were also under heavy pressure.

83. Vote of 14-33-7. See A VII/WP; A VII/9/1; A VII/WP.20. Brazil was solidly in support of the French position as it saw itself as the "Japan of Latin America."

MSC proposal failed by a single vote to carry the required two-thirds majority.[84] An adjournment produced a Japanese compromise proposal (drafted with a substantial contribution from a member of the IMCO Secretariat). Though it mirrored the MSC proposal for tankers up to 400,000 dwt, it did allow increasingly larger tanks for tankers above that size and thus helped keep the construction of the ULCCs economically viable. Despite continuing strong pressure from France, the countries of Brazil, Greece, the Soviet Union and Japan now agreed to support it.[85] This was a decisive shift and any further opposition collapsed. The resolution passed unanimously.[86]

One aspect of the new amendment is of particular interest. It would apply to vessels built after the date it formally came into force, but it also applied to all vessels delivered after January 1, 1977 and those delivered before this date whose building contracts were signed after January 1, 1972. This assured that almost all vessels built after the Assembly meeting (October 1971) would meet the amendment's technical requirements since their owners would want to assure their ability to use the ports of contracting states after the amendment's formal entry-into-force. This innovative provision reflected the interest of the participating states in promoting the quick implementation of the regulations. More importantly, however, the participants saw the provision as a way of preventing states from obtaining a competitive advantage for new flag vessels by delaying acceptance of the amendment. In fact, the shipowner associations in a number of countries had opposed the amendment prior to its passage for fear that it would be imposed on them by their governments but would not be imposed on rival shipping companies in other states.[87] These provisions were included so as to quiet their anxieties. It is noteworthy that at the 1973 Conference a resolution observed that almost all tankers ordered after January 1, 1972 had, in fact, complied with the 1971 amendment.[88]

84. Vote of 37-3-16. See VII/SR.9, pp. 18–19. A U.S. official noted that the U.S. delegation called the Chilean delegate and requested that he attend the session to vote for the proposal. His car was caught in London traffic and he arrived late. If he had arrived in time to vote for it, it would have passed.

85. The USSR changed its position several times in 1971. In March it backed the MSC proposal. In October it first acceded to French requests to oppose the MSC proposal. Then, it broke with France in backing the Japanese compromise.

86. Resolution A.346 (VII).

87. The technical grounds on which the International Chamber of Shipping opposed the MSC proposal are set forth in MSC XXIII/8/2. ICS officials cited the widespread fear of a loss of competitive advantage as having a strong influence on the shipowners' positions.

88. Resolution 11 of the 1973 International Conference on Marine Pollution. Regulation 24 of annex 1 of the 1973 Convention did alter subparagraph (1)(*b*)(*ii*) to read, "the building contract is placed after 1 January 1974, or in cases where no building contract has previously been placed, the keel is laid or the tanker is at a similar stage of construction after 30 June 1974."

THE 1973 CONVENTION FOR THE PREVENTION OF POLLUTION FROM SHIPS (MARPOL)

Almost twenty years had elapsed since the London Conference of 1954. It was a different world in 1973. The size of the trade, the size of the tankers, and the scope of the pollution were now of a new order of magnitude. But the law in force was only the 1954 Convention as amended in 1962. And that was a very deficient regulatory system—virtually unenforceable outside of ports and hence widely unobserved. The real control system in use throughout the world (load-on-top) was illegal because the discharges sometimes exceeded the standards in the 1962 Amendments, and the 1969 Amendments were not yet in force. The situation was anarchic. And pollution was getting worse.

Just as the initiative behind international pollution regulations shifted from Britain to the oil industry in 1962, it shifted again soon after the *Torrey Canyon* incident to the United States. More quickly there than elsewhere, a public environmental consciousness was beginning to take hold. Public interest groups and "doomsday" literature flourished. In 1968, the issue had already attained high political prominence through such individuals as Democratic vice-presidential candidate Senator Edmund Muskie. In 1970, the new Nixon administration sought to respond to the growing current of awareness by creating the Environmental Protection Agency. After years of unchecked technological expansion, it seemed as if the political establishment finally would give environmental protection its due recognition.

Vessel-source oil pollution was high on the list of concerns and, in a message to Congress in May 1970, President Nixon urged the secretaries of state, commerce, and transportation to press for effective multilateral action. He demanded "more effective international standards for both the construction and the operation of tanker vessels [to] materially reduce the potential hazard."[89] The 1973 Conference on Marine Pollution was significantly influenced by the American initiatives taken as a response to this directive. The specific impetus for the meeting was actually provided in 1969 by Iceland,[90] but it was the American government which generated the most important proposals and sought the necessary cooperation of the

89. "Marine Pollution from Oil Spills: The President's Message to Congress (May 20, 1970) Proposing Administrative and Legislative Actions," *Weekly Compilation of Presidential Documents* 6 (May 25, 1970), p. 661. The message also requested the ratification of the 1969 Amendments as well as the 1969 Intervention and Civil Liability conventions.

90. At the 1969 Assembly the particular concern of Iceland was oil pollution from seabed mining, but after discussions with the IMCO Secretariat it submitted a resolution calling for a conference to deal with other types of marine pollution as well. The resolution passed, although few delegates had views on specific proposals which might be considered three or four years hence. The 1973 Conference did eventually concern itself with pollution from offshore drilling platforms in areas under national jurisdiction, but it was the Third U.N. Conference on the Law of the Sea which dealt with pollution from mining on the deep seabed.

European maritime powers for the conference. In 1970, it lobbied intensively in a number of European capitals and, at a November 1970 meeting of the NATO Committee on the Challenges of Modern Society (CCMS), it was able to achieve the vague acceptance of a number of specific policy proposals. Indeed, the policies the United States suggested at the meeting significantly set the agenda for the negotiations leading to the 1973 Conference.

At the 1970 meeting, U.S. officials strongly criticized the entire LOT system (which had just been officially accepted at IMCO the previous year). It was already clear, they asserted, that LOT was not meeting the claims of its supporters. As it had turned out, the technique often produced an inadequate separation of oil and water in cargo tanks; a tanker's crew frequently could not judge the oil content of an effluent; and the system just was not appropriate on short or rough voyages. Most important, unconscientious crews could easily avoid compliance. The delegation certainly supported the development and installation of technologies on tankers to facilitate compliance with the 1969 discharge regulations, but they recognized the need for very diligent enforcement in port if the system was to work.[91] As it was, the American government now strongly preferred the installation of "segregated ballast" tanks which would never be used for the carriage of oil and would thus completely eliminate oily ballast water (but not tank cleanings) as a source of pollution. In addition, the delegation advocated the construction of nontankers which would never use fuel tanks for ballast (a practice then widely accepted) and the installation of "double bottoms" on tankers. This last proposition was intended to mitigate the impact of an accidental grounding, as it was believed that the existence of a double bottom would inhibit the breaching of the cargo tanks.

The American initiatives, although strongly supported by Canada, obviously threatened to undermine the great triumph of the industry in promoting the 1969 Amendments. Many European maritime states left the CCMS meeting very perturbed by these new environmentalists and especially by the diplomatic pressure they were under to accept the American policies.[92] IMCO, they hoped, would provide them with a more favorable forum for their negotiations, but it was not to be. At IMCO in 1971 the U.S.

91. The U.S. also supported the installation of reception facilities in ports which serviced tankers unable to employ the LOT system. Reception facilities, while necessary for vessels unable to use LOT, can lead to environmental damage in that the water extracted and discharged into ports does have an oil content of 15–30 ppm.

92. NATO doc. C-M(70)64 (revised), November 30, 1970; interviews with a number of European and North American government officials. Some European officials approached their oil and shipping companies after the CCMS meeting and told them that if they did not demonstrate the value of LOT to the Americans, they might be forced to accept some very expensive technologies.

continued to press its demands vigorously, and it obtained acceptance in 1971 of an accelerated work program for the conference preparations.[93] Nine major research programs, six bearing directly on oil pollution, were initiated at that time, and an ambitious normative goal was set for the conference.[94] It was to seek "the achievement by 1975 if possible but certainly by the end of the decade of the complete elimination of the willful and intentional pollution of the seas by oil and noxious substances other than oil, and the minimization of accidental spills."[95] From the time of these initial planning sessions until 1973, the Subcommittee on Marine Pollution (SCMP) with the assistance of other IMCO bodies produced five drafts of the convention. In all of the meetings the U.S. had tremendous influence, but its policies were in full consonance with the increased public recognition being given "the environment" throughout the world.[96]

The key decision for the preliminary negotiations was simply between improving the operation of LOT or demanding the introduction of segregated ballast tanks for all new tankers. The oil industry played an important role in the debate, as it had during the formulation of the 1969 Amendments. But now it was a defensive one. Indeed, until the submission in June 1972 to the Subcommittee on Marine Pollution of the American report favoring segregated ballast, the oil companies had flatly opposed the expensive segregated ballast proposal.[97] But with the American submission, the handwriting was on the wall, and the oil companies began intensive negotiations to consider its adoption.

The industry negotiations were conducted at the international level under the auspices of OCIMF, the nongovernmental organization in which

93. This accelerated program was adopted by the Assembly in October 1971 (A VII/Res. 241). The IMCO Secretariat was, in fact, very influential in setting the framework for the negotiations on the Convention. Following the failure of the SCOP meeting to agree on this matter in September 1970, the Secretariat put forward proposals to the June 1971 meeting of the Council. The Council then authorized the Secretariat to submit a first draft to the October 1971 session of the MSC. It strongly influenced the structure of the convention. (C XXVI/4/Add.1; MSC XXIV/3/1/Add.1, pp. 1-4.)

94. The research programs were initiated at the September 1971 meeting of the Subcommitee on Marine Pollution. Each study was directed by a country with a particular interest in the area of work: the U.S. on segregated ballast and double bottoms; the U.K. on LOT and "retention on board"; and France on short-voyage non-LOT systems. Needless to say, the conclusions of each of the studies submitted in June 1972 tended to support the prior positions of the states undertaking them. See MP XIII/2(*c*) /5 and PCMP 2/2 for the U.S. reports; MP XIII/2(*a*)/5 for the British; and PCMP 214 and 218 for the Norwegian and Japanese studies on LOT.

95. Assembly Resolution A VII/Res. 237, October 1971.

96. The United Nations Stockholm Conference on the Human Environment was held in June 1972 and substantially contributed to this growing priority. See in particular its Statement of Marine Principles (U.N. doc. A/CONF.48/14).

97. See the OCIMF submission of September 1971 to the Subcommittee on Marine Pollution (OP/X/2/4).

American companies had the dominant voice. Through it, a very rapid shift in international industrial policy occurred. In November 1972, the OCIMF Council, by a vote of 11-2, accepted the segregated ballast proposal with only the French and Japanese representatives still opposed.[98] Important European oil companies (most notably Shell and B.P.) as well as the American multinationals with tanker fleets registered in Europe now informed the European governments of their willingness to go along with the American initiative. As had happened a few years earlier, their acceptance led quickly to the conversion of many governments. Among others, Britain, the Netherlands, Belgium, and Italy had changed their policies by the time of the February 1973 preparatory conference.

The acceptance of segregated ballast did not, unfortunately, represent a new commitment to environmental values. For the American oil companies it was their desire to "stay on the good side" of the American government that prompted their transformation. In any event, the tanker-building boom was past its peak, so requirements for new tankers would not have much economic impact for many years to come. In addition, they calculated that continued opposition from them in this area could lead to more harmful policies on other matters.[99] In particular, they hoped that their acquiescence on segregated ballast would weaken the American commitment to the even more expensive "double bottoms" proposal. And there were some distinct advantages to segregated ballast tanks: they did prevent corrosion in the cargo tanks, and they reduced the likelihood of the explosions caused by mixing oil and water. Most importantly, the LOT system was not having the positive effects which the oil companies had anticipated, and they were bearing the brunt of the blame for continued oil pollution.

Considering their commitment to the LOT system, the realization of its inadequacy was not easy for the oil industry. The evidence is clear, however. In 1971 and 1972 ARAMCO (a consortium of Exxon, Texaco, Standard of California, and Mobil), Gulf, and Shell all undertook secret surveys at their oil-loading terminals in the Middle East and found surprising results. Only one-third of the tankers using their terminals were employing the system well, another third were employing it very poorly, and another

98. Our information on the activities of the oil companies in 1972 has been obtained from interviews with oil industry and government officials. It should be pointed out that the oil companies during this same time period were able to obtain a modification in the American proposal. Instead of the original proposal that 45 to 60 percent of a tanker's tanks should be employed solely for the carriage of ballast, they secured American acceptance of 30 to 35 percent. The exact specifications were drawn up in meetings of naval architects from the "seven majors" during the last half of 1972.

99. U.S. officials freely showed their displeasure toward industries which opposed them. At the Marine Pollution Working Group meeting in September 1972, American delegates obtained a ruling that the proposals of representatives of OCIMF or the ICS should not be recorded in subcommittee reports unless explicitly backed by a government representative. Interviews.

third were not using it at all.[100] As the managing director of the B.P. Tanker Co. Ltd., George King, noted in 1975, LOT is "a palliative and not a solution because it is subject to human frailty and external conditions such as weather. Segregated ballast tanks are the only real solution."[101]

Perhaps the most important motivation for industry was the certainty that, regardless of the outcome of the 1973 Conference, the U.S. government would require segregated ballast tanks for all tankers entering its territorial waters. The passage in July 1972 of the Ports and Waterways Safety Act gave the Coast Guard the power to require ship standards such as segregated ballast tanks and double bottoms even if they were not accepted internationally.[102] The threat of American unilateral action under congressional pressure was set forth very explicitly by the commandant of the U.S. Coast Guard, Admiral C. R. Bender:

> The legislative history of Title II [the Ports and Waterways Safety Act] dwells on the need for double bottoms and segregated ballast capacity for tank ships and certain barges (ocean and coast-wise). . . . In Title II of the Act Congress recognized the advantages of seeking multilateral agreements requiring that proposed U.S. regulations be submitted to the "appropriate international forums" such as the 1973 Conference. If the principles forwarded for consideration are not earlier adopted internationally, then we are required to effect them unilaterally not later than 1 January 1976. To comment at this time on either possibility would be pure conjecture. We are hopeful that the Convention will meet the principles of the Act. However, there is precedent for unilateral action. Examples are in our requirement for passenger vessel fire prevention construction standards (special fire safety measures for passenger vessels, SOLAS 1960) and the requirement for all vessels in our waters to comply with U.S. regulations (46 CFR, Subchapter 0) when carrying cargoes considered to be a particular or unusual hazard.[103]

100. These reports were secret and were never made available to the authors. Interviews with industry officials did reveal their existence and their basic findings. These findings prompted the ICS and OCIMF in 1972 to issue a booklet entitled the *Clean Seas Guide* which described procedures for using LOT effectively. The studies were briefly described by a Shell official in 1976. M. P. Holdsworth, "Loading Port Inspection of Cargo Residue Retention by Tankers in Ballast," SYMP/1. The best data on oil company surveys throughout the 1970s were provided to hearings of the Subcommittee on Government Operations and Transportation, Committee on Government Operations, U.S. House of Representatives, by the American Petroleum Institute on July 18, 1978. They will be published in an annex to the hearings. Despite improvements since 1972, probably 30 pecent of oil residues from the world fleet is still being discharged.

101. Interview.

102. Public Law 29-340, July 10, 1972. The new law, along with a commentary on it, was sent to IMCO in September 1972. MP(WG)/4.

103. Admiral C. R. Bender, "Ports and Waterways Safety Act," *Report on the Eighteenth Annual Tanker Conference, May 7-9, 1973* (Washington, D.C.: American Petroleum Institute, Division of Transportation), pp. 48-49.

This threat of unilateral action had its greatest influence on American oil companies, but it also influenced non-American oil and shipping companies whose vessels often entered U.S. ports.

During these preliminary negotiations, many other changes in the draft pollution regulations were also approved and then incorporated into the fifth draft of the convention at a twenty-five-nation preparatory conference in February 1973.[104] On those issues on which there were still significant differences among the participants, the draft described the divergences. Interestingly, all of the major provisions on which the participants at the preparatory conference agreed were later incorporated into the final convention at the October conference, and only one major provision which was rejected in February was accepted in October.

The 1973 Convention was intended both as a consolidation and as an expansion of the revised 1954 Convention. By the end of 1973, forty-eight states were parties to the 1954 Convention as amended in 1962. This was the only existing international law at the time, but few—if any—ships over 20,000 dwt were actually complying with its "general prohibition" provisions. Many were, however, supposedly complying with the 1969 regulations, but these had not yet entered into force, having received only twenty ratifications, well short of the required acceptance by two-thirds of the parties to the 1954 Convention. It was ironic that, despite the fact that the world's major maritime states representing at least 75 percent of registered tanker tonnage had ratified them, the prospects for their becoming law were growing increasingly slight as the number of contracting parties to the 1954 Convention steadily increased during the 1970s. (They finally did enter into force in January 1978.)

The International Conference on Marine Pollution was held in London from October 8 to November 2, 1973. It was attended by seventy-one states and, for the first time, the developing states constituted a majority.[105] With the advent of the Third United Nations Law of the Sea Conference, states had also now organized themselves into consulting and voting blocs—"coastal" states,[106] "maritime" states,[107] the developing world's "Group of 77,"[108] and the omnipresent Soviet bloc. The substantive scope of the

104. Considering that the developing states constituted the majority at the October conference, it is significant that only four such states attended the February meeting: Egypt, Chile, Brazil, and Liberia. The draft convention drawn up by the preparatory conference as well as states' comments on it can be found in MP/CONF/4, August 23, 1973.

105. There were eight Eastern European states, twenty-four noncommunist developed states, and thirty-nine Latin American, African, and Asian developing states.

106. Its members included both developing states and some developed states without large shipping industries. Canada, Australia, Spain, Mexico, Egypt, and India were the leaders.

107. Most of the Western European states as well as the U.S. and Japan.

108. This group had little coherence or organization in 1973, and they tended to voice their views through the coastal state grouping. Only four (including Liberia) had attended

task facing these states was truly enormous. Oil was only one, albeit the most important, of the many polluting substances for which a vast compendium of regulations was being compiled. The draft convention contained hundreds of provisions.[109]

The character of the proposed discharge regulations for existing ships still reflected the basic framework on the 1969 Amendments, and they were accepted with little dispute. The maximum discharge for *new* tankers was reduced from 1/15,000 to 1/30,000 of their cargo, and for nontankers a prohibition on all discharges in a twelve-mile zone was now substituted for the stipulation that they discharge ''as far from land as practicable.'' The U.S. sought some additional and more stringent restrictions, but they attracted insufficient support.[110] The most important addition to the discharge regulations was a provision allowing for the creation of ''special areas.'' The areas would be ones in which no discharges would be permitted except for ''clean ballast,'' that is, ballast with an oil/water mixture of less than fifteen ppm.[111] Such a restriction would, in practice, amount to a total prohibition and would be very costly to implement. But the opposition was minimal, and a number of such areas were agreed on for the enclosed seas surrounding Europe and the Middle East—the Mediterranean, Baltic, Black, and Red seas and the Persian (Arabian) Gulf. The lack of opposition by maritime states was easily understandable—a special area could be *implemented* only after the littoral states which would benefit from it had provided reception facilities for oily residues or ballast. It will likely be many years, therefore, before areas *declared* as ''special areas'' are actually *implemented* as such. At the same time, the maritime powers took great pains to ensure that the provision would not be taken as a precedent allowing other states to legislate unilaterally in ecologically sensitive areas. A ''special area'' can be declared as such only on the basis of thorough scientific information and only as a result of a formal amendment.

---

any of the meetings between 1971 and February 1973, and they had very little expertise on the technical issues.

109. The oil discharge regulations were discussed in Committee II and are included in Annex 1. Annexes 2 through 5, respectively, cover noxious liquid substances, packaged dangerous substances, sewage, and garbage. General provisions pertaining to jurisdiction and enforcement were discussed in Committee I and are contained as articles in the main body of the Convention. If a regulation obtained the backing of a simple majority in a committee, it went to the plenary where it required a two-thirds majority. Amendments also required a two-thirds majority in the plenary so very few changes were made there.

110. It wanted the tanker zones to be extended from 50 to 100 miles in width, the rate of discharge to be reduced from 60 to 30 liters per mile, and ''clean ballast'' to be defined as 10 ppm instead of 15 ppm (regulations 9 and 10). The U.S. obtained very little support for these proposals. In order to ascertain states' policies on these and other issues we examined the only record of the proceedings of Committee II which dealt with the oil pollution regulations. This was a single typed transcript of the meetings which was available to the authors at the IMCO Secretariat.

111. Regulations 9(4) and 10(4).

More important than the changes in the discharge regulations themselves were the ship and shore technologies which were accepted in order to enable ships to comply with the regulations. It was in these areas that previous international agreements had been so clearly deficient. Annex I sets out a whole series of technologies allowing improved retention-on-board and controlled effluent discharge. Regulations 14 through 18 include requirements for slop tanks, oil monitoring systems, oily water separators, filtering systems, oil/water interface detectors, special piping arrangements, and storage tanks. Having tried between 1971 and 1973 to promote the LOT alternative on the basis of these cheap and improved technologies, the European maritime powers now found it difficult not to accept the technologies that went along with LOT. Only in the case of an oil monitoring system was there major debate, and only then because of the lack of a reliable, well-developed monitoring system—particularly for refined or "white" oils. The Americans argued that requiring such a system would put pressure on industry to develop it, but traditional maritime states feared that such a maneuver could result in inefficient technological refittings that would impede ratification. As was common throughout these negotiations, the Scandinavian states submitted a compromise which allowed a flag state to waive compliance with the regulation if it felt that a reliable monitoring system did not exist.[112] This rather open-ended provision allowed an agreement to be reached but removed the incentive created by the original proposal. The lack of a monitoring system still stands as an impediment to gaining reliable pollution control.

By far the most important technical innovation at the conference was the inclusion of regulation 13, requiring segregated ballast tanks for all tankers over 70,000 dwt. Clearly an expensive proposition,[113] it was opposed to the end by states with large independent shipowning interests,[114] and by the two states anticipating the construction of the mammoth ULCCs.[115] But with enough tankers now constructed to provide capacity for another decade, the costs were a long way off. The oil companies accepted it, and most other states did so as well.[116]

112. Regulation 15(6).

113. The estimates of the increase in building costs—and hence shipping costs—ranged from 5 percent to 15 percent at the time. Oil industry officials indicated to the authors in 1975 that it will probably be around 8 percent to 10 percent. Others now estimate it at much less.

114. Norway, Denmark, Sweden, Germany, Greece, and the ICS. Independent shipowners are particularly worried about the problems of passing on extra costs, competitive advantage, and the time taken to recoup their investment.

115. France and Japan. The segregated ballast proposal reduces the economies of scale that a ULCC could otherwise produce. See PCMP/2/3 for the French position. The French decided to vote for the proposal at the time of the conference only when they realized that there was no chance of its defeat.

116. Even Liberia supported the proposal. A large proportion of its fleet is owned by the major oil companies, although much of that is hidden through dummy "independent" corporations.

Even the developing states supported the proposal, despite the indirect costs it would impose on them through its effects on the costs of shipping.[117] These states, even more than the others, opposed any *direct* costs being placed on their governments or fleets. At the same time, their delegations were preoccupied with the political issues—that is, the jurisdictional and enforcement questions—and they devoted little study to the *indirect* economic implications of technological innovation. Also, they argued that they could avoid the burdens of the convention simply by nonratification, ignoring the fact that their oil was imported in foreign ships.[118] Without strong environmental concerns and with a desire to avoid a diversion of resources from economic development, the support of many developing states for segregated ballast tanks was illogical. Only a few coastal African and Arab states had been suffering a rapid increase in operational coastal pollution since the 1967 closure of the Suez Canal, and the Arab states, with their numerous oil-loading terminals, were well aware of the impact of operational discharges. As a result they supported both segregated ballast and the application of the "special areas" regulations to their waters. Moreover, Egypt was one of the self-proclaimed leaders of the developing states at the conference, and it had its own compelling economic reason for supporting the proposal—namely, avoiding the alternative of reception facilities.[119] The Egyptian delegate, M. Fawzi, was thought to be the most knowledgeable developing state delegate on the technical ramifications of the oil pollution provisions, so the other developing states tended to follow his lead.

With regard to nontankers, segregated ballast tanks were not specifically prescribed, but regulation 14 in effect achieved the same thing by requiring all new nontankers over 4,000 tons not to use their fuel tanks for ballast. (In fact, most nontankers were then being so built.) There was no retroactive application of this regulation but, even so, it and the one for tankers could have a profound, if long-term, effect in eliminating intentional oil pollution.[120]

117. In the Committee II vote (31-7), approximately twelve developing states supported segregated ballast, and none opposed it. The Committee II transcript and interviews have allowed us to identify most of its supporters.

118. See particularly the Brazilian comment at MP/CONF/SR.8, p. 10.

119. Egypt was in the process of building a pipeline from the Red Sea to the Mediterranean, from whence a considerable amount of oil would be moved on short voyages to Europe. As vessels on these voyages would not be able to use LOT, especially in the Mediterranean "special area," all vessels would require expensive reception facilities at their loading terminals. However, if they had segregated ballast tanks, the reception facilities would not have to be as large, and hence the cost would be less. Regulations 10 and 12.

120. A weakness in the 1973 Convention is the lack of any "retrofitting" requirement. It will probably not be until the mid-1990s that most of the tankers (over 70,000 dwt) and nontankers without segregated ballast tanks will have been taken out of service and replaced by vessels with such tanks. In fact, given the drop in demand for new tankers after 1973, it will probably not be until the mid-1980s that a significant number of tankers will be built with

Other important technical innovations were also accepted. After twenty years of procrastination, the installation of reception facilities was finally to be required for ports where tankers and nontankers would have to discharge oil residues.[121] Because of its extreme costliness there had always been reasons in the past for postponing installation.[122] Now, its acceptance seemed to indicate a new willingness to deal with the problem. But the provision requiring it was misleading. First, the stipulation in regulation 12(1) that "the government of each Party undertakes to ensure the provision" of reception facilities has been interpreted by many states that supported its inclusion as not being legally binding.[123] This was made worse by the fact that many states felt able to approve the regulations only since there was little likelihood that their governments would ratify them. Furthermore, many of the governments which did view the regulation as legally binding did so assuming that they were not the party in their country which was obligated to build and pay for reception facilities. As one British government official commented, "We are sure to have the argument between Industry and Government as to who should pay. In my opinion . . . the industry and hence the public must pay in the end, and it is up to the industry to organize itself so that these facilities will be available when required."[124] Port authorities, repair yards, the shipping industry, and the various levels of government all disagreed as to who had ultimate financial responsibility. Thus, as has happened time and again, the paper agreement was but a prologue in the long negotiating process of procuring responsibility for, and perhaps implementation of, the regulations. Little had changed.

---

segregated ballast tanks. For a criticism of this aspect of the convention, see Eldon J. Greenberg, "IMCO: An Environmentalist's Perspective," *Case Western Reserve Journal of International Law* 2 (winter 1976), p. 137.

121. The ports that will require reception facilities include those servicing nontankers (for the residues from bilge water and oily ballast water) and oil/bulk ore carriers (OBOs), oil-loading terminals used by vessels on voyages of less than seventy-two hours or located within special areas, ports loading refined oils (at more than 1,000 metric tons per day), and ship repair yards or tank-cleaning facilities (regulation 12).

122. For Canada, which already has considerable reception facilities, the cost was estimated in 1974 to be $12 million (interviews).

123. Some delegates, especially those from developing countries, have stated that the provision was acceptable to them only because of its nonbinding character. They stated that their only obligation was to *urge* local authorities and industry to build facilities. Interviews.

124. Gordon Victory, "Discussion," *Proceedings of the Symposium on Marine Pollution*, Royal Institute of Naval Architects (February 27 and 28, 1973), p. 20. This viewpoint was challenged by the ICS, which feared that the costs would be imposed on port authorities and subsequently shipping companies in the form of port dues. Its memorandum stated, "The ICS respectfully submits the responsiblity for ensuring the provision of such facilities was part of the obligation which governments implicitly undertook to fulfill in adopting resolution A 175 (V) under which the 1969 Amendments were agreed" (MSC XXVII/5/2). There was a lengthy and confused debate on this issue in Committee II on October 24, 1973.

As ever, this debate on financial responsibility for the new regulations was even more intense between states than within them. For example, while France and Italy supported reception facilities in ports for tankers on short voyages, they argued that these would, of course, be installed at the loading terminals in the Middle East and not at the discharge ports in Europe. Indeed, France submitted a report to IMCO revealing the savings that would be made by so doing.[125] French support for the proposal is, therefore, almost meaningless in terms of its actual implementation—a problem now in the hands of the Arab oil exporters.

Apart from requiring new discharge standards and technologies, some significant changes in their scope were also made.[126] One such change was extending their scope to cover "fixed or floating platforms." It was American companies which operated the majority of off-shore drilling platforms, and the United States strongly opposed this new provision. Such platforms were very different from ships, and the United States—and others—opposed their control by this convention. But others were not so impressed with the American argument and the regulations were amended—largely as a consequence of an alliance between the Western and Eastern European states.[127]

A more contentious issue was the extension of the same regulations to "white" (refined) oils as are applied black oils.[128] The Americans took a strong position on this based on the findings of its scientists that refined oils had toxic, carcinogenic, and bioaccumulative properties and, therefore, posed a serious threat to marine life and human health.[129] The U.S. had included these refined oils in its own Water Quality Improvement Act of 1970. Their view was supported by the USSR, but the scientific evidence was not conclusive. The European states, and particularly the U.K., denied that there was scientific evidence to support the American claims,[130] but to this the United States retorted that any "doubt should be resolved in favour of protecting the environment."[131] Were this American assessment of the proper approach to environmental regulation accepted, it would have

125. MP XIII/2(*a*)/8 and PCMP/WP.22, p. 10.

126. Some minor changes involved the deletion of the exceptions for whaling vessels and vessels on the Great Lakes, the inclusion of all nontankers above 400 tons (instead of 500 tons) [regulation 9(2)], and the inclusion of the stipulation that the discharge of oil for purposes of combating oil pollution not be subject to the discharge regulations [regulation 11(*c*)].

127. The provision, article 2(4), was passed by a vote of 24-15-8 in Committee II on October 11, 1973.

128. Committee II Proceedings, October 10 and 16, 1973.

129. MP XIII/3(*c*)/5, pp. 5–10; PCMP/2/5 and 2/9; MP/CONF/15/8 and 15/9. There were long and even hostile debates in Committee II between the opposing sides (especially between the U.S. and the U.K.) on October 10 and 16, 1973.

130. PCMP/4/33; PCMP/4/17; MP/CONF/8/18.

131. Transcript of Committee II, October 16, 1973, p. 27.

constituted a profound development in the continuing dialogue between environmental and industrial priorities. But, not surprisingly, it was the Europeans who shipped more refined oil and who, therefore, had cause to consider the many practical problems of applying the regulations to light oils.[132] However, the United States did have broad support for its proposal: the Eastern European bloc, Canada and Australia (traditionally environmentalist allies), a number of developing countries, and, surprisingly, two conservative European maritime states (Greece and Italy) all rallied to the cause. By a narrow vote in Committee II, the American proposal on white oils squeeked through.[133] The "defection" of Greece and Italy (which were becoming visibly alarmed about the chronic ill health of the Mediterranean) was crucial to the passage of this important amendment.[134] The provision on refined oils has, therefore, been incorporated into the convention although, as the European states argued, the reception facilities and monitoring systems required to implement the change do not exist. With the convention needing numerous ratifications before it will enter into force, this provision may be one of the most serious stumbling blocks to achieving sufficient acceptances to bring the convention into actual operation.

The major focus of Committee II was operational or intentional oil pollution, but it also dealt with the control of pollution from accidents. The 1971 Amendments on tank-size limitation were integrated into the new convention;[135] the committee passed a rather noncontroversial regulation on the survival capability of a vessel;[136] and it rejected the very controversial proposal for building double bottoms on tankers over 70,000 dwt.[137] This

132. They raised a host of arguments: that it was impossible to use the LOT system with white oils since one cargo could not be loaded on top of another; that adequate reception facilities for their reception did not exist and might not exist for a long time; and that the application of the regulations to white oils would lead to much extra time being spent in ports by tankers carrying such products. They also argued that it was not sensible to impose the same regulations for oils with such different properties. Moreover, the aromatic fractions lethal to marine organisms are decidedly higher in refined oils produced in the U.S. and the Soviet Union than in those produced in Europe. Transcript of Committee II and interviews.

133. The vote was 23–19. The positions of all states are recorded in Chart 6.

134. Greece and Italy probably have the most polluted beaches on the Mediterranean, and their delegates were very troubled by the scientific claims of the Americans. It was also pointed out that the Arab states might refine more oil in the future with a dramatic effect on the volume of refined oil discharges in the Mediterranean (interviews). The extent of Italian concern is revealed in the comment of the Italian delegate to the Plenary that "if draconian measures were not taken at the earliest possible moment, the Mediterranean would become a source of desolation and death." MP/CONF/SR.8, p. 10.

135. Regulations 22–24. The dates in the 1971 Amendments were altered.

136. Regulation 25.

137. The proposal was defeated in Committee II by 9–22, and a second proposal that double bottoms be required on vessels between 20,000 and 70,000 tons was defeated by a vote of 5–21.

last proposal was one to which the American delegation was firmly committed. The Council on Environmental Quality, a sizable number of politicians (particularly Senator Magnuson), and the American environmental lobby groups all placed this demand high on their list of priorities. The installation of double bottoms would, it was strenuously argued, reduce substantially the likelihood of oil spillage from groundings.

But growing doubts persisted about the real merits of the idea. Indeed, despite the support for it in the American study done for IMCO in 1972,[138] there were nagging doubts about it even within the U.S. Coast Guard at the time of the conference.[139] And others flatly opposed it.[140] It would, they said, increase pollution, not reduce it. For one thing, the likelihood of explosions resulting from oil leaking out of the cargo tanks into the double bottoms would increase. More significant, as a result of unsymmetrical flooding after a puncture, double bottoms would destabilize the tanker and reduce the possibility of salvage. Flooding of the double hull would certainly tend to settle the ship so as to make the ship's—and cargo's—total loss more likely. And, of course, for major tankerowning states, the prospect of spending vast sums of money for a structural change which would increase the likelihood of the loss of a vessel was an abhorrent specter.[141] As the U.K. delegate commented, "One tanker which fails to get off a rock because it has had added weight to it and which thereby becomes a total loss could completely wipe out all the advantages of segregated ballast and double bottoms in respect of accidental pollution."[142] In the absence of more convincing technical arguments the American initiative lost all but its "diplomatic" supporters.[143]

Despite this "loss" and the general doubts concerning states' willingness

138. MP XIII/2(2).

139. Interviews. Although the delegation maintained a united front during the Conference, the difference of opinion was known to some other delegations. The Coast Guard formally came out against double bottoms in October 1975. It cited a study revising previous evaluations of the importance of double bottoms and suggesting alternatives. *Federal Register* 40 (October 14, 1975), pp. 48289–48290. See footnote 163 in this chapter.

140. PCMP/4/23 and ICS submission PCMP/4/13. See in particular the U.K. submission.

141. The cost depended on many features of the vessel, but could be over 10 percent. An American official who was familiar with the U.S. government's negotiations on double bottoms with its own oil and shipping industries and other governments said that industrial opposition was based solely on cost and that their other arguments were "window-dressing." In fact, there do seem to be substantial differences of opinion on its technical merits.

142. Transcript of Committee II, October 17, 1973, p. 124.

143. Even Canada and Australia opposed the U.S. on this issue on technical grounds. And the Soviet Union, which had generally voted with the Americans in the Conference, would support only the proposal for tankers over 70,000 dwt (in which it had little interest) and not the one for those between 20,000 and 70,000 dwt (in which it had a great deal of interest because it owned many small "coastal tankers"). The positions of the states are recorded in Chart 6.

or ability to ratify the convention, the 1973 Conference—especially from an historical perspective—was a landmark in international environmental regulation. For the first time the installation is required of those ship and shore technologies necessary for the retention on board and proper port disposal of oil residues. Moreover, in the long term, the possible sources of oil discharges will be reduced by requiring the installation of segregated ballast tanks on tankers and the construction of nontankers that will not use their fuel tanks for the carriage of ballast.

If the convention has a major flaw, it is that some of its provisions for oil *and* other substances may be too ambitious.[144] If so, their inclusion may impede the convention's entry-into-force. Success is not without its costs. It is, unfortunately, highly questionable whether the political will power exists to finance the installation of reception facilities or to force port authorities or industry to build them. This is true not only for the developing countries but for the developed ones as well. A particular problem exists in securing the building of reception facilities for voyages of short duration or in special areas, as this is the responsiblity of the Arab oil-exporting states on the Mediterranean (Syria, Egypt, Libya, and Algeria). In drafting the convention, many of its strongest backers admitted that it was unlikely that it would have a major effect on the oil pollution problem until the early 1980s. Even this may have been too optimistic a projection.

After many decades of indifference or even opposition to international oil pollution controls, this was the United States' conference. Three American proposals were the main focus of debate at the conference and, in retrospect, they provide a fair litmus test for the policies of a wide spectrum of participants. Chart 6 records the positions of the participating states on the application of the regulations to refined oils, the installation of segregated ballast tanks, and the construction of double bottoms. The strongest backers of the U.S. were the Eastern European and developing states. For the Soviet bloc, their location on semi-enclosed seas and their minor status in the world oil transportation industry undoubtedly had a strong influence. In addition, as many at the meeting perceived, the influence of "détente" was also potent. For the developing states, many influences helped shape their policies. Some—the Arab and African states—were concerned about the increasing pollution they were suffering from the shipping activities of the developed countries; most of this developing group had small fleets composed of older and smaller vessels that would not be affected

144. This applies particularly to the requirement for reception facilities for the residues of all existing nontankers and for all tankers either traveling on short voyages, to repair ports, or to ports in special areas, or for those carrying refined oils. Another major problem is that states were required to ratify annex I and annex II on chemical substances together. This was changed in the 1978 Protocol to the Convention.

Chart 6
State Positions on Major U.S. Technical Proposals at 1973 Conference

	Application of regulations to refined oils	*Mandatory segregated ballast on tankers over 70,000 dwt*	*Mandatory double bottoms on tankers over 70,000 dwt*
	Accepted Vote: 23-19	Accepted Vote: 30-7	Rejected Vote: 9-22
U.S.	X	X	X
Egypt	X	X	X
Bulgaria	X	X	X
USSR	X	X	X
Argentina	X	X	
India	X	X	
Australia	X	X	0
Canada	X	X	0
Italy	X	X	0
Romania	X		
South Africa	X		
Belgium	0	X	0
Finland	X	0	0
Greece	X	0	0
France	0	X	0
Liberia	0	X	0
Netherlands	0	X	0
New Zealand	0	X	0
U.K.	0	X	0
Denmark	0	0	0
Germany (F.R.)	0	0	0
Japan	0	0	0
Norway	0	0	0
Sweden	0	0	0

Key
X = Vote or statement in favor of proposal
0 = Vote or statement against proposal

NOTE: Votes were recorded, but a roll call was not taken. Delegation positions were ascertained by statements made in the Committee and by interviews. Although the information is therefore incomplete, it should be recalled that the largest number of states participating in any vote in Committee II was only forty-two. Indeed only four of the thirty-nine developing states participated in the debates on these issues (Argentina, Egypt, India, and, of course, Liberia).

Spaces between groups of states delineate those groups with overall silmilar policies.

by the new regulations; and few, if any, of the developing state representatives considered the indirect or long-term effects of the regulations. These states were supported, and often led, by several nonmaritime developed countries such as Canada and Australia. On the other hand, as was to be expected, the major opponents of the American initiatives were the developed maritime states which had substantial shipping interests or which imported large volumes of oil. It was their industries which would be faced with the problem of paying for the new technologies and passing on the costs. And it was their consumers who would ultimately have to pay for their ratification of the convention.

THE 1978 TANKER SAFETY AND POLLUTION PREVENTION (TSPP) CONFERENCE

At least on paper, the 1973 MARPOL Convention was a triumphant achievement for IMCO. The reality, unfortunately, was different. Existing law still rested on the antiquarian 1962 Amendments, and there was no indication that it would soon be superseded. In addition, the MARPOL Convention was itself filled with problems that would have to be resolved before the convention would be acceptable to most states.

To procure the entry-into-force of the 1969 Amendments was IMCO's most pressing challenge after the 1973 Conference, and it was to this that the new secretary-general, C. P. Srivastava, turned his attention. With the approval of the Council, he repeatedly urged members to accept the five-year-old regulations, a campaign that was eventually successful. His efforts were also supported by a vigorous diplomatic campaign by the United Kingdom. In January 1977, Nigeria became the thirty-eighth state to ratify the amendments, and they came into force one year later on January 20, 1978, over eight years after their adoption. Meanwhile the MEPC had begun work on the new 1973 Convention, setting specifications for the operation of oily-water separators and oil-monitoring systems[145] and formulating guidelines for the construction of reception facilities.[146] In addition, a special symposium on the implementation of the convention was convened in Mexico in 1976. But the situation was not hopeful. Some of the necessary technologies still had not been developed and others, such as reception facilities, were viewed by many as too expensive. Furthermore, annex I dealing with oil pollution was inseparably linked to that for hazardous chemicals (annex II). This further deterred ratification, as the latter annex imposed additional onerous burdens. By 1976, after three years, only three minor states had accepted the instrument. Certainly, no one was considering a new and more stringent one.

145. Res. A.393(X).
146. MEPC VI/8.

The single largest source of ship-generated oil pollution comes from intentional discharges such as the one pictured here. This vessel was caught by a Canadian surveillance aircraft as it deliberately pumped its oily waters into the sea off the east coast of Canada. Even with such evidence, few violators are successfully prosecuted by either the flag or coastal state. (Photo: Transport Canada.)

Ironically, the only proposal to amend (and strengthen) the MARPOL Convention was essentially unrelated to environmental considerations. Having heavily overbuilt during the tanker boom of the late 1960s, the tanker market was caught in a tremendous slump with the decline in oil demand following the 1973 OPEC price rise. Within a couple of years, approximately 15 percent of world tanker tonnage was laid up.[147] This situation presented an unusual opportunity for a solution to the operational pollution problem by ''retrofitting'' segregated ballast tanks. A proposal to this effect was put forward to IMCO by a nongovernmental source,[148] but the uncommonly welcome reception it was given by some states was motivated not by environmental concern but because its adoption would reactivate much of the laid-up tonnage by reducing current capacity. It was for this reason that those states whose owners had been hardest hit by the slump (the ''independent'' operators, especially those on the previously lucrative ''spot'' market) quickly seized on the suggestion. Not previously

147. By March 1976 almost 46,000,000 dwt was laid up. See ''World Merchant Shipping Laid Up for Lack of Employment'' (London: General Council of British Shipping, as of December 31, 1977) (mimeo.). In fact, since the industry was then using various methods of dealing with the tanker surplus (slow-steaming; trading part cargo, that is, not fully loaded; extended port time), unused capacity was much higher.

148. The original proponent was Arthur Mackenzie of the Tanker Advisory Center (New York), an outspoken environmental advocate on shipping matters. His letter to the secretary-general is found in MEPC III/17/2.

Despite their smaller contribution to the total quantity of oil entering the marine environment, it has been the dramatic effects of shipping accidents that have provided the continuing impetous for legislative change. Pictured above is the ship that started it all, the 119,000 ton *Torrey Canyon*. After waiting several days, the British Royal Airforce bombed the stricken ship. Its shattered remains are pictured here. (Photo: Press Association.)

renowned for their environmentalism, Norway and Greece in 1975 proposed to the MEPC that a study on retrofitting be done. With Italy, they were given the task of preparing one.

Their report was issued in March 1976 and, not surprisingly, it recommended that just such a course be adopted.[149] Greece, Norway, and Sweden supported the recommendation, as did a few Mediterranean countries (which had no tankers but much pollution) and a few oil exporters (which saw the idea as an alternative to reception facilities). In addition, these states were buoyed by a report of the OECD which also saw retrofitting as

149. MEPC V/7. See also the Greek submission, MEPC VI/6/3. Interestingly, Italy later dissociated itself from the report.

The 46,000 ton Liberian tanker, *Pacific Glory*, leaves a trail of oil behind it, after a collision with another Liberian tanker, *Allegro*, in the English Channel in October 1970. The qualifications of officers aboard both ships was later discovered to have been inadequate. This prompted a storm of protest against flag-of-convenience shipping and initiated a change in Liberian maritime policy. This accident also helped initiate the move to upgrade crew standards world-wide. (Photo: Keystone.)

the best answer to their grave economic problem.[150] On the other hand, those maritime states with tanker fleets largely owned by the oil industry were unsympathetic, as were the developing maritime nations. So too were most oil importers for whom the slump was, if anything, a boon in that the cost of transport had plummeted. Even "environmentalist" Canada opposed the idea fearing, with the others, that such a large expenditure and reduction in capacity would further raise the price of oil. The United States

150. Interviews. The OECD report is not public. Of the seventeen options considered by it, retrofitting was the most desirable, followed by trading "part cargo" and early scrapping.

Within hours of launching the gleaming supertanker *Olympic Bravery* founders on the rocks off the Brittany coast of France in 1976. The ship was totally lost causing heavy pollution from the discharge of its bunker fuel. Although newly-built, this tanker was on its way to Norway where it was to be laid-up. With the tanker tonnage surplus that resulted from the OPEC oil price rise of 1974 there was no market for it. (Photo: Ledruff, Sipa Press.)

(which was conducting its own study) was silent. In a vote of the MEPC in December 1976, the proposal was defeated.[151]

No sooner had the gavel fallen to end the MEPC session than nearby in the "City" the bell was ringing in Lloyd's. Another shipping disaster. The *Argo Merchant* was aground off Cape Cod. The bell was to ring many times in the next few months, as the *Argo Merchant* signaled the beginning of a spate of accidents—many in American waters—and a renewed offensive for higher environmental standards.[152] Once again, the United States led the way.

151. "Report from the IMCO Marine Environment Meeting," *Norwegian Shipping News*, February 11, 1977, p. 10.

152. Two days after the *Argo Merchant* grounded on December 15, one of the *Torrey Canyon*'s two sister ships, the *Sansinena*, exploded in Los Angeles harbor killing nine persons. These and other accidents (*inter alia*, *Oswego Peace*, December 24; *Olympic Games*, December 27; *Grand Zenith*, December 29) were widely reported at the time. For a list of fifteen casualties between

The *Argo Merchant* sinking off Massachusetts in December 1976 leaves a snaking trail of oil. This was the first in a series of tanker accidents occurring adjacent to the American coast in late 1976 and early 1977. Although a small tanker (carrying 27,000 tons) the threat of pollution to the popular resort areas of New England and to the rich Georges Bank fishing grounds caused a sensation in the United States. The slick emanating from the ship stretched 100 miles. (Photo: UPI.)

Within weeks, Senator Magnuson convened hearings of the Senate Commerce Committee to discuss the tanker pollution problem and to consider new legislation to combat it.[153] The Coast Guard was already under legal challenge for its alleged failure to fulfill its environmental mandate, and at the hearings it was broadly attacked for environmental conservatism.[154] With

---

December 15, 1976 and March 27, 1977, see Hans-Frederick Grorud, "Tanker Safety," *Veritas* 23 (July 1977), pp. 2–3. The pattern of events that followed was strangely reminiscent of the circumstances surrounding the 1966 IMCO amendments to the Safety of Life at Sea (SOLAS) Convention. These amendments followed on a threat of unilateral American action after many American lives were lost in tragic fires on board the *Yarmouth Castle* and *Viking Princess* (IMCO doc. A/ES.III/SR.1–5, February 6, 1967).

153. *Recent Tanker Accidents: Hearings Before the Committee on Commerce, United States Senate, January 12–13, 1977* (Series 95-4) (Washington, D.C.: Government Printing Office, 1977). Magnuson's bill, S. 682, received Senate approval as the Tanker and Vessel Safety Act. After the TSPP Conference, House of Representatives bill H.R. 13311 replaced the Magnuson bill as the legislation most likely to pass. Although it does contain some proposals at variance with the 1978 Protocols, environmentalists have criticized it and its Coast Guard supporters as having "collapsed" to IMCO standards. One focus of criticism is the failure to require collision avoidance systems of the high caliber available on the market.

154. The Washington-based Center for Law and Social Policy had in February 1976 instituted an action on behalf of seven environmental organizations to force the Coast Guard

The bow of the oil tanker *Sansinena* (71,763 dwt) protrudes from the waters of Los Angeles harbor after an explosion and fire destroyed the ship. This accident in December 1976 was one of the rash of accidents occurring in the United States soon after the grounding of the *Argo Merchant*. The explosion helped prompt demands for "inert gas systems" to be installed in tankers to combat the threat of hydrocarbon gas buildup. Coincidentally, the *Sansinena* was one of two sister ships of the *Torrey Canyon*. (Photo: AP.)

the change in administration, the Department of Transportation (of which the Coast Guard is one part) was under new leadership, and there appeared to be a distinct difference in the orientations of the outgoing secretary of transportation, William Coleman, and the incoming Carter appointee, Brock Adams. Adams, like Magnuson from the State of Washington, struck a stronger environmental stance.[155] Combined with pressure from the "Hill," it was evident that significant changes were in the wind.

---

to promulgate additional regulations under the *Ports and Waterways Safety Act*. See *Natural Resources Defense Counsel et al. v. Coleman et al.* (District Court, Washington, D.C., Civil Action 76-1081). The action was later stayed by mutual agreement. As the Coast Guard had promulgated some regulations and had rejected others only after written evaluations, the outcome of the case was not at all certain.

155. Coleman had opposed unilateral action soon after the *Argo Merchant* disaster as

The changes happened fast. On taking office, Adams immediately appointed a Marine Safety Task Force under Rear Admiral Sidney Wallace, long one of the United States' most respected representatives at IMCO meetings. Although the task force was certainly involved in the ensuing discussions, the major impetus for the new administration's policy came from elsewhere, most notably from the Council on Environmental Quality (CEQ). Created under the National Environmental Policy Act (1969), the CEQ functions as a continuous prod to all governmental departments on environmental matters. Operating in this case under the overall guidance of another White House body, the Office of Management and Budget, the CEQ worked in close conjunction with environmentalists outside the bureaucracy and with other agencies (most notably with State and the EPA).[156]

The result was "the Carter Initiatives," announced on March 17, 1977, barely three months after the *Argo Merchant* had foundered.[157] In his initiatives, President Carter promised action to prevent further accidents by requiring within five years collision avoidance aids, inert gas systems, improved steering standards, and double bottoms on all tankers over 20,000 dwt. Rules for liability for pollution damage, enforcement, and certification were also to be improved.[158] To prevent operational pollution, President Carter endorsed the installation of segregated ballast tanks on new *and existing* tankers above 20,000 dwt and the ratification of the MARPOL Convention. Flexibility was retained, however, as alternatives would be

---

making him "almost sick." *Seatrade* 7 (January 1977), p. 11. Earlier, he had allowed the Coast Guard publicly to oppose the Environmental Protection Agency's support for double bottoms. Adams, as a former Congressman from Washington, was no stranger to the tanker issue. His home state had passed legislation barring any tankers in excess of 125,000 dwt from entering its waters, and he now testified of his firm commitment to bring the situation under control.

156. Interviews. There are clear differences of opinion as to the extent of the Coast Guard role in the formulation of the initial policy. Many participants argued that the policy was virtually imposed on it. Certainly the Coast Guard's policy of 1975 which opposed the requirement of double bottoms was reversed in this period. The "conservatism" of the Coast Guard is understandable, however. It had had experience with "unilateralism" before and knew, for one thing, that double bottoms were not going to be acceptable internationally. It was also the Coast Guard that would be given the task of developing any position in detail after the initial flurry of interest had subsided, and it was the Coast Guard that would have to take it to IMCO and defend it. Its officials were naturally concerned to prevent the government at home being locked into unrealistic standards.

157. "Tanker Safety Text," *Congressional Quarterly Weekly Report* 35 (March 26, 1977), p. 568. An excellent presentation of the American position, its projected effects on pollution, accidents, and safety as well as estimates of its costs can be found in *Draft Environmental Impact Statement: International Conference on Tanker Safety and Pollution Prevention* (Washington, D.C.: U.S. Coast Guard, February 1978).

158. On the matter of pollution damage liability, a House Merchant Marine and Fisheries Bill, HR.6803 (the "Super-Fund" legislation), was approved by the House of Representatives in mid-1977 and if passed finally would create a $200-million liability fund. See Chapter 5 for more discussion of this legislation.

considered where they "can be shown to achieve the same degree of protection against pollution." The President pledged "cooperation with the international community," but the United States would act alone if it had to. Two months later the Coast Guard began to move in just that direction as it issued a "notice of proposed rule making" under the Ports and Waterways Safety Act.[159] Once again, the United States was on the move.

At the IMCO Council meeting in May, the new secretary of transportation, Brock Adams, delivered the news. The United States, he said, retained its "commitment to international solutions" and noted that the Carter initiatives and proposed rules "did not imply unilateralism on the part of the United States" or a "breach of faith with IMCO."[160] But he added unequivocally that only "if IMCO tailors its moves to suit and protect the U.S., we will accept; if not, we reserve the right to impose our own rules."[161] The Council, clearly shaken by yet another threat of U.S. unilateralism, agreed to convene another conference on marine pollution—the Conference on Tanker Safety and Pollution Prevention. The date was set for February 1978, scarcely nine months away.

To prepare a basic working document for the conference, an intersessional working group met in May, June, and July, and a joint meeting of the MSC/MEPC was held in October.[162] During these sessions it was agreed that all new regulations would be formulated as protocols to the 1973 MARPOL and 1974 SOLAS (Safety of Life at Sea) conventions. After reviewing the American proposals to control accidental pollution, the October meeting also gave tentative approval to the proposals, except in two cases. Although the desirability of collision avoidance aids was accepted (these assist in plotting courses in traffic), the dearth of specifications for them led most countries to recommend that the issue be handled through the regular IMCO committees for future acceptance by the Assembly. The second proposal that was rejected was, of course, double bottoms. This was now replaced with a suggestion that all future segregated ballast tanks would be "protectively located" to provide for maximum protection against the breaching of cargo tanks in the event of an accident. The policy change that the Coast Guard had itself undergone on this issue in 1975 facilitated this result.[163]

On the major issue of debate, the requirement of segregated ballast tanks

159. Enforcement was to be strengthened by the introduction of annual inspections (a "tanker boarding program"), and the certification and training of crews was to be improved by expediting the 1978 IMCO Conference on the subject (see IMCO doc. STW X/7, October 6, 1977 for the draft convention).

160. C. XXXVIII/SR.1, p. 10.

161. *Seatrade* 7 (June 1977), p. 21.

162. The summary report is contained in MSC/MEPC 10.

163. This change resulted from a Coast Guard study showing that collisions were as important a source of oil outflow as groundings. Lt. Cmdr. James C. Card, Paul V. Ponce,

on all ships above 20,000 dwt, delegations were polarized. Most were unenthusiastic about building segregated ballast tanks on *new* vessels above 20,000 dwt (the 1973 Convention demanded them only above 70,000 dwt), but they were extremely hostile to the demand for retrofitting on *existing* ones. As a result, in addition to the American proposals two alternative packages were drafted for possible inclusion into a protocol to the MARPOL Convention.[164] Package #1 was quite close to the American proposals requiring segregated ballast tanks and inert gas systems for *all* crude carriers (new and old) above 20,000 dwt and for *all* product carriers above 50,000 dwt. This package was supported by Norway, Sweden, and Greece. Package #2 was backed by a sizable group of countries led by the United Kingdom in collaboration with the oil industry. In place of the segregated ballast tanks, this package offered a new system of tank washing—crude-oil-washing (COW)—for *all* crude carriers above 70,000 dwt. This process utilized crude oil to wash out the cargo tanks during the cargo discharge process. It was estimated that 80 to 90 percent of the oil residues and sludge could be removed in this way. Only those tanks that would be taking on ballast would still need to have a water wash, and even they would already have had much of the oil removed. The process was ingenious and seemed to offer an easy and inexpensive solution.

Like the introduction of load-on-top a decade earlier, COW was a product of the commercial dexterity of the oil industry. The system originally had been developed by British Petroleum and Exxon in the late 1960s as a way, not of reducing pollution, but of saving cargo. Technically feasible, the system had nevertheless been rejected as economically unprofitable—until the 1973 oil price rise! At that time, B.P. and Exxon decided to perfect the system and, when its commercial utility was demonstrated, its use was quickly expanded.[165] Though raised at IMCO in 1976 in response to the MEPC study on retrofitting, it was dropped along with the retrofitting proposal. Later with the introduction of the Carter initiatives it was quickly resurrected, and the environmental advantages of this "new" approach were now widely touted.[166]

---

and Lt. Cmdr. Warren P. Snider, "Tankship Accidents and Resulting Oil Outflows, 1969-1973," *Proceedings of the 1975 Conference on Prevention and Control of Oil Pollution, March 25-27, 1975* (sponsored by American Petroleum Institute, Environmental Protection Agency, and U.S. Coast Guard), pp. 205-213. The time period of this study is limited to 1969-1973. A larger period would likely paint a very different picture, especially if recent large losses from groundings were included (*Olympic Bravery, Argo Merchant*, and *Amoco Cadiz*). The arguments against double bottoms are not decisive. Some environmentalists have condemned the study just cited as "number juggling." Interviews.

164. MSC/MEPC 10, annex XXIV.

165. One Norwegian shipowner told the authors that COW was providing him with an 18 percent return on his investment. However, the economic return is less and may actually impose an economic burden on smaller tankers.

166. Descriptions and critiques of crude oil washing can be found in MSC/MEPC/

In choosing between crude-oil-washing and retrofitted segregated ballast tanks, two issues were of paramount concern: their respective environmental benefits and their relative costs. Unfortunately, discussion of these issues was so clouded by differing and often politically biased technical and economic studies that many states found it difficult to make dispassionate, rational assessments.

Supporters of retrofitting argued strenuously that only a *structural* solution to pollution prevention would work given the sorry experience with *operational* solutions like LOT and, by extrapolation, COW. Segregated ballast tanks would completely remove the routine problem of dirty ballast water, although it would not affect the necessity to clean cargo tanks. However, even here it was suggested that by not using the cargo tanks for ballast, sludge buildup would be reduced by about 30 percent.[167] In any event, COW could always be employed on these tanks *in conjunction* with segregated ballast tanks. As the primary method of pollution control, COW was strongly attacked by its opponents. It was an operational technique requiring highly skilled crews who would invariably be tempted to take short cuts to save time in port. Therefore, it required stringent and expensive enforcement. In addition, for those tanks that would be used to carry ballast, a final water wash and use of the load-on-top procedure was still necessary. Moreover, if mishandled, it could be dangerous (the system necessitated the use of sealed tanks employing inert gas systems to prevent explosions) and could cause substantial air pollution if the highly contaminated fumes were allowed to escape during unloading and cleaning operations. There was such limited experience with its operation that it was impossible to say that it would work.

In response, the backers of crude-oil-washing (endearingly calling themselves the "Friends of COW") criticized segregated ballast tanks for being only a partial answer to the problem, and they pointed to the many obvious advantages of the COW system.[168] If operated effectively (which they alleged it would be, as it would be profitable to do so, and, unlike LOT, would occur under inspection while in the discharge port), it would substantially reduce *all* the tanker residues without use of a salt water wash. Certainly it would require stringent operational guidelines, but its potential

---

INF.17, INF.22, INF.23, INF.35, and MSC/MEPC 10/Annex XIX. The debate was carried on at the eighth MEPC meeting in December 1977. See MEPC VIII/WP.6 and MEPC VIII/14/2 and 3. OCIMF had submitted descriptions of the system at the earlier MEPC sessions that discussed retrofitting. See MEPC VI/6/3 and the British submission MEPC VI/6/2. Also see the following Conference documents, TSPP/CONF/INF.2 and INF.3 (submitted by OCIMF) and TSPP/CONF/7/3, 7/19, 7/20, 7/21, and 7/22.

167. Interviews.

168. For a summary of the arguments, see MSC/MEPC/10, annex XVII. There was a plethora of other papers arguing the many pros and cons of each system. Only the major arguments have been cited here.

effectiveness was so significant that it was to perfecting these guidelines that attention should be turned, not to disputing its desirability vis-à-vis the outlandishly expensive retrofitting proposal. This was the real crux of the matter: cost.

Even more than environmental desirability, arguments about costs varied widely. Perhaps the only consistency was that the size of a state's estimate of the cost for retrofitting the world fleet increased with its opposition to the proposal (and decreased with its support). Differences were often accounted for by differing assumptions underlying their calculations, but these were so numerous and so deeply buried that they were virtually unfathomable.[169] These cost variations were one of the central issues of the entire conference, and they made rational analysis very difficult.

Before the conference began, the OECD had itself undertaken a study of the implications of the retrofitting proposal. Although the OECD's Maritime Transport Committee had been in existence since 1947, its major focus was the commercial dimension of shipping. Hence, it had not concerned itself with IMCO's work, but rather with the liner conferences and UNCTAD.[170] The submission of its large study on retrofitting was its first foray into maritime environmental affairs. Unfortunately its conclusions varied so widely, were so complex and so disputed that the paper only caused confusion. A Norwegian/Swedish/Greek study analyzed in it estimated the total cost of conversion of the world tanker fleet at only $2.59 billion, while the French-commissioned "Micro" study estimated the costs at $5.09 billion.[171] The French later became more sympathetic to retrofitting, and another study by them brought the costs down to $3.35 billion.[172] Other estimates varied as widely: OCIMF put the conversion price tag at $6 billion, the United States at $2.93 billion.[173] When combined with varying estimates of the impact on shipbuilding, on the loss of cargo-carrying capacity, and on the reemployment of laid-up tonnage, estimates of the impact of retrofitting on the final cost of the price of oil varied from 0.8 percent to 2.0 percent.[174]

169. For guidelines to identify the cost assumptions of retrofitting segregated ballast and COW, see MSC/MEPC/10, annex XVIII.

170. Its earlier work on flags of convenience and substandard ships was motivated not by environmental concerns but by opposition to "unfair competition."

171. The OECD paper was submitted as TSPP/CONF/7/4. It noted "significant differences" resulting from "the number of different parameters." Because of the "major uncertainties" it concluded that "one must accept *all* estimates," as none is "likely to be '*truer*' than the others" (p. 39).

172. The differences in these estimates are nicely summarized in a Swedish paper, TSPP/CONF/7/6. It notes that the cost of corrosion protection by coating of ballast tanks varies from $3.3 billion in the Micro study to $1.4 billion in the Swedish estimates (p. 3).

173. OCIMF estimate, TSPP/CONF/INF.3; U.S. estimate, TSPP/CONF/INF.17.

174. See TSPP/CONF/7/1, p. 16, TSPP/CONF/7/4, p. 48. The United States paper noted that "the cost to the consumer would be marginal as . . . the impact on freight rates (12

Well before the conference opened, it was clear that the participants were polarized. At one extreme were the oil companies. At the 1973 Conference, with enough tankers already constructed to provide service for many years to come, the prospect of segregated ballast tanks on *new* tankers was a far-off investment. With the present American proposals, retrofitting and loss of capacity meant immediate and sizable expenditures. For the first time what was at issue were environmental costs with no commercial side benefits, and the industry was fiercely opposed.[175] Although OCIMF was active throughout the preparatory meetings and at the conference itself, the chief spokesman for the argument of the oil industry was the United Kingdom.

The position of the U.K. had been developed in very close consultation with OCIMF and the oil industry, and it was adamantly opposed to the U.S. position. Within the British civil service the lead actor was the Marine Division of the Department of Trade, whose major nongovernmental ties were those with the shipping and oil companies. There were certainly no important formal interdepartmental or public consultative mechanisms, though final instructions were circulated to other departments prior to the conference. As a result, the interests of the environmentalists and of the shipbuilders (which would greatly benefit from the business in a period of little construction) were scarcely represented.[176] The British-based environmental organization Friends of the Earth did get the support of the national shipbuilders association for retrofitting and it did promote a debate on British policy in the House of Lords, but these had no noticeable impact on government policy.[177]

At the other end of the spectrum were the United States and its allies, Norway, Sweden, and Greece. Although not the lead actor in formulating the Carter initiatives, the Coast Guard was now firmly in charge of the American delegation. As with the 1973 Conference, the American position was a product of wide consultations among many governmental agencies

to 13 percent) is less than the annual average variation in those rates" (TSPP/CONF/INF.17, p. 5). Meanwhile, the OCIMF paper saw freight rate variations as being "between a low of 18 percent for large ships up to almost 50 percent for small ships" (TSPP/CONF/INF.3, p. 25).

175. One knowledgeable delegate informed the authors that the industry had actually offered to pay the costs of building reception facilities in the Arab states if the U.S. proposals were defeated. Interviews. If so, it will be interesting to see if this promise is fulfilled.

176. The Department of the Environment had one junior officer on the twenty-nine-man U.K. delegation but, despite the fact that the conference was held in London, not one representative of the Department of Industry (which is concerned with shipbuilding) was represented on the delegation. In contrast, six representatives of the private British Council of General Shipping were accredited and, throughout the discussions, the delegation openly conferred with representatives of the ICS and OCIMF, from which much of their information was obtained.

177. Interviews and *Hansard* 388 (House of Lords Official Report), January 25, 1978, pp. 344–361 and 370–400.

and private interests.[178] The final preparation of instructions was left to the Coast Guard, which had to prepare a package balancing the demands of Congress and the environmentalists with their own conception of technical feasibility and with the more conservative cost demands of the Treasury.[179] Especially with President Carter's personal support for stronger environmental controls, the result was a firm commitment to segregated ballast tanks—with COW as an alternative only where it could be shown to provide equivalent protection. To advance this position, a tour of a dozen carefully selected countries was carried out in the months preceding the conference.[180]

More extreme than the United States in advocating the retrofitting of segregated ballast tanks were Norway, Sweden, and Greece. Unlike the U.S., they did not entertain any alternative to retrofitting, as their interests were obviously commercial, not environmental. In mid-1977, 39 percent of Sweden's tanker tonnage, 28 percent of Norway's, and 21 percent of Greece's was laid up.[181] Other countries had fared far less badly. Indeed, for the first time in the history of Nordic participation at IMCO, Denmark broke with its Scandinavian allies on an important issue.[182]

It was amidst such polarity and with grave fears of failure that the TSPP Conference opened in London on February 6, 1978. Fifty-eight states were in attendance—twenty-two from the developed Western group, four from the East European bloc, and thirty-three developing countries. With the numerical majority clearly on its side, the United Kingdom had intimated that it might seek an immediate vote, a move that would certainly provoke an American walkout and a collapse of the meeting.[183] From the beginning, therefore, it was the earnest concern of the secretary-general and of the conference chairmen to avoid such a possibility. Indeed, Mr. Srivastava had

178. Representatives of all spectrums of the American bureaucracy commented on the lack of special treatment given to the oil industry.

179. The composition of the American delegation reflects this broader based participation. Of twenty-seven representatives, eleven were accredited from the Congress (only a few attended), four were environmental representatives from inside and outside the government, and two were from the industry.

180. The tour included Canada, the four Scandinavian states, the USSR, Netherlands, France, Japan, Kuwait, Egypt, Ivory Coast, Brazil, and Argentina.

181. Sweden's situation was exacerbated by its idle shipyards and Greece's by lay-ups in its nationals' Liberian-registered tonnage as well. "World Merchant Shipping Tonnage Laid Up for Lack of Employment" (London: General Council of British Shipping) (mimeo.).

182. There is only one large Danish tankerowner, A. P. Muller, and he strongly opposed intergovernmental interference into the commercial aspects of the tanker industry. Also, he owns oil-drilling rigs which are leased to the oil companies and so is highly sensitive to their views. Finland too opposed retrofitting but kept silent at least during the public conference sessions out of deference to its Nordic partners and to the United States delegation that had visited it. Interviews.

183. An exchange on this matter had taken place between the U.K. and the U.S. before the conference convened. Interviews.

quietly urged the key states to avoid any confrontation for fear that a U.S. walkout and unilateral legislation would destroy the organization. In addition, he had sought to ensure that competent and conciliatory chairmen were selected to run the meetings. In this regard, the central figure was the chairman of Committee II (also chairman of IMCO's Maritime Safety Committee), L. Spinelli of Italy.

After the opening plenary session, the action quickly shifted to Committee II where the alternative packages were debated.[184] As state after state spoke out against retrofitting, it was soon apparent that the U.S. and its three key allies were becoming increasingly isolated. Some states with heavy coastal pollution (such as Portugal, Spain, and Morocco) supported the American position, as did a few other countries (for example, Cyprus and Venezuela). Most, however, strongly preferred COW.[185] The urgent task was, therefore, somehow to get the delegates talking about a compromise.

The opportunity for compromise certainly existed. Some key states were truly cross-pressured and saw the need for an accommodation. Japan tended to oppose retrofitting because of its huge dependence on foreign oil carried over a long distance, but it was also feeling pressure from its mammoth but slumping shipbuilding industry and from its environment ministry, which had responded quite favorably to the American delegation's visit to Tokyo. Liberia too was in somewhat of a quandry, not wanting to oppose publicly its largest client states, the United States and Greece, but feeling strong pressures from the oil industry which had much of its tonnage registered there as well.[186] These two states remained remarkably silent but clearly favored a compromise. On the whole, the USSR was also quiet but did state strongly in the opening plenary session its desire to achieve a consensus.[187] Unlike previous conferences, the Soviets were not well disposed toward the American proposals, for they would have felt the costs more than with previous measures (especially as they had begun in recent years to expand their shipping into non-Comecon markets). They

184. The other committees were concerned with the legal aspects of the Protocols (Committee I) and with the proposals on safety and accidental pollution (Committee III).

185. Among those opposing it were the U.K., Netherlands, West Germany, and the developing countries Brazil, Argentina, India, Indonesia, Nigeria, and Thailand. The Netherlands, in particular, has close links to the oil industry and, through Shell, to the United Kingdom.

186. The Liberian policy was directed to a significant extent by President Tolberg and the minister of finance, who worried about possible U.S. congressional legislation against flags of convenience. Interviews. Although the delegation was silent throughout the conference, it made a surprising but shrewd gesture in the final session. It abstained on the vote and, as a political sop to both Greece and the American environmental lobby, it bemoaned the fact that there was not more retrofitting in the final package.

187. TSPP/CONF/SR.1/p. 7.

CHART 7
State Positions on Retrofitting Segregated Ballast Tanks vs. Crude-Oil-Washing, 1978

	Retrofitting Segregated Ballast Tanks	*Uncommitted*	*Crude Oil Washing*
Western Europe and others	Cyprus	France	Australia
	Norway	Japan	Belgium
	Portugal		Canada
	Spain		Denmark
	Sweden		Finland
	U.S.		Germany (F.R.)
			Italy
			Netherlands
			New Zealand
			Turkey
			U.K.
Soviet bloc			Germany (D.R.)
			Poland
			Romania
Developing	Egypt	Kuwait	Argentina
	Kenya	Liberia	Brazil
	Morocco		Chile
	Venezuela		India
			Indonesia
			Mexico
			Nigeria
			Thailand
			Tunisia
			*Most of the other 17 developing countries

NOTE: The "uncommitted" states (except France) did tend to lean toward crude oil washing. The preferences of states were obtained from statements of and interviews with delegates. The authors attended the conference.

did not, however, wish to oppose publicly their partner in "détente," although their position was evident from the public opposition of Poland, Romania, and East Germany. In addition, despite the numerical imbalance, all states recognized the power and determination of the United States.

On the second day of discussions, Dr. Spinelli sought to break the growing freeze and to channel the discussions into considering a compromise. France too was visibly cross-pressured and had already put forward a compromise proposal.[188] Although it was clearly too early for a compromise actually to be considered, the Italian delegation at Spinelli's instigation also suggested one.[189] At the time some delegates considered that Dr. Spinelli had made a serious tactical error, but he achieved his real intention. Soon the negotiations moved out of the committee room and into private discussions. For two days each side met separately to consider what elements they could accept in a compromise. By Thursday, the fourth day, discussions between the opposing forces began in earnest. As Dr. Spinelli had told the negotiators that he would not favor alternative proposals being brought back to his committee, strong pressure existed to achieve a single package.

The progress of the negotiations on a segregated ballast/COW package was obviously the major interest of national delegations, but there were many other important discussions going on at the same time. For example, a working group had been convened to draw up guidelines for the operation of crude-oil-washing, and the outcome of this was crucial if COW were ever to be acceptable to the United States as an alternative to segregated ballast. The work here was highly politicized, therefore, especially as Norway (which did not want COW accepted) was visibly trying to sabotage the working group. Another crucial working group was that on the "protective location" of segregated ballast tanks. This new measure too would be acceptable as a substitution for double bottoms only if a suitable formula could be devised giving sufficient protection against oil outflows in the event of an accident.

188. French policy was controlled by the Ministry of the Merchant Marine, with little input from other ministries. With its shipping well protected under a 1928 cargo preference law (whereby two-thirds of oil carried to France must come in French ships), it had little need to support retrofitting. Yet it did give qualified backing to the American proposal. France had suffered a major pollution casualty, the *Olympic Bravery*, off its coast two years earlier, and this had caused quite a row in the Assembly. Moreover, in perennial competition with the U.K., France hoped to be a peacemaker at the conference, get the U.S. and IMCO off the hook, and upstage the British. (Interviews.) Its compromise proposal (to have retrofitting only on tankers below 70,000 dwt) would have had little impact on the French fleet, which utilized much larger vessels. Barely a month after the conference ended, France was, with the grounding of the *Amoco Cadiz*, to experience the worst oil spill in history just a few kilometers from where the *Olympic Bravery* went down.

189. TSPP/CONF/C.3/WP.1. In fact, the Italian proposal bears a remarkably close resemblance to the final product.

But the main action was out of public view. Interestingly, as the United Kingdom and the United States confronted each other, a new mediator surfaced, W. O'Neil, the commissioner of the Canadian Coast Guard. Despite its radical "environmentalist" stance at the 1973 Conference, the Canadian government (particularly the Department of External Affairs) had lost much of its interest in IMCO affairs once it no longer offered a promise of jurisdictional gain. As a result, even though this was clearly a critical conference for IMCO and for international environmental affairs, the Canadian government had left its policy almost entirely to its Ministry of Transport. The policy it adopted was not now a particularly aggressive or visibly environmentalist one (it opposed retrofitting), but it permitted the delegation to take a very constructive mediatory role through its maritime ministry representatives. One official chaired the working group on crude-oil-washing, while Mr. O'Neil worked closely to facilitate the discussions on the major issue.[190]

As the negotiations continued, the developing countries in attendance also played a constructive role. In large measure, they opposed the high costs of the American initiatives but were also convinced of the necessity of achieving a compromise acceptable to the United States president and Congress. Their balanced position had certainly been assisted by the American tour to key developing states, but the secretary-general had also played an important role. Especially in his discussions with the de facto leader of the developing countries, India, he was able to emphasize the value of preserving the organization by forestalling American unilateral action. The developing nations only caucused once, but many did make clear in private talks that they would accept a compromise if the costs for themselves were not high.[191]

After four days of intense private discussions, a compromise proposal

190. The COW working group was chaired by R. Parsons, also from the Ministry of Transport. Only one representative from the Department of External Affairs attended this conference, and he was exclusively concerned with particular legal issues concerning the protocols. The Canadian position was, like that of most other opponents of retrofitting, dictated both by opposition to the extremely high costs it entailed and by the belief that COW, properly operated, would be as effective environmentally.

191. India was clearly the most important of the developing states and, in general, their spokesman, although Egypt also played an active role. A common position was never advanced publicly, although informally most did communicate their opposition to the more costly proposals. In particular, the developing states were concerned that their smaller tankers be exempted from some regulations. Only once was a "double standard" sort of issue raised. All ships on certain "special trades" (for example, those operating between ports with reception facilities) were to be exempted from some of the new requirements, and some countries suggested this be extended to some ships traveling on international voyages to the ports of the developing countries. This suggestion was quickly rejected. The "special trades" regulations can be found in TSPP/CONF/11, annex I, reg. 13C. The Conference also decided that IMCO should examine the regulations and perhaps revise them and extend them to other voyages. TSPP/CONF/12, res. 16.

was ready for submission to the conference.[192] In recognition of his constructive mediatory role, Mr. O'Neil was asked to present the package, and he brought back to Committee II what seemed to be significant advances. First, for all *new crude* carriers above 20,000 dwt, segregated ballast tanks "protectively located," crude-oil-washing, and inert gas systems were all to be required. The 20,000-dwt figure was a reduction from the 70,000-dwt limit set by the 1973 Convention, and the other requirements were new. The same regulations were to apply to all *new product* carriers above 30,000 dwt. Both of these regulations are to take effect for any carrier for which the construction contract is agreed to after mid-1979, regardless of whether the protocol has entered into force by that time. For new tankers the U.S. secured virtually everything it wanted. Second, for *existing crude* carriers above 40,000 dwt, either crude-oil-washing or the retrofitting of segregated ballast tanks was required—signifying, of course, that COW would become the operational system.[193] On *existing product* tankers above 40,000 dwt, retrofitting was required—but most product tankers are below this size. Although no fixed data for the acceptance of these regulations for existing tankers was set in the Protocol, a separate resolution recommended dates and urged states to implement them "to the maximum extent" possible even before their entry-into-force.[194] This legitimizes unilateral action on these regulations, something on which the United States had firmly insisted. The agreement, of course, contained many more detailed provisions, but these were the most important ones.[195] Combined with stringent new guidelines for the operation of crude-oil-washing, the package was satisfactory to the United States. By rejecting mandatory retrofitting, most maritime nations and the oil industry were satisfied as well.

Agreement was also reached on new measures to control accidental pollution through a protocol to the SOLAS Convention and some additions to the MARPOL Protocol. The requirement for inert gas systems, while necessary for tankers using crude-oil-washing, was also seen as useful for

192. TSPP/CONF/C.3/WP.11. The major elements of the package were then accepted as part of the MARPOL Protocol (TSPP/CONF/11, regs. 13, 13B). An integral part of the final accord was the specifications for crude oil washing (TSPP/CONF/12, res. 15).

193. The figure of 40,000 dwt was a concession to the developing countries by the United States in exchange for their willingness to accept the package. The United States would have preferred a figure of 20,000 dwt.

194. TSPP/CONF/12, res. 1.

195. Two other measures which will expedite the entry of the regulations into force were the "decoupling" of annex II (TSPP/CONF/11, art. II) and the linkage of the new protocol to the MARPOL Convention so that they have to be ratified as one instrument (TSPP/CONF/11, art. I). Unlike the MARPOL Protocol, the SOLAS Protocol and SOLAS Conventions can be accepted separately (TSPP/CONF/10, arts. I, II, and IV). This was agreed because in February 1978 ten states with 40 percent of world shipping tonnage had already accepted the 1974 SOLAS Convention. Participants hoped it would secure sufficient ratifications to enter into force by 1979.

preventing explosions on all tankers.[196] Such systems were viewed as highly desirable following the explosion on the *Sansinena* in December 1976. Although many shipping states wanted them installed only on larger tankers (above 70,000 dwt) and opposed their installation on product tankers, they bowed to American insistence on this matter. The requirement for back-up radar for vessels over 20,000 dwt was also accepted, but the acceptance of collision avoidance aids was postponed until after the MSC and Assembly approved specifications in 1979.[197] The U.S. was somewhat reluctant to leave the obligation to employ such aids out of the SOLAS Protocol but accepted the opposition of the European maritime countries to imposing such demands before operational specifications had been agreed to. On the very important question of emergency steering standards, the United States was also forced to compromise. The American delegation had wanted not only duplicate steering controls but a requirement for the manning of steering control spaces in certain situations as well. Most maritime powers objected strongly to the latter stipulation, and the U.S. reluctantly retreated while reserving its right to enact manning requirements for its internal waters.[198] The American willingness to compromise did not, however, extend to the formula for the protective location of segregated ballast tanks, which was directed at minimizing oil outflows resulting from collisions. American officials argued that they had already backed down in approving this structural provision as a substitute for double bottoms, and they remained adamant throughout the conference. They made it clear that acceptance of their formula for maximum surface protection was a prerequisite for their approval of the protocols. There was extensive discussion on just what the stipulations on this would be but, as the last remaining obstacle to agreement, an acceptable formula was achieved.[199] Finally, as they did with the MARPOL Protocol, the participants passed a resolution inviting governments to implement the new regulations ''by'' certain dates or soon thereafter—thus legitimizing their adoption by the U.S. soon after the conference.[200]

The results of the TSPP Conference are testaments to the powers of both the United States and the maritime community. Despite the strong reluctance of most states to consider any changes to the 1973 MARPOL and 1974 SOLAS conventions, the threat of American unilateralism forced

196. TSPP/CONF/10, chap. II-2, reg. 60.

197. TSPP/CONF/10, chap. V, regs. 12, 19, and TSPP/CONF/12, res. 13. A primary radar for all vessels above 1,600 grt was also required.

198. TSPP/CONF/10, chap. II-1, reg. 29. The U.S. probably will unilaterally impose the manning requirement. Interview.

199. TSPP/CONF/11, reg. 13E. The participants did, however, agree that IMCO should study other possible formulae which could be substituted at a future time.

200. TSPP/CONF/12, res. 2.

them to do just that—and in record time. In revising these conventions, the United States was able to achieve most of its objectives. Yet the oil industry and the maritime states also wielded influence. The United States is more easily capable of threatening unilateral action than actually undertaking it. It, like all other states, has a strong attachment to uniform international rules and does not want to undermine the efficacy of the central maritime regulatory agency, IMCO. As a result, unilateral action would be extremely costly, especially as the European states still control substantial maritime resources. As a result, and considering the ability of shipping and oil interests to provide an alternative to retrofitting segregated ballast tanks, maritime interests did have an important impact. Even here, however, the compromise was achieved only with the adoption of stringent new regulations for the operation of crude-oil-washing, regulations that could be implemented by the United States as soon as it was able.

At a time when the 1973 MARPOL Convention and to a lesser extent the 1974 SOLAS Convention were languishing for lack of support, the TSPP Conference not only generated more stringent regulations but also greatly enhanced the possibility for their application and entry-into-force. The "decoupling" of annex I and II of the MARPOL Convention will encourage ratifications of that instrument but, more importantly, the commitment of the U.S. to legislate the regulations should spur ratifications by other maritime states concerned to protect the competitiveness of their own shipowners.[201] It would be foolish to predict "clear sailing" for the 1978 protocols and their mother conventions or to judge that their regulations constitute "the answer" to oil pollution. To do so would be to ignore the past history of false hopes and undiscerned impediments. With a serious commitment, however, real progress can be achieved.

201. The U.S. Coast Guard set out its plans for implementing various aspects of the new agreements between 1979 and 1985. The dates conform to the recommendations in the conference resolutions. "Tanker Safety and Pollution Prevention: Information and Regulatory Implementation Plan," *Federal Register* 43 (April 20, 1978), pp. 16886–16890.

Chapter V

Coastal State Rights: Intervention and Compensation

By dawn the next morning it was clear that something had gone wrong. At 7:30 A.M. small amounts of oil could be seen in the sea around the ship. Half an hour later, a fisherman called to say oil had appeared in nearby Gearhies Harbour. A sea valve had been left open during the night. The spillage was 2,597 tons or 650,000 gallons.

In Bantry Bay itself, where attempts to clean up 22 miles of affected coastline continued yesterday, large oil slicks can still be seen and the smell is appalling. The rocks are black and seaweed dredged up is a tangled mass of sludge.

("The Blunders of Bantry Bay,"
Sunday Times [London],
November 6, 1974, p. 3)

Whether by accident or design, oil pollution is a statistical certainty that will exist for as long as oil is moved by sea. Frequently, the pollution attacks only the environment with its assorted, unrepresented inhabitants. Such cases are not reported nor the victims recompensed. But frequently the victim is also another innocent bystander, the coastal state. Not just a shipper or receiver, the coastal state is a special party and one with its own interests and rights to be protected.

In the last decade, the cause of the coastal state has been paramount. Two areas, in particular, have been the focus of industry and intergovernmental initiatives: the scope of a coastal state's power to intervene in a maritime casualty that threatens pollution, and the right of that state to be compensated for any damage that results. Unlike the conventions and amendments on discharge regulations,[1] a high degree of success has been achieved on these issues: all of the five agreements concluded have entered into force.[2] It was, however, only the shock of a major disaster, the *Torrey*

1. It will be recalled that until 1978 the international law in force concerning shipping discharges was that of the 1954 Convention as amended in 1962. In January 1978 the 1969 amendments entered into force. As of 1978, the 1973 Convention had not entered into force.

2. Two nongovernmental arrangements on coastal state compensation have entered into force: TOVALOP on October 6, 1969, and CRISTAL on April 1, 1971. Of two intergovernmental agreements on coastal state compensation, International Convention on Civil Lia-

Canyon, that provoked states, in the face of woefully inadequate legal rules, to the successful formulation of the new accords. As with the development of new discharge controls for flag states, the history of these new laws for coastal states is the history of the push and pull of competing economic and political interests.

The beginning of this story is the wreck of the *Torrey Canyon*, a "jumboized" tanker, lengthened to keep pace with the ever-rising demands for imported crude oil. When, on March 18, 1967, it ran aground off the coast of Cornwall, it was carrying 119,328 tons of Kuwaiti oil. The accident was a product of extreme human negligence,[3] but that was in itself almost commonplace. It was in its impact that the incident was unique: 35 million gallons of heavy black oil were spread over a hundred miles of British and French beaches in Cornwall, Normandy, and Brittany. Thousands of birds were destroyed while contamination continued day after day despite drastic but incompetent attempts to limit, burn, or neutralize the oil outflow.[4] And, most importantly, media coverage was given worldwide to a new type of man-made spectacle, the environmental disaster. That of the *Torrey Canyon* was one of unprecedented proportions.[5]

Burrows, Rowley, and Owen have established approximations of the *quantifiable* costs of the incident, although "it proved impossible . . . to attribute monetary valuations to the ecological damage."[6] They concluded that the quantifiable costs were £14.24 million. These costs are listed in Table 6. Excluding the ship and cargo losses, the prevention and control costs alone were estimated to have been about £7.70 million ($18 million), which "(perhaps) for the first time in maritime history . . . substantially exceeded the value of the ship and cargo."[7]

---

bility for Oil Pollution Damage entered into force on June 19, 1975 and the International Convention on the Establishment of an International Fund for Compensation for Oil Pollution Damage entered into force on October 16, 1978 following France's ratification. The other intergovernmental agreement, the International Convention Relating to Intervention on the High Seas in Cases of Oil Pollution Casualties, entered into force on May 6, 1975. These agreements are discussed in this chapter.

3. See the report of the Liberian commission of inquiry. C/ES.III/3/Add.6.

4. Ninety thousand gallons of detergent were poured onto the oil every day in an attempt to break it up and dissolve it. In fact, the toxicity of the dissolved detergent was even more damaging to sea life than was the oil floating on the surface. Furthermore, the late but drastic action of bombing the vessel did burn off some of the oil left in the tanker, but the use of napalm to ignite the oil on the water largely failed since the oil had become thinly spread and many of the volatile fractions of the oil had already evaporated.

5. Numerous accounts of the incident have been written but for a dramatic and detailed exposition, see E. Cowan, *Oil and Water: The Torrey Canyon Disaster* (New York: Lippincott, 1968).

6. P. Burrows, C. Rowley, and D. Owen, "The Economics of Accidental Oil Pollution by Tankers in Coastal Waters," *Journal of Public Economics* 3 (1974), p. 258.

7. Ibid.

Prior to the incident few were aware or concerned about the possibility of such a massive environmental disaster, as the total inability of the British government to deal with the incident revealed. Of the few rules applicable to the incident, many had been written by the industry (through such bodies as the nongovernmental Comité Maritime International, CMI) and these were largely directed toward the regulation of responsibilities among those involved in shipping (shipowner, cargo-owner). The legal responsibilities of these interests to third parties—to coastal interests—were only of incidental importance.

Two aspects of the restricted nature of the international regulations were highlighted in the disaster. First, existing admiralty laws provided that there was to be a waiting period before anyone other than the shipowner could intervene in a high-seas shipping casualty. This was intended to give the shipowner a chance to exercise his right of salvage and prevent further loss. Only when he had failed to do so could another interest take over, it never having been anticipated that this could seriously affect coastal interests. In good maritime tradition, the British government did wait several days before taking serious action against the *Torrey Canyon*.[8] Second, in maritime law there exists the principle of "limitation of liability," whereby a shipowner may limit to a set amount his obligation to provide compensation for personal or property damage. By the widely accepted 1957 Convention on the Limitation of Liability, the amount set for property damage was equivalent to about $67 per ton of the ship's tonnage. (This figure and other provisions were substantially changed in 1976.) The convention was the product of an agreement to adjust costs for damage occurring largely between participants in the trade. Indeed, the origin of the concept of limiting one's liability is founded in an earlier era when it was necessary to protect embryonic merchant industries against ruinous losses and when it was likely that only other venturers were apt to be affected by such losses. But times had changed.

Therefore, apart from the obvious need to review rules for accident prevention, the *Torrey Canyon* pointed to the need for a reconsideration of the coastal state's right both to intervene for its own protection in a shipping casualty and to be fully compensated for any damage suffered. Simple as it may seem, it was the first time that it was widely realized just how the shipping industry could have "significant repercussions on the coasts."[9] Within a month of the accident, the United Kingdom submitted a paper to IMCO requesting an immediate international meeting to consider the

8. See G. W. Keeton, "The Lessons of the *Torrey Canyon*," *Current Legal Problems* (1968), p. 96. See also Cowan, *Oil and Water*, p. 71.

9. U.K. submission to the IMCO Council, C/ES.III/3.

TABLE 6
Cost Estimates for the *Torrey Canyon* Spill

	£ million	
Internal cost (to shipowner)		
(*a*) hull of Torrey Canyon	5.90	
(*b*) cargo	0.60	
(*c*) salvage operations	0.04	6.54
External cost of prevention and control (U.K.)		
(*a*) cost of avoiding coastal pollution	2.00	
(*b*) cost of cleanup	2.70	4.70
External cost of control (France and Guernsey)		
(*a*) minimum estimate based on compensation claims	3.00	
(*b*) external cost of damage, extensive but unquantifiable	--	3.00
Total quantifiable cost		14.24 million

SOURCE: P. Burrows, C. Rowley, and D. Owen, "The Economics of Accidental Oil Pollution by Tankers in Coastal Waters," *Journal of Public Economics* 3 (1974), p. 258.

problems raised by the incident,[10] and at an "extraordinary session" of the Council in May 1967 a list of eighteen items was established which required study.[11] The majority of these items dealt with such things as the establishment of navigation lanes and shore navigational installations, standards of training and qualifications of crew, design and construction of ships, mapping of hazards, multilateral pollution combat ("contingency") plans, and pollution research and information exchanges. Five other issues recommended for study were legal questions dealing with general rights of salvage, coastal state rights of surveillance and control, participation in official enquiries, and

> The extent to which a State directly threatened or affected by a casualty which takes place outside its territorial sea can, or should be enabled to, take measures to protect its coastline, harbours, territorial sea or amenities, even when such measures may affect the interests of shipowners, salvage companies and insurers and even of a flag government . . . and . . . all questions relating to the nature (whether absolute or not), extent

10. Ibid. The French government also submitted its own proposals, C/ES.III/3/Add.4.
11. See C/WD.III/5.

and amount of liability of the owner or operator of a ship or the owner of the cargo (jointly or severally) for damage caused to third parties by accidents suffered by the ship involving the discharge of persistent oils or other noxious or hazardous substances and in particular whether it would not be advisable:

(*a*) to make some form of insurance of the liability compulsory;

(*b*) to make arrangements to enable governments and injured parties to be compensated for the damage due to the casualty and the costs incurred in combatting pollution in the sea and cleaning polluted property.[12]

While the other questions were channeled into the appropriate technical subcommittees, these last two commanded high political priority and the Council sought an immediate resolution of them. The task which IMCO set for itself in 1967 was a large one, made even greater by the virtual absence of law in many areas and disagreement over what the law was in others.

THE LAW OF COASTAL STATE INTERVENTION—1967

A coastal state's "right" under general international law to intervene when an accident occurred within that state's territorial seas was, in 1967, not really open to question. Subject to the right of innocent passage and to the usual legal rules of reasonable conduct, the territorial sea was, and is, an area of full coastal state sovereignty. Although a state might be liable for tortious conduct or for interference with passage which "is not prejudicial to peace, good order, or security of the coastal state,"[13] it is otherwise free to act.

On the high seas, the issue was not nearly so clear. Under the rules of international law, a coastal state's right of intervention was subsidiary to the right of free, unimpeded usage by all. Article 2 of the High Seas Convention (1958) proclaims this basic freedom: "The high seas being open to all nations, no State may validly purport to subject any part of them to its sovereignty. Freedom of the high seas is exercised . . . by all States." This customary free use of the oceans has imposed a heavy burden on those interfering with it. Indeed, it may even be argued that coastal state intervention was permissible only to the extent that it was specifically allowed. No treaty provisions proclaim otherwise. Article 24 of the High Seas Convention does impose a duty on states to draft rules to prevent pollution of

12. Ibid., p. 5.

13. Geneva Convention on the Territorial Sea and Contiguous Zone (1958), art. 5(2). On the impact of tort liability in the area of coastal pollution, see Lance C. Wood, "An Integrated International and Domestic Approach to Civil Liability for Vessel-Source Pollution," *Journal of Maritime Law and Commerce* 7 (October 1975), p. 1.

the seas, but such rules are stated to be subject to "existing treaty provisions" and they are directed toward discharge or construction standards and not to any self-proclaimed right of intervention.[14] Moreover, the "existing treaty provisions" to which it refers, the 1954 Convention on the Prevention of Pollution of the Sea by Oil, themselves give exclusive control powers over ships on the high seas to the *flag* state.

In the absence of any treaty law supportive of a coastal state "right" of intervention, recourse would have to be made to customary international law in incidents such as the *Torrey Canyon*. The principle of self-protection provided coastal states with the major jurisdiction for intervention beyond territorial seas. In the classic United States Supreme Court case *Church* v. *Hubbard*, Chief Justice Marshall supported Portugal's seizure of an illegal trading ship on the high seas as an exercise of its

> right to use the means necessary for its protection. These means do not appear to be limited within any certain marked boundaries, which remain the same at all times and in all situations. If they are such as unnecessarily to vex and harass foreign lawful commerce, foreign nations will resist their exercise. If they are such as are reasonable and necessary to secure their laws from violation, they will be submitted to.[15]

With the ever-mounting threat posed by maritime activities to coastal interests, this principle has been increasingly supported by some authors,[16] as has the similar principle of "self-help." Of the latter, Professor L. F. E. Goldie has written,

> Coastal states enjoy, in general international law, a right of self-help, the condition of which was well expressed by Secretary of State Webster in the *Caroline Case* well over a hundred years ago when he said that there must be a necessity which is instant, overwhelming, and leaving no choice of means and no moment for deliberation. Countries situated as

14. The article reads, "Every state shall draw up regulations to prevent pollution of the seas by the discharge of oil from ships or pipelines or resulting from the exploitation and exploration of the seabed and its subsoil, taking account of existing treaty provisions on the subject."

15. *Church* v. *Hubbard* (1804) 6 U.S. (2 Crank) 187, pp. 234–235.

16. Hydeman and Berman have written, "a rule of customary international law seems to be emerging with respect to the right of a coastal State to take necessary action beyond territorial limits for its self-protection or self-preservation. While publicists seem somewhat reluctant to admit such a right, a number of them have done so. The reluctance appears to stem from concern about the potential abuse of the right and, more specifically, the possible establishment of permanent zones in the high seas contiguous to territorial seas for defensive purposes." L. M. Hydeman and W. H. Berman, *International Control of Nuclear Maritime Activities* (Ann Arbor, Mich.: University of Michigan Law School, 1960), p. 216. See also Donat Pharand, *The Law of the Sea of the Arctic* (Ottawa: University of Ottawa Press, 1973), pp. 235–244.

France and Britain were in the *Torrey Canyon* casualty are clearly in that position.[17]

Even should such a principle provide the justification for coastal state intervention, the occasions and the manner in which it might be "reasonably" exercised remained unclear in 1967. As it was, the *Torrey Canyon* disaster was an extreme situation and, even there, the United Kingdom, by waiting as long as it did before bombing the ship, fulfilled its traditional maritime obligations.

THE LAW OF COMPENSATION FOR POLLUTION DAMAGE—1967

Before a state can claim compensation for pollution damage, it must first be able to identify a suitable party against whom it can bring a legal action. Following the *Torrey Canyon* incident, both the British and French governments found that simply locating the party could pose as great a problem as obtaining the compensation from it.

The *Torrey Canyon* was owned by a Bermuda corporation, the Barracuda Tanker Corporation, and it was on a full "bareboat charter" to the Union Oil Company of California.[18] At the time of the accident, it was also on a "voyage charter" to British Petroleum.[19] Beyond this contractual relationship, however, a Union Oil spokesman said that there was "no corporate relationship between Union Oil and the Barracuda Tanker Corporation," thereby leaving the latter to bear full liability as the "owner" of the ship.[20] In fact, Barracuda was a corporate creation of Union Oil (if not a legal subsidiary).[21] Its incorporation meant that Union Oil, as a separater corporation, could avoid responsibility for damages caused by the ship. This is a notoriously common arrangement. Often these shadow corporations are

17. L. F. E. Goldie, "Principles of Responsibility in International Law," *Hearings*, Subcommittee on Air and Water Pollution of the United States Senate Committee on Public Works, 91st Congress, 2nd Session, July 21 and 22, 1970, p. 99. This position has been disputed. See, for example, E. D. Brown, "The Lessons of the *Torrey Canyon*," *Current Legal Problems*, 1968. See also Dennis M. O'Connell, "Reflections on Brussels, IMCO, and the 1969 Pollution Conventions," *Cornell International Law Journal* 3 (1970), p. 1. O'Connell supports Goldie in that he argues that a right of intervention is "reasonably well-grounded in current customary international law."

18. The owner of a ship may lease out the ship in a number of ways. The shortest charter is for a single "voyage." It may also be leased for a set period of "time." In either of these cases the shipowner provides the crew and maintains close supervision of the ship. On the other hand, the owner may charter the ship for such a long period of time, say twenty years, that all control effectively passes to the charterer. This is a "bareboat charter."

19. See Cowan, *Oil and Water*, pp. 1–7. Union Oil had granted a voyage charter to B.P., not having required the *Torrey Canyon*'s use at the time.

20. *New York Times*, July 18, 1967, p. 39.

21. See Cowan, *Oil and Water*, pp. 20–23.

only "one-ship companies" in which, after an accidental loss, there are no other assets except for limited insurance coverage to provide for *all* the ship's liabilities. In this case, the *Torrey Canyon* did have two sister ships: the *Lake Palourde* and the *Sansinena*. As a result, if either were discovered in a territory where Britain or France could serve a writ, it could be seized as an asset to cover the corporation's potential liability.

It was not until four months after the accident that the *Lake Palourde*, in urgent need of two rolls of wire rope, put into Singapore harbor. Immediately, it was met by lawyers working for the British government who, following long-standing maritime tradition, affixed a writ to the ship's mast.[22] This compelled the ship to post a bond for £3 million before it could depart from the harbor. French officials were not so prompt, allowing the ship to slip out of the harbor and escape their writ, although they did give chase in a police motor launch. It was not until April 1968, thirteen months after the disastrous pollution incident, that the French government finally was able to secure a bond—and again only as a result of seizing the *Lake Palourde*, this time in Rotterdam.

Such a confused state of affairs again stems from the desire for self-protection in an industry of an unrivaled competitive nature. In order to protect the shipowner from the claims of other parties, artificial legal barriers are erected so as to limit one's liability to that which is recoverable from the company actually involved in the incident. Although this is the nature of all legal "corporations," the practice has been taken to an unreasonable extreme in the maritime field through the creation of the "one-ship company." Being totally undercapitalized, the new corporation buys its vessel with a guarantee (a "bankable charter") from a major oil or shipping company, repayment being in the form of an assignment to that company of much of the revenue to be derived from the charter services of the ship. The oil or shipping company itself then contracts to charter the boat on a long-term "bareboat" or time charter. As the legal shipowner has no capital and often just that one ship as an asset, should the ship be destroyed in an accident, no funds are available from him for compensation. The fact that the owning company is a separate corporate entity protects the other company by means of the "corporate veil." Not only can that company not be sued—often it can not even be identified! With such large potential damages to innocent third parties, this was a problem to be rectified.

Even when contact is made with a financially sound owner, full recovery of damages depends on establishing the shipowner's liability and on his ability to "limit his liability." Liability for damage can be established in a

22. For a fascinating account of this legal cat-and-mouse game, see ibid., pp. 194–196.

number of ways. First, a party's liability can be based on *fault*; that is, he may be liable only when the claimant can prove that the accident resulted from his negligence. Second, it can be founded on *fault with a reversed burden of proof*. In this system it is the party from whom compensation is being claimed who must prove that the accident was *not* due to his negligence. This is, therefore, more favorable to the claimant. An even more rigorous regime is *strict* liability, where responsibility for compensation is imposed on the party causing the damage whether or not he was at fault. Some specific exceptions to his liability (such as an "act of God") are allowed. Finally, liability can be *absolute* so that the party causing the damage is always liable, regardless of circumstances. Traditionally in most legal systems and in maritime law, liability has been based on *fault*. In the case of the *Torrey Canyon* there was no dispute about the liability of the shipowner, as the Liberian Board of Investigation found negligence on a number of grounds.[23]

The impetus for change from this traditional fault regime was already developing in domestic English law at the time of the *Torrey Canyon* incident. At common law two cases, *Southport Corporation* v. *Esso Petroleum*[24] and *The Wagon Mound*,[25] supported the proposition that damage resulting from an oil spillage is not compensable without proof of fault, but another case, *Rylands* v. *Fletcher*,[26] held a person "strictly liable" for damage caused by the escape of harmful materials from an "extra-hazardous activity." Considering the damage caused by massive releases of crude oil, it was reasonable that the transportation of such oil should be considered such an extrahazardous activity. Whatever the position of the domestic English common law (which might not, in the event, carry great weight internationally), justice would seem to require a liability more rigorous than one based on fault. Indeed, the great English jurist, Lord Devlin, argued this position when, as chairman of the International Subcommittee of the Comité Maritime International, he submitted his report on the *Torrey Canyon*. Needless to say, his suggestions were unacceptable to the CMI and he was later, as British representative to the IMCO conference, to put forward a different view. However, he wrote at the time,

> Thus the imposition of strict liability in the case of a marine casualty would be a marked departure from what has hitherto been accepted but it may be said that it is a departure which is to some extent overdue. . . .

23. See C/ES.III/3/Add.6.
24. [1954] 2 Q.B.182; [1956] A.C.218.
25. [1961] A.C. 388 (P.C.). See also J. D. Dunn and J. A. Hargrave, "Oil Pollution Problems on the Pacific Coast," *University of British Columbia Law Review* 6 (June 1971), p. 139.
26. (1968) L.R. 3, H.L. 330.

> It would be a recognition of the change in economic and legal conditions that has been taking place over the last half century. . . . The law of torts [which is based on the need to prove fault] is framed and administered on the assumption, which has now become quite unrealistic, that the defendant will pay out of his own pocket. . . . It is much easier for the cargo to be insured than for a number of small people to seek property insurance independently. It is also fair that the cost should be borne by the cargo; it can be reflected in the price of oil and so paid by the users for whom the cargo is being brought and not by those whose property is damaged in the course of carriage.[27]

With regard to hazardous activities within state boundaries, public international law would seem to impose strict liability for damage caused outside the state by such activities. The *Trail Smelter Arbitration* held Canada strictly liable for the tortious acts of its citizens and ordered Canada to pay $350,000 in compensation for damage in the United States caused by fumes emanating from Canada. The tribunal held that

> Under the principles of international law . . . no state has the right to use or permit the use of its territory in such a manner as to cause injury by fumes in or to the territory of another or to the property of persons therein, when a case is of serious consequence and injury is established by clear and convincing evidence.[28]

Although this case would not apply directly to pollution from ships on the high seas, Professor Goldie has argued that this case and others (*Corfu Channel*[29] and *Lac Laroux*[30]) "clearly point to the emergence of strict liability as a principle of customary international law. In none of these cases did the issue of fault prevail."[31] There is, however, no certainty about this point, and with the prevalence of fault liability in other areas of maritime law, there has been a strong inclination toward a continuing reliance on it in international pollution incidents. It was, in fact, on this issue that the 1969 IMCO Conference almost foundered.

The second issue affecting recovery for pollution damage is the principle of "limitation of liability," a peculiar, albeit well-accepted practice. This principle allows a shipowner to restrict his liability to an agreed limit. It has evolved with only incidental consideration of its possible effect on innocent third parties. As previously mentioned, the prevailing rules of limitation at

27. LEG/WP(II).I/WP.1, pp. 6–7.
28. 3 U.N. Rep. International Arbitration Awards, p. 1905. (Hereinafter cited as U.N. RIAA.)
29. [1947] I.C.J.4.
30. 12 U.N. RIAA, p. 281 (1957); *American Journal of International Law* 53 (1959), p. 156.
31. Goldie, "Principles," p. 92.

the time of the *Torrey Canyon* incident were those set by the 1957 Convention on the Limitation of Liability. The convention sets a limitation on liability for property damage at 1,000 gold francs (about $67) per ton of the ship's "limitation tonnage." This "limitation tonnage" is equivalent to about 40 percent of the deadweight (cargo-carrying) tonnage of a tanker and about 10 percent less than its gross registered tonnage.[32] Therefore, although the *Torrey Canyon* carried 119,000 tons of oil (its approximate deadweight tonnage), its gross registered tonnage was only 61,224 tons and its limitation tonnage, 59,308.[33] On this basis the limitation of liability for the *Torrey Canyon* would be set at about $4,746,000. Any damage above that level—and there was plenty—would be borne by the victim. However, the 1957 Convention also provides that where the shipowner (and not just the ship's master or crew) is negligent, no limitation is permitted. With the *Torrey Canyon*, despite the finding by the Liberian Board of Investigation that the master was alone at fault, a settlement in excess of the limitation figure was agreed on in order to avoid a complicated legal battle. The agreed compensation for both the United Kingdom and France was slightly over $7 million, still far below the total costs and damage incurred by these states (see Table 6).

Considering the pitfalls of the limitation system, these governments were fortunate to recover as much as they did. For one thing, the 1957 Convention applies to all property damage, so that claims other than those for pollution damage would share proportionately with the pollution claim whatever amount was available after the shipowner had limited his liability. This could substantially reduce the compensation available for pollution damage. In addition, the laws limiting liability in many states are even more restrictive than those set out in the 1957 Convention and applied in the United Kingdom. For example, in Japan, France, or in Union Oil's domicile, the United States, local laws fixed the limits of liability according to the value of the ship and cargo *after* the incident. In the case of the *Torrey Canyon*, this would have come to about $50, the value of one salvaged lifeboat![34]

Despite these glaring anachronisms, the revision of the rules was not to be an easy process. While the United States and others demanded that the

32. The deadweight tonnage is the weight of cargo, fuel, and so forth that will be required to push the ship down to its loadline limit in the water. The gross registered tonnage (grt) is not a measure of weight at all but of cubic capacity of the ship. One hundred cubic feet of enclosed space is equivalent to one gross registered ton.

33. Statement of James T. Reynolds, president, American Institute of Merchant Shipping, *Hearings*, Subcommittee on Air and Water Pollution, Committee on Public Works, United States Senate, 91st Congress, 2nd Session, July 21 and 22, 1970, p. 31.

34. Cowan, *Oil and Water*, p. 200.

figure should be set as high as $450 per ton to accord with their own domestic experiences, other states and commercial interests argued that a higher limitation figure was not necessary.[35] The Norwegian insurance group, the Assuranceforeningen Skuld, in a letter to the 1969 IMCO Conference asserted that of the 850 pollution cases handled by the London and Scandinavian group of twelve Protection and Indemnity (P & I) Clubs, only two settlements would have exceeded the $67 per ton figure. The American figures were attacked as "vastly overshooting" the mark.[36] At a later IMCO conference, the Oil Companies International Marine Forum (OCIMF) submitted a paper looking at both costs and settlements.[37] The paper did recognize that costs could run to $350 per ton but only with *very* small spills (where the unit costs would be higher than for large spills) and only in heavily populated harbor areas where meticulous cleanup procedures would be employed. In general, the paper argued that virtually all spills would be covered by a $150 per ton limit.

These projections are, of course, based on limited and arbitrary definitions of "cost," "damage,"[38] and "cleanup." But two things were clearly established: the old 1,000 franc per ton ($67) limit was inadequate; and any new limit would, in the acerbic words of the OCIMF paper, be acceptable only if it was not "arbitrarily raised on the basis of theoretical calculations founded on the extrapolation of figures which are irrelevant to the hypothetical cases cited."[39]

FROM CATASTROPHE TO CONFERENCE: 1967–1969

Background

It was in this setting of consternation and complacence that the IMCO Council met in extraordinary session in May 1967. To bring some order out of the chaos, one of the Council's first actions was to create a special ad hoc Legal Committee open to all members. Charged with the tasks of elucidating the nature of the existing legal norms, the new committee was to

35. See, for example, LEG/WG(FUND)/IV/2/Add.1. At an estimated $16 million in control and cleanup costs *alone*, the per ton costs for the *Torrey Canyon* incident would be $266. Wood ("Integrated Approach," p. 36) says American studies estimated costs at between $450 and $2,250 per ton. It is difficult to rely on industry cost quotations as they usually use the costs of *settlement* rather than the full environmental costs and losses to the victims. Moreover, all cost figures vary according to one's definition of "damage."

36. LEG/CONF/6/Annex III, *Officials Records*, International Legal Conference on Marine Pollution Damage (1969) (London: IMCO, 1973), p. 53 (hereinafter cited as *Official Records*).

37. LEG/WG(FUND)V/2 Add.1.

38. For a discussion as to what "damage" should cover, see Wood, "Integrated Approach," pp. 29–37. See also the case of *Burgess* v. *The Tamano*, 1973 AMC 1939 (D.Ne 1973).

39. LEG/WG(FUND)V/2 Add.1, p. 5.

determine "whether intergovernmental action is required and what form it should take."[40] With its broad membership, its ability to call on expertise in all fields of international law, and its continuing liaison with other IMCO organs and outside agencies such as the CMI, the Legal Committee quickly produced a plan of action. In the crisis atmosphere of the time, the Legal Committee was effective, and it soon evolved from an ad hoc emergency body into the most important permanent U.N. legal body dealing with marine pollution.

Traditional commercial interests were, however, not at all happy to see IMCO's role expanding into the realm of "private" legal liability. While the major maritime states were willing to have IMCO draft a "public law" convention outlining a state's rights and responsibilities regarding intervention in polluting casualties, they steadfastly opposed its consideration of the "private law" issues of liability and compensation, issues which, they argued, were of private commercial and economic significance only. It was only after a long and acrimonious debate at the Legal Committee's first session that IMCO supporters succeeded in keeping the development of a new environmental compensation scheme within the public domain and out of the private precinct of the CMI.[41] Private industrial representatives and the CMI were, however, invited to participate, particularly in the working group set up to draft the Private Law Convention.

Under the Legal Committee's direction, two working groups on intervention and compensation met between 1967 and 1969 and each produced a draft convention. The groups worked largely from skeleton drafts and comments submitted by national governments and the CMI. Minor differences were resolved in the working groups, although most differences of any substance were referred to the Legal Committee. In fact, the Legal Committee itself rarely resolved issues of any consequence. For example, although in its second meeting the Legal Committee decided to restrict the scope of both conventions to "oil" and in its fourth meeting to impose the primary liability on the ship- (not cargo-) owner, these issues continued to be debated and became major topics at the conference itself. But progress was made in both the working groups and the Legal Committee, and it was decided in June 1968 to convene a conference for the following year.

The CMI had been thwarted in its attempt to preempt IMCO, but it did participate actively in both the private law working group and the Legal Committee, submitting a proposed draft as well as reports on various technical aspects of the convention. At the same time, having already created its

40. C/ES.III/5.

41. The participants that strongly favored leaving the drafting of the Private Law Convention to the CMI were the United Kingdom, Norway, Ireland, France, Germany, Denmark, and the International Chamber of Shipping.

own "International Subcommittee" on questions arising from the *Torrey Canyon* incident, the organization continued to work independently of IMCO as well. In the same manner as it had conducted the drafting of its commercial conventions for over sixty years, the CMI relied on submissions from its many member "maritime law associations" to create its own draft convention on oil pollution liability. These private submissions often did not reflect the positions of the associations' governments,[42] but they formed the basis of a private draft convention which was accepted at the CMI's nongovernmental conference in Tokyo in April 1969 and presented to the sixth session of the IMCO Legal Committee, the last session before the convening of the intergovernmental conference. It was a disappointing document, its approach little changed from traditional practices. Without continuing pressure, it was now clear that this commercial organization would not act to protect coastal interests.[43] Except for the reluctant inclusion of a provision requiring liability insurance, the CMI convention showed scant evidence of the changes being demanded by many national representatives at IMCO. With the completion of an intergovernmental alternative at the 1969 IMCO Conference, the CMI draft convention ceased to be of any further interest.

Industry activities during this time were not, however, restricted to the proposed convention. In a familiar pattern, it was to the seven major oil companies themselves (particularly Shell International Ltd.) that the initiative now fell. Together they produced the "Tanker Owner's Voluntary Agreement on Liability for Oil Pollution" (TOVALOP). Although unquestionably intended to preempt or at least influence IMCO,[44] TOVALOP did provide a timely and much needed, albeit limited, improvement in international pollution compensation. It continues to do so and as such must be considered an important document in its own right.

42. For a comparison of the position of the American government and of its maritime law association, see Allan Mendelsohn, "The Public Interest and Private International Law," *William and Mary Law Review* 10 (summer 1969), p. 783.

43. In the past, the CMI would hold a conference attended by governmental representatives after it had privately prepared the working draft. This arrangement had worked well in areas of technical/commercial interest, but it was not suitable to the politically highly charged issue of pollution liability. Their "new" convention still provided for shipowner liability based on fault, although now with a reversed burden of proof. Strict or absolute liability was rejected. The limitation figure was maintained at 1,000 francs per ton, but with a new provision for compulsory insurance for that amount.

44. Various industry officials admitted in interviews this rather obvious motivation. TOVALOP was introduced by the CMI to IMCO at the fourth session of the Legal Committee and came into effect one month before the 1969 Conference. When it first came into effect it covered 50 percent of all Western tanker tonnage and it now represents about 99 percent. Combined with its later partner, CRISTAL, it was for years the only comprehensive and widely accepted international compensation regime.

TOVALOP

It was a time when change was needed, and many radical options were being propounded. IMCO was planning to hold its intergovernmental liability conference in a few months, but some states had shown signs of impatience with the slow pace and cautious trend of negotiations. A convention was imminent, but unilateral national action was also possible—with all the attendant unreasonableness and lack of control that entailed.[45] Could governments fail to notice the creation of a practical, functional compensation scheme so soon before the IMCO meeting?[46]

Although it was initiated by the seven major oil companies, TOVALOP is a guarantee by *any* "participating tankerowner" that he will reimburse national *governments* for *preventive and cleanup expenses* incurred as a result of the spillage of crude oil from his tanker. TOVALOP is, therefore, a voluntary agreement among shipowning interests for the benefit of a third party, the government of a polluted coastal state. It is not a contract with such a government. No states are party to the agreement, and no questions of jurisdiction arise.

TOVALOP was the first in a series of complex compensation arrangements that were to be developed over the next few years. Each agreement differs from the others and, to simplify our discussion of them, the central provisions of each are summarized in Chart 12 (see pp. 194–195, following). As can be seen from that chart, liability under TOVALOP is imposed on the tanker "owner," now defined to include the bareboat charterer.[47] Reacting to one of the greatest sources of annoyance after the *Torrey Canyon*, the TOVALOP parties had moved to remedy the deceptive corporate barriers encountered with the Barracuda Tanker Corporation and Union Oil.

The nature of liability is little changed, however. It extends in scope only to governmental "cleanup costs" (not damage) and, even then, is incurred only on the basis of the "fault with a reversed burden of proof" test. The effect of this is to create a *presumption* of negligence against the tankerowner

45. Compliance with uniform international requirements concerning the nature and limits of liability, as well as any mandatory insurance coverage for such liability, would be vastly cheaper and more efficient than would compliance with a variety of national provisions. Fears about burdensome legislation were justified when one compares the provisions of TOVALOP with those of the Canadian Arctic Waters Pollution Prevention Act, the Canada Shipping Act, part XX, and the Florida Oil-Spill Prevention and Pollution Control Act.

46. The fact that TOVALOP covered only governmental cleanup expenses (and not losses by private parties) indicates the parties the agreement was designed to influence. In fact, its effectiveness in influencing the conference was limited by the fact that governments had had so very limited exposure to it. In any event, TOVALOP clearly represented a type and limit of liability that many states had already rejected.

47. Clause I(*b*). Where there is a bareboat charter, the charterer is considered as the "owner."

which he can rebut if he "can prove that the discharge of oil from his Participating Tanker occurred without [*any*] fault on the part of the said Tanker."[48] Actually rebutting such a presumption and proving no fault may prove difficult, but where it can be done, there can be no claim under TOVALOP. Interestingly, a double standard is applied here. Under clause V, the tankerowner is himself to be reimbursed for "reasonable expenditures" incurred by him in taking his own "preventive measures" after an incident, whether or not the incident was caused by his negligence. The TOVALOP memorandum states that this provision was "designed to encourage a tankerowner to take prompt action," although the same incentive was not felt to be necessary for governmental authorities which would presumably act for their own self-protection.

In keeping with traditional maritime practice, even where the liability of the shipowner is established, under TOVALOP he may still limit his liabilities: $100 (U.S.) per gross registered ton with a new maximum liability of $10 million (U.S.). The per ton limitation set out here represents an increase of about 50 percent over the international limitation figure, and it applies to pollution damage only. But the interests responsible for TOVALOP balanced this benevolence with the introduction of the new overall limit of $10 million, a figure of some advantage to the very large tankers. In fact, for tankers exceeding approximately 150,000 grt, the tonnage limitation is actually *lower* than it was under traditional law for other liabilities.[49] Interestingly, this innovation of a new overall limitation eventually found its way into the 1969 IMCO Convention, even though it had not been included in the draft convention. This provision is one instance in which the industry's early initiative later produced dividends at IMCO.

In order to ensure full compensation to the limits of coverage under TOVALOP, clause II (*c*) of the agreement requires that every participating tankerowner "establish and maintain its financial capacities to fulfill his obligations under this agreement." Existing P & I coverage of pollution risks did not extend as far as the new TOVALOP system, and this new provision was a source of much dissension within the ranks. The oil industry foresaw the need for mandatory insurance, but the independent shipowners and the Protection and Indemnity Associations opposed giving pollution special treatment from other claims. As a result, to implement the agree-

48. Clause IV(B).

49. Clause VI(C). With the global limitation of $10 million, any tanker over approximately 150,000 gross tons (about 300,000 dwt) would be subject to pollution liability of about $67 per ton. This is equal to the amount set under the 1957 Convention for all liabilities. Although now treated separately, the pollution liability is of no larger order of magnitude for these large tankers than for other liabilities.

ment, the oil companies created a mutual insurance company, the International Tanker Indemnity Association (ITIA), to provide coverage for tankerowners who would not otherwise be able to obtain it.[50]

TOVALOP has proven to be a very useful guarantee to governments. Today over 99 percent of the world's privately owned tanker tonnage is party to it,[51] and many hundreds of claims have been settled under it.[52] It is, however, a limited remedy. It is limited in that while the per ton liability is increased by 50 percent, overall liability is now limited. It extends only to governmental cleanup expenses, excluding all indirect or ecological damage and all actual losses. Strict liability is avoided and fault liability is retained although with a "reversed burden of proof." Finally, being a voluntary agreement *among* tankerowners, TOVALOP "does not create any rights against the federation, and the federation shall have no liability hereunder or otherwise to any government."[53] TOVALOP does not therefore give governments any means of enforcing the provision for mandatory financial responsibility, although this has not proven to be a problem.

But its creation was a subtle strategy. Given the general absence of unilateral action in the wake of the *Torrey Canyon*, TOVALOP achieved its primary objective. Moreover, at least one of its provisions found its way into the 1969 Civil Liability Convention, a feat the CMI convention did not achieve. In addition, incorporation of the new overall limitation manifestly redounds to the benefit of the owners of the *very* large tankers—the large independents and Western oil companies—and reduces the costs of the scheme for them. Finally, by expanding TOVALOP's coverage to the rest of the world fleet, the oil industry ensured that all ships were covered and not just those run by their own, more responsible tanker divisions. This benefits the oil industry (as well as the coastal states) for it is the oil industry, the *cargo*-owner, and not the shipowners, that have borne the brunt of public reaction to oil spillages—especially to those for which there was no available compensation.

50. Neither of the British companies, B.P. and Shell, felt free to leave their P & I Clubs to get the new pollution coverage from ITIA. Under pressure from the American oil companies, however, both these companies entered into a "subsidy agreement" with ITIA where they agreed to contribute the difference between the ITIA charge and their P & I charge into the ITIA premium pool. This agreement lasted until about 1975. Interviews.

51. *Report and Accounts* (1975), International Tanker-Owners Pollution Federation. This percentage amounted to about 156 million gross registered tons in 1975.

52. Interviews with TOVALOP officials revealed that ITIA alone paid out over two hundred pollution claims in 1974. As most of the TOVALOP coverage is now provided as part of the broad P & I liability insurance, cumulative statistics on the tankers in all the P & I Clubs do not exist. (Interviews.) See also Gordon L. Becker, "A Short Cruise on the Good Ships TOVALOP and CRISTAL," *Journal of Maritime Law and Commerce* 5 (July 1974), p. 609.

53. Clause VII(H).

The 1969 IMCO Conference

Amidst these legal and political developments the International Legal Conference on Marine Pollution Damage opened on November 10, 1969 in Brussels at the Palais de Congrès. Although an IMCO-sponsored conference, it was held in Brussels in recognition of the long-standing contribution of the Brussels-based Comité Maritime International to the development of international maritime law. Forty-eight states attended as voting participants, and six more states, six nongovernmental bodies, and four U.N. and intergovernmental agencies were present as observers.[54] The conference was organized into a central plenary body and two "committees of the whole" to consider the two draft conventions.

The Intervention (Public Law) Convention. As is common with most IMCO draft conventions, the draft Public Law Convention incorporated the more conservative provision developed at regular IMCO sessions by the maritime states. With the broader attendance at the conference, the balance usually shifts. But the new recruits are at a disadvantage. They are unfamiliar with the draft and often inexperienced in the subject matter; yet they have the onus of showing the need to change it. In Committee I, where the Intervention Convention was discussed, there were three major areas in which attempts were made to improve the position of the intervening coastal state: the definition of the maritime incident justifying intervention; the obligations and safeguards to which a state should be subject when intervening in a maritime casualty; and the relationship between coastal state and shipowner obligations in a maritime casualty.

Early negotiations focused on the definitions of the right of intervention and of the "incident" which would justify an intervention. To states with substantial shipping interests, it was crucial that the rights of the coastal state be kept clearly limited. No legal excuse should be given for unnecessary interference with their very costly operations. On the other hand, for states that remembered the helpless plight of the British government in 1967, it was not wise to be left too dependent on the flag state. Six amendments to the draft provisions were submitted and two were accepted.[55] The

54. Of the fifty-four states in attendance, twenty-two were Western developed states from Western Europe, North America, and Australasia, six were from Eastern Europe, twelve from Africa, nine from Asia, and five from Latin America.

55. The first Canadian amendment would have deleted all reference to the "consequences" of a casualty [art. I(1)]. Canada argued that this was redundant and restrictive of coastal interests in light of the reference already to "grave and imminent danger to their coastlines." Canada also sought to delete the requirement that a coastal state consider the "material damage to a ship or cargo" before acting [art. II(1)]. Although it did get substantial support (see Chart 8), this was rejected for obvious reasons—it was precisely their condition that was relevant to the need to intervene.

results of these negotiations are set forth in the following extract. The italicized sections indicate the areas of contention and of change where this occurred. The basic right of intervention was set out in article 1(1):

> Parties to the present Convention may take such measures on the high seas *as may be necessary* to prevent, mitigate or eliminate *grave* and imminent danger to their coastline or related interests from pollution or threat of pollution of the sea by oil, following upon a maritime casualty or acts related to such a casualty, *which may reasonably be expected to result in major harmful consequences.*

The "maritime casualty" giving rise to this right is set out in article 2(2): "*Maritime casualty* means a collision of ships, stranding or other incident of navigation, or other occurrence on board a ship or external to it resulting in material damage or imminent threat of *material damage to a ship or cargo.*"

The two amendments accepted were minor ones proposed by West Germany and the United States. They both related to article 1(1). Germany's amendment resulted in the substitution of the word *grave* for *disastrous*, while the American amendment resulted in the replacement of the phrase *major or catastrophic* consequences with the phrase *major harmful* consequences. Four other proposals were rejected. The American suggestion to require only *imminent* danger to the coastline rather than *grave and imminent* danger was unacceptable, as was the Indonesian suggestion to give the coastal state full freedom to decide what measures were necessary. This was to have been accomplished by substituting the phrase *as it deems necessary* for the phrase *as may be necessary*. Needless to say, maritime states adamantly opposed the substitution of a "subjective" for an "objective" standard. Two contentious, but not significant, Canadian amendments were also rejected.[56]

The outcome of these debates reveals a clear tendency toward community control of coastal state actions. Despite the recognition of coastal state interests, its control of activities which, though occurring outside its jurisdiction, would have effects within it was limited in favor of a balance of control with the flag state, which feared unjustified interference with its shipping interests. A number of coastal states—in particular, Canada and its followers—felt that the restrictions in the convention were even greater than those applicable under general international law.

The second area of contention centered on the obligations of a coastal state before, during, and after intervention. As it was drafted, the convention required an intervening state to notify and consult with the parties

56. Although it might seem that the changes are small, it was repeatedly pointed out that the vague theoretical obligations agreed to at the conference could be important factors in the sudden choices that a coastal state would later have to make in situations of great uncertainty.

A signpost points the way to the location of the wreck of the *Torrey Canyon*. The billowing column of smoke results from the bombing and burning of the tanker by the British Royal Air Force, six days after it ran aground in March 1967. (Photo: Jane Bown/Camera Press.)

French navy helicopters dropped depth charges on the *Amoco Cadiz* in March 1978 in order to dislodge any pockets of oil remaining in the vessel. With the investigations made after the disaster, it has become apparent that procedure for dealing with tanker casualties continue to remain most inadequate. Within weeks of this incident, another crippled tanker, *Eleni V*, was blown up by British Royal Navy divers. (Photo: UPI.)

affected by the intervention, to restrict its intervention to measures proportionate to the damage threatened, and to pay compensation for excessive measures. Although these proposals represent realistic expectations from the viewpoint of the shipowner, the third proposal in particular proved to be a major source of disagreement.

Canada proposed that the duty to notify and consult interested parties be placed on the flag state and not on the coastal state and that, in any event, the exemption from such an obligation in cases of urgency should be made the primary element of any article dealing with the issue. Although these proposals were supported by Canada's coastal allies at the conference—

New Zealand, Indonesia, Syria, and the Philippines—as well as by Italy, they represented a clear threat to flag state preeminence over their vessels on the high seas and were rejected. The French delegate succinctly expressed the traditional viewpoint: "It was again a question of balance. The Convention provided for an exception to the normal freedom of the seas, but that exception should be limited as much as possible."[57] On the other hand, a French proposal that would clearly have infringed on the coastal state "right" of self-defense was also rejected. Despite the support of the strong maritime states of Greece, Liberia, and Belgium as well as of the former French colony, the Ivory Coast, this French proposal to require coastal state consultation with independent experts before intervening was also rejected. Such consultation was made optional, and the coastal state was now obligated, under article 3, only to notify and consult with those affected parties whose interests were known to it and to "take into account" their views. The result was a fair balance.

The second obligation on coastal states, that measures taken "shall be proportionate to the damage actual or threatened to it,"[58] was accepted with little debate as it was already implicit in existing international law. Furthermore, it was agreed that, in considering whether the measures were proportionate to the damage, account should be taken of "(*a*) the extent and probability of imminent damage if those measures are not taken; and (*b*) the likelihood of those measures being effective; and (*c*) the extent of the damage which may be caused by such measures."[59]

The final and most significant obligation imposed on the coastal state is found in article 6, which provides that

> Any Party which has taken measures in contravention of the provisions of the present Convention causing damage to others, shall be obliged to pay compensation to the extent of the damage caused by measures which exceed those reasonably necessary to achieve the end mentioned in Article I.

This obligation was accepted with little debate, as it too was implicit in the requirement of reasonable and proportionate conduct under existing international law. However, in light of the inadequate accountability of shipowners for the damages *they* inflicted, there was substantial debate on the question of whether the coastal state's obligation should be separate from or

57. Statement by M. Jeannel (France), Committee I, *Official Records*, pp. 336–337. The effect of the defeat of the Canadian amendment, concerning an exemption from notification in cases of emergency, was simply to place the exemption further back. It became article III(*d*).

58. Art. V(1).

59. Art. V(3).

reciprocal to that imposed on the shipowner in the Civil Liability Convention (then being negotiated in Committee II). Canada, the major protagonist in this debate to "link" the two conventions, argued, "The consequences of preventative measures are inseparable from the consequences of a pollution incident itself and should not be made the subject of separate proceedings."[60] New Zealand's single delegate put forward the argument that it "would be unreasonable to require a coastal state to pay compensation to those responsible for a maritime casualty when they are not under a corresponding obligation to pay damages for any catastrophic damage which a casualty might cause to the coastline and related interests."[61]

Canada's proposal was, therefore, to "link" the two conventions by making the obligation of the coastal state to pay compensation to the shipowner conditional on the shipowner's payment of his pollution damage costs. To many coastal states this proposal was seen as crucial both to ensure their own compensation for pollution damage costs and as a test of the good faith of the shipowners. The provision was vigorously debated at length in both Committee I and in plenary and was introduced in varying forms four times—by Canada, Spain, the Netherlands, and New Zealand. However, the maritime states seemed to distrust the world's coastal states, for they argued that any article relieving the coastal state of its duty to pay for excessive intervention would lead to irresponsibility by that state. Such an argument was quite spurious, however, as any failure by the coastal state to pay compensation would occur only after the intervention and would, in any event, have been conditional on the failure of the *shipowner* to pay for his damages. Each time the proposal was reintroduced with greater safeguards but each time it was defeated, the last time in plenary and only then because it did not achieve the required two-thirds majority.[62]

Considering the orientation of the draft prepared by the IMCO Legal Committee, the major maritime states were clearly on the defensive throughout the debates on the convention. Both sides had their extreme

60. Canadian Submission on the Draft Convention, LEG/CONF/C.1/1, *Official Records*, p. 212.

61. Statement in Plenary by Mr. Beeby (New Zealand), *Official Records*, p. 95.

62. Canada's proposal was to make the existence of the coastal state obligation to pay damages conditional on its receipt of compensation from the polluting ship under the Liability Convention. This was the original "link" provision, and it was defeated by a vote of 10-24-4. Subsequent proposals were all rejected. The Spanish proposal (which would not have removed the coastal state obligation but only its duty to undertake compulsory dispute settlement on the matter) was rejected by 12-19-8. The Dutch proposal was defeated by 14-14-10, even though it would have authorized only a temporary withholding of payment by the coastal state until the other claim was settled. The New Zealand proposal in plenary was a tighter version of the Dutch amendment, and although it was supported by a majority vote of 18-14-14, it failed to achieve the required two-thirds majority and was therefore not accepted.

advocates. On the one hand, some states, notably Poland, the USSR, Liberia, Norway, and Greece, were almost uniformly opposed to any change (see group 3 in Chart 8). The Soviet bloc, especially, maintained its traditionally defensive posture about interference with their flag ships on the high seas and introduced additional proposals to prevent it.[63] On the other hand, coastal states such as Canada, Syria, New Zealand, Indonesia, and Spain strongly supported virtually every amendment put forward (see group 1 in Chart 8). In fact, this conference marked the emergence of Canada as a radical proponent of international environmental rights and as a leader of the coastal nations—most of whom were developing countries. Canada was the only government, apart from the host state, to be represented by a cabinet-level delegation, and it was the major protagonist in the debates.[64] While the whole thrust of the work on the Intervention Convention was theoretically to achieve a "balance" between maritime and coastal interests, the Canadian attitude was, according to the British representative, unfairly biased: "The Canadian representative should rather ask if his country was not in a particularly favourable position, because it could devote its main concern to the protection of its coasts."[65] Despite the overwhelming opposition to its proposals, the Canadian position was widely appreciated, especially among the poorly represented coastal and developing states. Defending Canada against the British remarks, the New Zealand delegate commented,

> Canada's situation was not unique; several countries—for example, the Philippines, Indonesia, Syria and New Zealand—had long coastlines and were not large shipowners. In fact the situation of most countries which were members of the international community, and in particular the developing countries, was precisely that. Even though many of those

63. The USSR sought to include an article *requiring* a coastal state to return the crew of a disabled tanker to the state of the ship's registry. This amendment was supported only by other socialist states but was later accepted as article III(3) when it was worded so as to request the coastal state only to "facilitate the repatriation" of the crew. A Romanian proposal to require notification of the flag state before any action was taken even in cases of extreme urgency [that is, to delete article III(*d*)] was also defeated. The only support for this proposal came from Bulgaria, Poland, U.K., USSR, Yugoslavia, Liberia, and Portugal.

64. The Canadian delegation was headed by then Minister of Transport Donald Jamieson. Indeed many of the speeches of the delegation were perhaps intentionally extreme and inflexible. Many maritime delegates went so far as to assert (in interviews) that Mr. Jamieson's Plenary speech at the start of the conference actually removed Canada as a credible bargainer on both conventions for the rest of the conference. For an analysis of the Canadian policy, see R. Michael M'Gonigle and Mark W. Zacher, "Canadia Foreign Policy and the Control of Marine Pollution," in Barbara Johnson and Mark W. Zacher, eds., *Canadian Foreign Policy and the Law of the Sea* (Vancouver: University of British Columbia Press, 1977).

65. Statement by Mr. Simpson (United Kingdom), *Official Records*, p. 381. For the text of the Intervention Convention, see 9 *International Legal Material* 25.

countries were not represented at the Conference, it was essential that their interests should be taken into account.[66]

The policy orientation of most of the coastal and developing states was consistently in favor of the more relaxed coastal state obligations proposed by Canada and its allies. In contrast, many of the maritime states (including the developing maritime states), while opposing any fundamental changes in the control of flag vessels on the high seas, frequently took positions which would lend balance to the convention. This was illustrated by the near unanimous support for the American amendment to article 1(1) (see Chart 8, column 2) as well as by the diverse positions of those states found in group 2. One of the explanations for this situation was the strong position coastal states were in with regard to existing international law. Maritime states argued that the right of intervention was being *conferred* by the convention, but coastal states did have a strong case for the argument that the convention merely *recognized* a preexistent right. As a result, all but a few maritime states took care to create a balance consistent with coastal state expectations of existing law. Without it, they foresaw that many states would prefer to continue to rely on "self-help" or "self-defense" in preference to the convention. Even so, these maritime states opposed granting any "greater additional rights to coastal states than were urgently needed."[67] In retrospect, an examination of the convention produced can only lead one to the assessment that, in light of the concerns at the time, a sound and reasonable balance of rights and obligations was indeed created.

The Civil Liability (Private Law) Convention (CLC). The tasks set for Committee II were not easy ones. The preamble to the Civil Liability Convention put it, in deceptively simple terms, as being both "to ensure that adequate compensation is available" and to adopt "uniform international rules and procedures" to achieve it. These two objectives touched distinct and crucial concerns of the participants. Reconciling them was nearly an unattainable task.

The issue of compensation for pollution damage involved the solution of three distinct questions: who should be made liable for the damage; what should be the basis of the liability; and what should be the limit of liability. The alternatives available were fairly clearcut. First, liability could be placed on the shipowner or the owner of the cargo which caused the damage, on the oil industry, or on both. Second, liability could be based on the

66. Statement by Mr. Beeby (New Zealand), *Official Records*, p. 384.

67. Statement by Mr. Breuer (Germany), *Official Records*, p. 303. Generally throughout the committee meetings, the positions of the German delegation were models of reasonableness. Although the convention seemed fairly balanced at the time, the *Amoco Cadiz* disaster pointed up some critical omissions, particularly the lack of a duty on the tanker captain to notify the authorities when an emergency arises and on the salvor to assist the tanker without being concerned with contract.

Chart 8

Public Law Convention: Proposals and State Positions

	Definition of right of intervention						*Consultation/ notification*			*Link proposal*			
	American amendment to art. I(1)—accepted	*German amendment to art. I(1)—accepted*	*American proposal to* delete grave—*defeated*	*Indonesian subjective standard—defeated*	*Canadian proposal to* delete consequences—*defeated*	*Canadian proposal to* delete material damage—*defeated*	*Canadian proposal to put consultation onto flag state—defeated*	*French proposal on mandatory consultation with experts—defeated*	*Canadian proposal on emergencies—defeated*	*Canadian proposal—defeated (roll call)*	*Spanish proposal—defeated (roll call)*	*Dutch proposal—defeated*	*New Zealand proposal—defeated*
I													
Canada	x	x			x	x	x	x	x	x	x	x	x
Syria				x	x	x	x		x	x	x	x	x
Indonesia			x	x	x	x			x	x	x		x
New Zealand					x	x	x		x	x		x	x
Spain	x		x		x	x					x		x
Philippines	x				x		x			x			x
Guatemala	x	x	o							x	x		x
India						x				x	x	x	
Australia										x	x		x
Yugoslavia										x	x		
Cameroon						x							
II													
U.S.	x		x		x	x				o	o		o
UkSSR	x									o	x		

Ivory Coast					x	o		o					
Ireland						o				o			x
Brazil	x						o				o		
Germany		x	o			o				o	o	x	
Italy	x				o	x	x		o	o	o	o	
Ghana	o					o			o	x			
Netherlands							o		o	o	o	x	x
III													
Romania										o			
Sweden		x			o					o	o		
Japan	x		o							o	o		
Switzerland										o	o		
U.K.										o	o		
Bulgaria			o	o	o		o	x		o	x		
Belgium				x			o	o		o	o		
Denmark				o						o	o		
Finland						o				o	o		
France	x	x		o			o	o	o	o	o		
Israel						o			o	o	o		
Liberia								o	o	o	o		o
USSR	x	o	o		o	o	o			o	x	o	
Poland				o	o	o	o		o	o	o		x
Norway	x	o			o	o	o		o	o			o
Greece	x	o			o	o		o	o	o	o	o	o

Key
x = Vote for coastal position
o = Vote for maritime position

existing maritime law requiring fault to be proven, it could impose a regime of fault liability with a reversed burden of proof (as TOVALOP had just done), or it could impose strict or even absolute liability. Finally, the limits of liability could remain at the level of 1,000 gold francs per ton or be raised or even abolished. Committee II attempted to go through these issues individually, but it was soon discovered that the answer to any one of the questions was so dependent on the resolution of the others that only an overall "package" could be considered.

The debates in Committee II opened with long speeches by a variety of delegations telling of the important shipping, coastal, and/or consumer interests they had to protect and introducing the best compensation scheme for these interests. The more radical coastal states favored a system of strict or absolute liability, with both shipowners and cargo interests *jointly* responsible. Many others, including some major shipping states (Germany, France, and the United States), argued that *strict* liability imposed on the shipowner alone would suffice. Unlike those few states supporting joint liability, the proponents of strict shipowner liability were often also major oil consumers for whom joint liability meant either the imposition of financial burdens on their oil companies or an increase in the cost of oil.

The majority of European maritime states adamantly opposed placing all of the responsibility on the shipowner and interfering with their traditional legal responsibilities based on fault. Their shipping industries would bear the brunt of such changes, so it was only the incorporation of *existing* maritime law into the convention or the complete shifting of liability onto the *cargo*-owner that was acceptable to them.[68] Strict shipowner liability was anathema. It raised fears of reduced bargaining leverage in settling claims, of vastly increased costs for the shipping industry, and of a precedent being set for all other areas of shipping liability. Belgium, ever mindful of the interests of the CMI, opposed any consideration of strict liability, as it "constituted a serious attack on the fundamental principles of maritime law and created a dangerous precedent for all sectors of transport other than that of maritime law."[69] Similarly, Denmark argued, "The notion of liability based on fault was a principle of law from which there should be no departure without valid reason."[70]

Linked to this was the opposition to any higher limits than the traditional 1,000 francs per ton. Speaking for the United Kingdom, Lord Devlin now stressed the counterproductive effect that imposing strict or higher

68. If *cargo*-owner liability were accepted, it would necessarily have to be strict as it would be rare, once the cargo were on board the ship, for an incident ever to result from the cargo-owner's fault.

69. Statement by Mr. Cuvelier (Belgium), Plenary, *Official Records*, p. 117.

70. Statement by Mr. Philip (Denmark), Committee II, *Official Records*, p. 628.

liabilities would have. In fact, he argued, either would result in *lower* compensation because the funds available through the P & I Clubs and on the world insurance market would be insufficient to meet such a sudden increase in demand. Replying to the Canadian minister of transport, who labeled such arguments "specious and out of place,"[71] Lord Devlin could

> not see what was to be gained by appearing sceptical about the terms which the London marine insurance market . . . would be prepared to offer. The market . . . was adamant in their view that the figures were the limit. They had plenty of other agreeable risks to insure and had not sufficient capacity to meet all. Unless delegations knew of other markets which were willing to quote that risk, other solutions would have to be sought.[72]

Such conservative commercial arguments seemed only to make the non-maritime states more stubborn in their demands. The American delegate represented a country whose shipping fleet had steadily declined but where public environmental interest was increasing greatly. He commented,

> The most important principle for the United States delegation was the avoidance of oil pollution to the detriment of innocent countries. The Convention should not be dictated by a small group of insurers. It should be drafted in the interests of the victims and the insurers should be expected to adapt themselves to such needs.[73]

Furthermore, these states insisted on the need to incorporate the equitable principle of "the polluter pays." As the Canadian Transport Minister put it,

> shipowners had not asked coastal states to share in the responsibilities which they shouldered when they undertook to operate their mammoth oil tankers. It was hard to see, therefore, why fishing interests . . . or taxpayers in coastal states should be asked to bear the financial burden of marine pollution damage, as that would amount to subsidizing the oil industry on a world scale.[74]

71. Statement by Mr. Jamieson (Canada), Plenary, *Official Records*, p. 86.
72. Statement by Lord Devlin (United Kingdom), Committee II, *Official Records*, p. 86.
73. Statement by Mr. Neuman (United States), Committee II, *Official Records*, pp. 652–703.
74. Statement by Mr. Jamieson (Canada), Plenary, *Official Records*, p. 86. The Canadian demands were, as Chart 9 indicates, the most extreme of all states at the conference. The delegation sought absolute liability to the limits of "maximum credible damage." This meant, to all intents and purposes, absolute unlimited liability. The German government put the "polluter pays" argument in less inflammatory terms (indeed, in terms similar to Lord Devlin's own earlier submission to the CMI). They argued for absolute liability as being "in line with the rules governing responsibility in case of other modern technical risks which must be accepted by the public but which are economically to be borne by the persons creating those risks." Observations on the Draft (West Germany), LEG/CONF/4/Add.1, *Official*

This division between the protectors of environmental interests and those of the shipowning interests was compounded by the reluctance of some of the environmentalists to have the liability burden shifted to the oil industry. At a time of rising domestic demand for foreign oil, the United States delegation failed to carry the "polluter pays" principle to its logical conclusion, instead finding "little justification for the desire shown by some delegations to shift liability for pollution of the seas onto the oil companies."[75] On the other hand, coastal states such as Canada were willing to support any solution that provided adequate coastal compensation,[76] and the shipowning states such as Denmark argued,

> Maritime transport was not dangerous in itself: it was only dangerous if the goods carried were dangerous and it was therefore normal to impose liability on the cargo for any damage caused to a third party. The industry which made a profit from that business should also accept the risks entailed.[77]

These arguments flared for two weeks without progress being made.

One very radical proposal had been simmering in the background throughout the entire period. But it was too new and unexplored to be capable of detailed discussion or drafting at a formal diplomatic conference. Advanced by Belgium, the proposal would impose pollution costs directly onto the oil industry through the creation of an international fund by those companies importing oil by sea. The fund would pay for pollution damage directly, although it would have recourse against any shipowner that caused a pollution incident through his negligence. In the absence of negligence, the fund alone would be strictly liable. It seemed an intelligent compromise, but not one that could be accepted on such short notice despite its appeal to many coastal and maritime states.[78]

Records, p. 505. Although the German submission refers to "absolute" liability, the government actually favored "strict" liability at the conference, the difference being the inclusion in the latter of a limited list of exceptions to liability.

75. Statement by Mr. Neuman (United States), Committee II, *Official Records*, p. 642 (9 *International Legal Material* 45). Many oil-consuming states—United Kingdom, France, Germany, Italy, Japan as well as the United States—were reluctant to accept a transfer of the costs directly to the oil industry.

76. Canada supported a number of alternative proposals when its own joint ship/cargo-owners proposal was defeated. See *Official Records*, p. 731.

77. Statement by Mr. Philip (Denmark), Committee II, *Official Records*, p. 628. This was logically a very reasonable position. As the Indian representative, Mr. Rajwar, argued (p. 624), "The maximum ability to shoulder the burden of liability was possessed by the international oil companies which had the resources to build up a fund to meet the claims for oil pollution."

78. In the absence of any interest in governments becoming directly involved, it also promised to be a relatively cheap alternative, as Mr. McGovern (Ireland) pointed out: "there would be no serious financial burden on the oil industry. On the basis of the figure of 1,050 million metric tons for total tanker cargoes in 1967 (British Shipping Statistics 1968–69,

The committee was in a quandry and faced a very uncertain future, as an intervention by the head of the IMCO Secretariat's legal division made forcefully apparent. Technical, administrative, and budgetary considerations made it impossible, he said, for another conference to be convened for at least two years. Unless decisions were made soon, the meeting would collapse, making unilateral action a real possibility. It was mandatory that an immediate compromise be negotiated. Discussions thereupon moved "into the corridors," and at virtually the eleventh hour (two days before Committee II was to have concluded *all* its work) a solution was found. The United Kingdom produced a package imposing *strict* liability on the *shipowner* to a limit of 2,000 francs per ton but with a new *overall* limitation (not found in the draft or in the old 1957 Convention) of 210 million gold francs ($14 million) per incident.[79] (For a comparison with TOVALOP, see Chart 12.) This solution was tied implicitly to an agreement to create, as soon as possible, an International Fund such as Belgium had proposed. The purposes of the fund would be both to provide additional compensation to that provided by the present convention and to "indemnify" the shipowner for part of the burden assumed under it. The *motivation* for creating the fund thus came from the maritime states that hoped to see the shipowner's new liability reduced in the future. The *need* for it was because of the new *limitation* that had been placed on the shipowner's liability. This limitation was, like the TOVALOP limitation, favorable to the supertankers of the Western European states and placed the smaller tankers at a further competitive disadvantage.[80] But its more immediate impact was the ceiling it would put on coastal state compensation in the event of an especially large spill. A supplement was necessary.

Having resolved the basic differences, the details of the convention followed easily. Liability was to be imposed only on the shipowner.[81] As it was only to be "strict" and not "absolute," the traditional exceptions to liability were included. These would apply where damage resulted from an act

---

p. 75, Table 32), a levy of 1 cent per ton would produce $10.5 million. Such a levy would be well within the capacity of the oil industry and would meet the objections to a government levy on imports discussed in the IMCO Legal Committee" (*Official Records*, p. 635). Throughout the meetings of Committee II, a working group on the proposal did meet in order to consider just how an international fund might relate to the specific proposals being discussed in the committee.

79. The official value of 2,000 gold francs in 1969 was approximately $134. In 1973, it was $144. In 1978, its value was about $160.

80. This point was particularly irksome to the developing states, which had to rely on smaller tankers for their national fleets. See the statement by Mr. Rajwar (India), Committee II, *Official Records*, p. 733. Canada, which sought unlimited liability, was infuriated by the higher but restricted limits. In the operation of TOVALOP and CRISTAL, however, these limits have made no difference and, in fact, both arrangements have benefited the small tankers more (by providing for more compensation for smaller spills from small ships).

81. Art. III(1), International Convention on Civil Liability for Oil Pollution Damage (1969).

of war; a "natural phenomenon of an exceptional, inevitable and irresistible character"; an intentional act of a third party; governmental negligence in the maintenance of navigational aids; or, finally, a situation where the damage resulted from an intentional act of the person who suffered the harm.[82]

Article 5(1) outlined that part of the agreement dealing with the extent or limits of liability: "The owner of a ship shall be entitled to limit his liability under this Convention in respect of any one incident to an aggregate amount of 2,000 francs for each ton of the ship's tonnage. However, this aggregate amount shall not in any event exceed 210 million francs." Important qualifications of this right of limitation were provided. If the incident occurred as a result of the fault or privity of the *owner* (as in the 1957 Convention, not just the master or crew), then the ability to limit liability was removed.[83] Article 5(3) stated further that the owner must provide a "fund in court" equal to the total amount of his potential liability after an incident in order to invoke his right of limitation. Other paragraphs of article V outline in detail the practical operation of these general obligations.

A point of some debate concerned the territorial scope of the convention. It was restricted "exclusively to pollution damage caused on the territory including the territorial sea of a contracting state, and to preventive measures taken to prevent or minimize such damage."[84] This article should be interpreted to mean that preventive measures taken on the high seas to control *damage* which could occur in the *territorial sea* are subject to compensation, as are preventive measures and damage inside the territorial seas. Furthermore, in order to encourage prompt action, the expenses of the shipowner incurred in trying to minimize the pollution damage are, as in TOVALOP, to rank equally with other preventive measures and damage claims.[85]

The substantive scope of the act was restricted to damage caused by "persistent oil" such as crude, fuel, or lubricating oils.[86] The parties potentially liable are *tanker*owners only, as the convention extends only to "ships" defined as vessels "actually carrying oil in bulk as *cargo*."[87] Furthermore, in the convention, unlike TOVALOP, the *owner* is narrowly

82. Art. III(2)(3).
83. Art. V(2).
84. Art. II. See Chapter 7 for a discussion of this point.
85. Art. V(8).
86. Art. I(5).
87. Art. I(1). Note that should the oil causing damage come from the fuel tanks, the Convention would still apply—so long as the ship concerned was a tanker. At no time was it ever suggested that the Convention should extend to cargo ships. Warships or other state-owned noncommercial vessels also are excluded from the scope of this Convention, as they are from the Intervention Convention.

defined so as not to include the "bareboat" charterer, except in exceptional circumstances.[88] Unlike the situation at the time of the *Torrey Canyon* incident, it is not important that the charterer be made liable, as article 7 now requires that a shipowner maintain insurance or other financial security to the full limits of his liability. This requirement, restricted to ships carrying over 2,000 tons of oil, is to be enforced by requiring that each contracting state apply the article "in respect of any ship, *wherever registered*, entering or leaving a port in its territory."[89] The effect of this is to force *all* ships trading with that contracting state to be insured (regardless of registry) to the limits of their liability as if they were parties to the convention. Furthermore, any contracting state "shall not permit a ship under its flag . . . to trade unless a certificate" of insurance has been issued.[90]

Finally, despite a solid vote in favor of the British compromise, the "Fund Resolution" was eventually accepted only in modified form.[91] The original resolution which had formed an implicit part of the British compromise contained a clear promise to return the shipowner's liability to a regime based on fault.[92] The resolution as accepted by the conference provided only that the proposed international compensation fund ensure that "(1) Victims should be fully and adequately compensated under a system based upon strict liability. (2) The fund should in principle relieve the shipowner of the additional financial burden imposed by the present Convention."[93] Both shipowning and coastal states accepted the *principles* of the resolution. But the vague wording was a result of demands by the large oil importers—France, Germany, and the United States—that the extent of shipowner relief be left undecided. An interindustry battle was beginning to take shape.

88. Article I(3) reads, "*Owner* means the person or persons registered as the owner by the ship or, in the absence of registration, the person or persons owning the ship. However, in the case of a ship owned by a State and operated by a company which in that State is registered as the ship's operator, *owner* shall mean such company." Therefore, in the unlikely event the ship is not registered, the Convention would invite recourse to the legal tests of ownership including possession and control.

89. Art. VII(11).

90. Art. VII(10). There are obvious problems of enforcement with this article, however, as the Convention makes no provision to hold the government responsible in the event a ship's insurance is faulty.

91. The vote was 35-3-3. Canada voted against the compromise (and Australia and the Philippines abstained) because the compensation provided was felt to be inadequate. Greece and Liberia also opposed the compromise (and Japan abstained) because they felt the burden on the shipowner to be too great. All six states expressed their concern that the proposed fund might not successfully augment the Civil Liability Convention.

92. The original resolution recommended that the new Convention base "the liability of the shipowner . . . on fault and limited to the amounts provided for in the Convention of 1957 on limitation of the liability of shipowners." See LEG/CONF/C.2/WP.38, *Official Records*, p. 599.

93. Resolution on Establishment of the International Compensation Fund for Oil Pollution Damage, *Official Records*, p. 185.

The pattern of national alignments in Committee II was quite predictable (see Chart 9).[94] The majority of both developed and developing shipowning countries supported the retention of traditional shipowner fault liability. The four largest—Liberia, the United Kingdom, Japan, and Norway—opposed any change at all from traditional practices, while many other large shipping states sought to maintain the shipowners' customary liabilities by shifting pollution liabilities to the cargo-owner. The fierce opposition of Norway and particularly the United Kingdom to the possibility that major costs would be imposed on the shipowners was also a reflection of their roles as centers for the placement of P & I insurance. These insurers they hoped to protect from excessive demands which might be beyond the limits of "insurability." Likewise, Belgium's role was significantly affected by its links to the Comité Maritime International. Indeed, in proposing the creation of an International Fund, the Belgian delegate, himself president of the CMI for most of its recent history, commented that: "the aim it [the delegation] had set itself at the start of the Conference . . . [was to get] the best possible safeguard of maritime interests."[95] Notably, it was Belgium that initiated the fund proposal and Norway that introduced and steered the Fund Resolution through the committee.

At the other end of the scale were the supporters of strict liability, largely coastal states including a significant number of developing nations with negligible shipping interests. Canada led the grouping here, proposing strict and *unlimited* liability imposed *jointly* on the ship- and cargo-owners. Canada's extreme demands remained unaffected by the negotiations and compromises going on around it and, in the end, Canada cast the only negative vote against the convention.[96] A number of states with maritime interests—in particular, the United States, France, Germany, and Italy—also supported strict shipowner liability throughout the conference. The American position was expected in light of its environmental interests,

94. The best indicator of states' preferences is given in two "indicative votes" taken in committee. These votes were taken to indicate the participant's first and second preferences concerning liability regimes. Their votes corresponded in all cases to their previous statements and submissions. Where states indicated their preferred level of limitation, this is noted as well.

95. Statement by Mr. Lilar (Belgium), *Official Records*, p. 118. The Swiss representative, Dr. Muller, also had very close CMI ties.

96. Delegates from other states commented that this defiant vote severely damaged Canada's relations with other IMCO members and weakened its subsequent contribution to the work on the proposed Fund Convention. There was not unanimous condemnation, however. Allan Mendelsohn, legal counsel in the Department of State prior to the conference, has written, "Had I controlled the United States vote at the 1969 Conference, I would have joined Canada and not permitted that great environmental-trail-blazing neighbour of ours to have been the sole vote opposing adoption of the final draft." "Ocean Environment and the 1972 United Nations Conference on the Environment," *Journal of Maritime Law and Commerce* 3 (January 1972), p. 389.

CHART 9
Indicative Votes on Nature of Liability and Party Liable, and Limits of Liability

States favoring strict liability				*States favoring shipowner fault liability*			
	Vote				*Vote*		
	1	*2*	*Limits*		*1*	*2*	*Limits*
Joint (strict):				*Traditional:*			
Canada	Js	C	unlimited	Japan	F	A	
Ghana	Js	S	1,000 +	Liberia	F	A	1,000 +
Indonesia	Js	S		UkSSR	F	A	
Yugoslavia	Js	S		USSR	F	A	1,000 +
				U.K.	F	A	2,000
Shipowner (strict):							
France	S	A	3,000	Norway	F	C	1,000
Germany	S	A	1,000 +	Brazil	F	C	1,000 +
Cameroon	S	A					
China	S	A		Korea	F	S	
Egypt	S	A					
Italy	S	A		*Cargo Liability:*			
Libya	S	A		Belgium	C	A	
Monaco	S	A		Greece	C	A	
Malagasy	S	A		India	C	A	1,000
Poland	S	A					
Singapore	S	A		Denmark	C	F	1,000
Spain	S	A		Finland	C	F	
				Switzerland	C	F	
U.S.	S	Js	2,000				
Venezuela	S	Js		Sweden	C	Jf	1,000 +
Bulgaria	0	S					
Australia	S	C	2,000 +				
Netherlands	C	S					
Portugal	C	S					
Ireland	C						

Key
A = Abstain
F = Fault liability
C = Liability on the cargo-owner
S = Strict liability on the shipowner
Js = Joint strict liability
Jf = Joint fault liability—primary liabilities on cargo-owner with recourse to shipowner based on fault.

but it is surprising to note the vehement French position incorporating, for example, a demand for a 3,000 franc per ton limitation of liability. By his repeated references to the *Torrey Canyon* incident it was evident that the French delegate was much more affected than his British counterpart by that experience than by the country's immediate commercial interests. Italy's participation in the conference was minimal and, due to the language barrier, it tended to support the one delegation with which it could easily communicate, France.[97]

SUPPLEMENTS TO THE 1969 ACCORD

CRISTAL

Just as the intergovernmental discussions prior to the 1969 Conference prompted the major multinational oil companies to produce their own liability scheme (TOVALOP), so too the results of that conference led these companies to formulate their own supplementary agreement, CRISTAL. In each case the actions of the oil industry were prompted by the hope that they would engender goodwill and that states would, therefore, refrain from unilateral action and follow the industry's precedent at the ensuing conference. This was particularly important in the creation of the international fund, as the shipping and oil industries would obviously be urging divergent solutions. The Fund Resolution sought to shift the shipowner's burden onto the oil industry and to impose still further compensation liability on it. The costs of these measures could run quite high, as would any international administrative arrangements to implement them. The oil companies hoped that an effective example would serve as a useful guide to the conference.

With this in mind the oil industries felt the need to create their own lobbying body independent of the shipowners in order that they might most effectively influence the development of the new convention.[98] The Oil Companies International Marine Forum (OCIMF) was the result. Its two immediate goals were to create a functioning industry fund and to participate fully in the drafting of the intergovernmental alternatives at IMCO. Through the very active participation of OCIMF and with the assistance of many individual companies, OCIMF successfully accomplished both goals. The Contract Regarding an Interim Settlement of Tanker Liability for Oil Pollution (CRISTAL) was concluded among thirty-eight oil companies in

97. Delegates from both France and Italy pointed this out to the authors.

98. Previously, oil industry input was only through the ICS in its capacity as tanker-owners. This was no longer sufficient as the ICS had, after all, strongly supported the Fund Resolution, the idea of primary cargo-owner liability, and the full indemnification of the shipowner to his traditional position.

January 1971 and was in force on April 1, 1971, a full seven months before the intergovernmental conference. With this tangible achievement and with its continuous liaison with IMCO and privately with key delegates, OCIMF was able to have a significant impact on the outcome of the 1971 Conference—far greater than that of the oil industry through TOVALOP two years earlier.[99]

Considering that its primary purposes were to influence the IMCO Conference and to forestall unilateral action by coastal states, it is not surprising that CRISTAL completely neglected to deal with the indemnification of the shipowner. As a supplement to TOVALOP (where the liability of the shipowner was only $100 per ton), the need to reduce the shipowners' liability was not, in any event, as crucial for CRISTAL as it might later be for the international fund. The thrust of the contract was to provide compensation for cleanup expenses beyond those available under TOVALOP as well as for pollution damage itself—the latter having been completely omitted from the TOVALOP scheme. In keeping with the goal of forestalling unilateralism, such compensation would not be forthcoming where alternative domestic arrangements were available to a pollution victim. Its application was limited in other ways as well.

Unlike the 1969 IMCO Convention, CRISTAL provided compensation for *direct* damage only, explicitly excluding "any loss or damage which is remote, or speculative."[100] It covered only oil owned by an oil company participating in the contract[101] and also only if the ship involved was registered in TOVALOP. As it served as a link between the earlier private scheme and the new needs of the IMCO community, CRISTAL accepted liability according to the application of both TOVALOP and the Civil Liability Convention (CLC).[102] But TOVALOP/CRISTAL is a completely separate industrial arrangement having no contractual or legal connection to the intergovernmental agreements, although, like the CLC, CRISTAL

99. Close contact was maintained with the work of the Legal Committee and with the principal drafter of the IMCO Fund Convention, Ulf Nordenson of Sweden, as well as with the very influential maritime lawyer, Walter Muller of Switzerland. Through its member countries, OCIMF also kept track of the development of policies within many of IMCO's key member states. For OCIMF's submissions to IMCO throughout this period, see LEG/WG(Fund)II/4; III/2/Add.3; V/2/Add.1 and LEG/CONF.2/INF.2.

100. Clause I(E).

101. Clause IV(B). The effect of this then is to provide compensation additional to that available from a shipowner who is party to TOVALOP in cases where the oil spilled was owned by an oil company party to CRISTAL.

102. Clause IV(A). By referring to the 1969 CLC as being part of the basis on which the liability of CRISTAL is determined, the contract simply "incorporates by reference" some of the terms of the CLC into its own operation. As CRISTAL was intended as an interim measure until the proposed intergovernmental fund could come into existence, this reference to the CLC was intended to aid in the transition.

In setting regulations for pollution compensation, the insurance industry exercises a profound influence. The floor of Lloyd's Underwriting Room with its numerous insurance syndicates is seen above. When a shipping casualty is reported, the Lutine Bell is rung once as a notice of bad news. Two strokes are rung to signify the announcement of good news to the Underwriters. The bell was recovered from the bullion ship *HMS Lutine* which sank off the coast of Holland in 1739. (Photo: Popperfoto.)

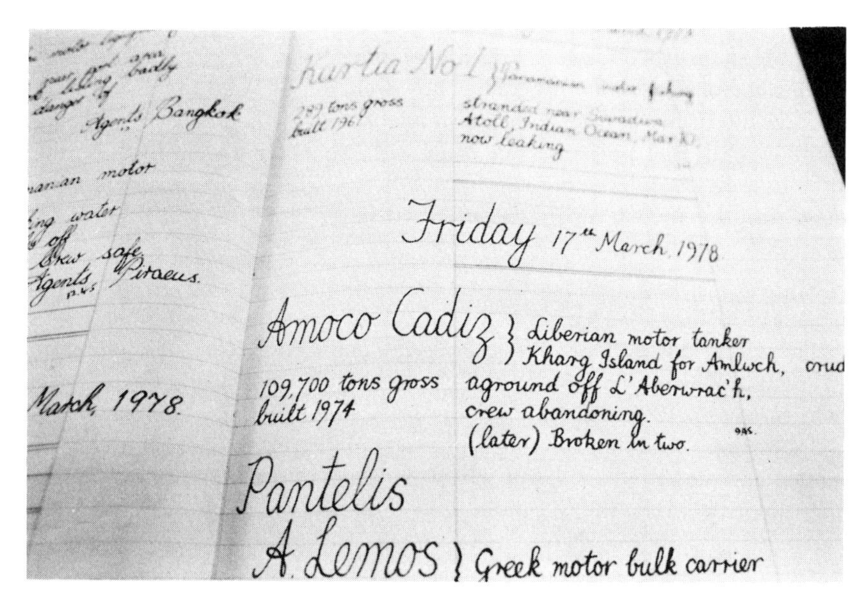

After a casualty is announced a notation is made in the record book situated across the floor from the bell. Above is the notation for the loss of the *Amoco Cadiz*. (Photo: Topix.)

was to operate on the basis of strict liability and subject to the same defenses as those available under the CLC.

The maximum liability of the CRISTAL fund was set at $30 million per incident less the liabilities under TOVALOP for cleanup measures.[103] The money was to be provided by a levy imposed on all participating oil companies in amounts proportional to their movements of oil by sea. The initial operating fund was set at $5 million, with further money to be raised by "periodic calls" only as required to cover expenditures.[104] CRISTAL has been in operation since April 1971.

While TOVALOP alone was an insufficient remedy, combined with CRISTAL, many inadequacies had been overcome. The limitation on compensation was raised to $30 million per incident; pollution "damage" and private claims were covered as well as governmental cleanup expenses; and the "fault liability with a reversed burden of proof" was replaced by the strict liability test of the CLC. On the other hand, although it did increase the amount of coverage available over that of the 1969 Convention, it did not broaden the scope of that coverage, being restricted to

103. Clause IV(C).
104. Clause V.

"direct" provable damage (excluding "ecological" damage) and incorporating the same exceptions as in the 1969 Convention. Other criticisms have been made of it as well. In general, however, the TOVALOP/CRISTAL package has been a simple, efficient, and effective method of covering some additional costs of pollution not dealt with by either TOVALOP or the 1969 Liability Convention.[105] As a "voluntary agreement" it was clearly a major improvement over the very limited scope of TOVALOP.[106]

The 1971 International Fund Conference

The 1971 IMCO Conference capped almost five years of diplomatic wrangling. Thirty-one months had elapsed between the *Torrey Canyon* incident in March 1967 and the creation of the Civil Liability Convention in October 1969. And that meeting had ended in discord: discord between the oilmen and the shipowners over the allocation of the oil pollution liabilities, and discord between the coastal interests and everyone else over how great the liabilities should be. Now, in November 1971, another draft convention was ready. This, it was hoped, would resolve the remaining problems.

The benchmark for the IMCO Legal Committee's preparatory work on the International Fund was an eleven-point proposal produced by an ad hoc working group meeting throughout the 1969 Conference to discuss the original Belgian submission. With the benefit of this proposal and of detailed written submissions from governments, a long and complex series of provisions setting out basic obligations and a detailed administrative structure to implement them had been drafted over two years. As in earlier negotia-

105. Some have argued that the system of sharing pollution costs on the basis of one's proportion of the oil transportation market removes the direct economic incentive (premium payments) that encourages the avoidance of spills and so "should be scrapped in favour of an extended insurance arrangement of the TOVALOP kind." This argument ignores one of the greatest advantages of the "compensation fund" approach—its efficiency. Rather than paying annual premiums on the potential liabilities above $10 million, only the direct costs of such infrequent liabilities are incurred as they arise. This is a far less expensive arrangement, and although insurance costs are certainly an important incentive to improved ship standards and operation, the additional incentive that would have been achieved for the higher magnitude of spill should a straight insurance approach have been used would not have been significant. (In fact, insurance premiums do not seem to decline with an increase in the shipowner's antipollution measures, as these measures themselves enhance the value of the ship. See "Marine Anti-Pollution and Economic Reality," *Fairplay International Shipping Weekly*, January 9, 1975, p. 125.) Any such benefit would, in any event, have been offset by the lower cost, higher efficiency, and greater reliability of the fund alternative. Finally, the CRISTAL system of mutually shared costs will likely give rise to substantial corporate pressure should the more responsible shipowners find they are subsidizing the less careful operators as a result of significant and disproportionate pollution cost burdens.

106. In its first seven years of operation (to January 31, 1978) CRISTAL received notice of sixty-five incidents, of which twelve resulted in payable claims. These claims averaged $401,000. At that time six claims totaling about $2 million were pending.

tions, no major differences were resolved in the years before the conference. They were instead incorporated into the draft as alternatives to be dealt with at the diplomatic conference.

The four-week conference was again held in Brussels at the Palais de Congrès. Forty-nine states were represented by delegations, with five more states, seven nongovernmental and three intergovernmental organizations sending observers.[107] The primary division at the previous conference between the coastal and shipping interests was now compounded by one between shipping and oil interests. Three important issues divided those groups and provided the fulcrum for debate: the nature of the incident for which the fund should be liable; the extent of relief that should be given to the shipowner and the conditions on which it would be given; and the extent of the fund's role in providing such relief.

That the fund should provide compensation additional to that given by the Civil Liability Convention was not disputed. Unquestionably an element in the 1969 compromise, an article implementing the obligation was now easily agreed upon: "the Fund shall pay compensation to any person suffering pollution damage if such person has been unable to obtain full and adequate compensation for the damage under the terms of the Liability Convention."[108]

The issue of the extent of the fund's total liability was more controversial but not significantly so. Without major disagreement, it was set at 450 million francs per incident (about $35 million), with a proviso that the "Assembly of the Fund" (when constituted) should be empowered to raise the limit to 900 million francs ($70 million) if experience showed the need for it.[109]

It was on the question of the types of incidents for which the fund should provide compensation that the debate really began. For many states, no satisfactory conclusion was ever reached. Under the 1969 Convention,

107. Of the forty-nine states sending delegations, nineteen were from Western Europe, North America, and Australasia, twelve were from Africa and the Middle East, eight from Latin America, five from Eastern Europe, and five from Asia (including Japan). Of the five state observers, three were from Asia and two were from Africa and the Middle East. One of the observers, the Philippines, had been a vocal advocate of the coastal position in 1969, but now it did not have a vote.

108. Art. 4(1), International Convention on the Establishment of an International Fund for Compensation for Oil Pollution Damage. This compensation is to be paid should no liability arise under the 1969 Convention, should the damage incurred exceed the limits of liability set down in that Convention, or should the "financial responsibility" provision fail to produce adequate compensation [art. 4(1)(*a*)(*b*)(*c*)].

109. Art. 4(6). There was little debate on the limitation, despite Canadian and American suggestions that the liability be unlimited. There was some debate on the proviso to raise the limit. Britain felt that such a high limit was unnecessary, and some federal states felt that they would encounter constitutional difficulties in later amending the Convention. Three proposals to delete, reduce, or restrict the provision were rejected.

there were a number of exceptions to the shipowner's "strict" liability and a shipowner would, of course, be liable only for pollution damage for which he could be identified as the source. With the creation of a fund providing compensation for damage caused by the oil industry's transportation of its own oil, it is not surprising that many coastal states felt that the fund should now accept the costs of all such damage, however caused—there would be no exceptions to its liability so that the fund would, in fact, be "absolutely liable." Compelling as the logic of this position might seem, it was bitterly —and, in large measure, successfully—resisted. Two minor exceptions available to the shipowner were deleted, making the Fund liable for them.[110] But two other similarly minor exceptions were retained,[111] and yet two more were accepted only after long, heated, and acrimonious debate. Of these last two exceptions, the first was the question of liability for pollution from incidents caused by "natural phenomena." (For state positions, see Chart 10). Under the Civil Liability Convention, the shipowner was exempted from liability for such phenomena only if they were of an "exceptional, inevitable and irresistable character." This obviously did not exclude such regular occurrences as hurricanes, tropical storms, or other normal weather patterns. The shipowner and fund were already liable for these. What, then, if coverage were now extended under the fund to the very rare and exceptional disaster such as an earthquake or resulting tidal wave? Fears were expressed that these incidents, when they did occur, would be so catastrophic as to be financially devastating or, at least, to consume all the resources of the fund before they could be applied toward the shipowner's indemnification.[112] These worries were much aggravated by the fact that the extended liability was a very unsettling precedent for all facets of maritime transport. It was the shipowners and not the oil industry that were concerned by the proposal. Indeed, three amendments that would have retained the exception for the fund were submitted, and all of these emanated from states with only minor oil imports but major shipping interests.[113] But coastal states were insistent in their demands that the

110. The Fund was thus made liable for damage wholly caused by an act or omission done with intent to cause damage by a third party (for example, sabotage) and damage wholly caused by the negligence of a government or governmental authority in the placing or maintenance of lights or other navigational aids.

111. The Fund will still be exempted from liability for damage resulting from an intentional or negligent act of the party suffering the damage. Another exception to the Fund's liability is damage resulting from war or hostilities. [Art. 4(3) and 4(2).]

112. Of course, the alternative would be that innocent coastal victims would bear these costs. Coastal states argued that the party taking the risk should bear the responsibility, especially for damages which are a hazard of the trade beyond anyone's control. On the nature of the natural phenomenon exception, see Wood, *Oil and Water*, p. 66.

113. Proposals were made by Greece (defeated: 14-19-8), Norway (defeated: 10-13-6), and Belgium (defeated: 2-21-11).

industry help shoulder this great potential burden. Finally, a compromise was put forward by Belgium's delegate, Albert Lilar. As president of the CMI he was quite familiar with the legal implications of the decisions of the conference and his suggestion was accepted. The fund would indeed be liable for "catastrophic" damage, but to a limit of 450 million gold francs ($35 million) per natural disaster, regardless of the number of distinct shipping incidents arising from it.[114] Clearly this was a "special" case.

Of tremendous significance was the debate on liability for pollution from unidentified ships. (For state positions, see Chart 10.) It was, after all, the continuing accumulation of oil on British beaches from unidentified passing ships that provided the momentum for the first international convention on oil pollution, and it is still true today that well over two-thirds of ship-source oil pollution comes from intentional discharges whose sources seldom can be identified. Yet, because of their very magnitude, the imposition of liability for these discharges aroused great fears. After all, if the fund were made liable for all this, it really would be forced to internalize *all* the identifiable environmental costs of its operations. Anyway, was not the fund *intended* to deal only with large spills? Should it have to concern itself with numerous petty—but cumulatively expensive—claims? And wasn't it the shipowner and not the oil industry that benefited from such intentional discharges? The oil-importing states flatly rejected the inclusion of unidentified spills. It was a totally unwarranted expansion of the role of the fund, and it amounted to a subsidy of the shipping industry and especially of the less responsible shipowners.[115] To coastal states such as those in Africa, whose shores are regularly fouled by the passing traffic, the industry concerns were niggling.[116] These states certainly were not benefiting from the shipments *at all* and yet were paying the entire pollution costs. Moreover, detecting the violators was almost impossible, especially for the developing states, since the oil slicks often drifted ashore months after they had been discharged. To retain the exclusion would, in the words of the Algerian delegate, "be tantamount to penalizing the developing countries."[117] But the majority of shipowners and oil importers were unmoved, and an attempt by the United States to have the exception deleted was defeated.[118] Instead,

114. Art. 4(4).

115. This argument was only partially true, however. Most oil companies are themselves major tankerowners and "bareboat charterers," and their own secret investigations in 1971 and 1972 had revealed that the primary method of pollution control, load-on-top, was not being utilized efficiently or at all by two-thirds of the world's tankers.

116. Since the closure of the Suez Canal, these states have been the worst affected by such chronic pollution.

117. Statement by Mr. Ait-Chaalal (Algeria), LEG/CONF.2/C.1/SR.25.

118. As shipowners did not now pay for such pollution, they had no reason to support deleting the exception to liability. On the contrary, they were wary of the precedent of anyone

Chart 10
State Positions on Deletion
of Exceptions to Compensation

	Natural phenomenon	*Unidentified ships*
Australia	x	x
Canada	x	x
Ghana	x	x
Singapore	x	x
Spain	x	x
U.S.	x	x
Algeria		x
Indonesia		x
Kenya		x
Poland		x
Portugal		x
Ireland	x	
Sweden	o	x
Germany	x	o
Brazil	o	
Denmark	o	
USSR	o	
Liberia		o
Romania		o
Belgium	o	o
France	o	o
Greece	o	o
Italy	o	o
Japan	o	o
Norway	o	o
U.K.	o	o

Key
x = Delete exception
o = Retain exception

it was retained as article 4(2)(*b*). As a result, any claimant against the fund must show "that the damage resulted from an incident" involving a ship carrying oil *as cargo*, a requirement that would necessitate proving the identity of the ship as a tanker. This was not only a defeat for the coastal states, but also for environmentalists who had hoped to establish an effective method to force genuine preventative action by the industry in all parts of the oceans.

Any hope that the fund would introduce some broad industry accountability for the condition of all the oceans was dashed. Together with the coastal states' demand that the fund be liable to the coastal state for damage beyond the territorial sea (see Chapter 6), these always remote environmental ideals were now dead. Both shipping and oil interests—and the states defending them—had been united in their opposition. In one area, however, the commercial interests were not in such unison, and the environment was to benefit as a result. On the issue of shipowner indemnification, the oilmen and the environmental states were united against the shipowners. (For state positions on this and related issues, see Chart 11.)

CRISTAL had made no plans for reducing the liability burden on the shipowner, and many at the IMCO meeting now felt that, despite the 1969 Fund Resolution, the IMCO Convention should avoid this function as well. The preparatory work of IMCO's Legal Committee had, however, acted in accordance with the Fund Resolution and had included this as a mandatory function of the fund. Indeed, the draft articles provided that the shipowner's liability would be reduced right back to its original limit of 1,000 francs per ton with the fund as "automatic guarantor" of the balance. This would, in effect, return the shipowner to his traditional maritime legal position and shift all extra liability onto the fund. Two areas of debate emerged in connection with this issue: should there be shipowner indemnification at all, and if so, on what conditions; and what should be the role of the fund in this indemnification?

Realizing the integral part that the Fund Resolution had played in the success of the 1969 Conference, most states agreed that indemnification of the shipowner was at least necessary, if not desirable. Some did oppose it—notably the United States and, in a clear reversal of its adamant stance in 1969, the United Kingdom.[119] Others, such as Canada and Singapore, supported indemnification only if it were tied to certain conditions. After

being liable for such damage. An American proposal to delete the exception was narrowly defeated by a vote of 18-20-1. An earlier Scandinavian compromise to delete the exception for unidentified spills over 15 million gold francs in value ($1 million) was rejected 17-18-3.

119. This opposition was doomed to failure and was, in any event, likely a bargaining tactic for the imposition of conditions on indemnification. In fact, it was the American delegation to the Legal Committee's working group that had first raised the idea of conditions as a prerequisite to its country's acceptance of indemnification.

CHART 11
State Positions on the Role of the Fund in Indemnifying Shipowners

	In principle, should there be indemnification?	*Should conditions be attached?*	*French proposal for high indemnification and Fund as guarantor*	*Scandinavian compromise*	*Statement as to preferred level of shipowner liability*
I					
U.S.	x	x	x	x	+ 1,500
Singapore	x	x	x	x	+ 1,500
Australia		x	x	x	1,500
Germany		x	x	x	+ 1,000
Canada	x	x	x		1,500
U.K.	x	o	x	x	1,700
Netherlands	x	o	x	x	1,500
Portugal	x	x		x	
II					
Belgium	o	x		x	+ 1,250
Sweden	o	x	x	x	
Finland	o	x		x	1,000
Japan	o	o		x	+ 1,000
Denmark	o		o	x	1,000
Norway	o	x	o	x	1,000
III					
Italy	x	x	o	o	1,500
Poland	x		o	o	1,250
Ireland	o	x	o		1,000
Egypt	o		o	o	
USSR	o	o	o	o	1,000
Bulgaria			o	o	
India	o	x		o	1,000
Liberia	o	o	o	o	1,000
Greece	o	o	o	o	1,000
France	o	o	o	o	1,250

Key
x = Statement or vote unfavorable to shipowner
o = Statement or vote favorable to shipowner

all, blanket indemnification would reduce the incentives to safety provided by the higher limitation and would be a much greater drain on the fund than would the compensation provision.[120] Some major shipowners such as Greece and Liberia attacked anything less than a full return of the shipowner's liability to 1,000 francs per ton as a betrayal of the promise of 1969.[121]

The issue was profoundly divisive. Independent shipowners complained that they had been cheated and that any suggestion of putting conditions on the indemnification was a terrible infringement on their *right* to indemnification. The oil industry, though, found this suggestion an imaginative way to reduce the indemnification of the less responsible shipowners. And to the environmentalists the proposal was an incentive to shipping safety and an unexpected boon for environmental protection. But the atmosphere in the meeting hall was tense, filled with accusation and recrimination, and a total breakdown in the conference loomed as a possibility. To avoid this disappointing conclusion, the issue was referred to an *in camera* "conciliation group" where the differences could be hammered out.[122] Unfortunately, the lack of records of the discussions limits our knowledge of national policies, especially as the conciliation group's recommendation, being the delicate result of embattled negotiations, was accepted almost without debate.[123] The result was a welcome and imaginative bonus for environmental interests. The fund would provide indemnification for a portion of the shipowner's liability but only where the shipowner had complied with IMCO's four most important maritime conventions:

(*i*) the International Convention for the Prevention of Pollution of the Sea by Oil, 1954, as amended in 1962;
(*ii*) the International Convention for the Safety of Life at Sea, 1960;
(*iii*) the International Convention on Load Lines, 1966;
(*iv*) the International Regulations for Preventing Collisions at Sea, 1960;
(*v*) any amendments to the above-mentioned Conventions.[124]

120. Indemnification would occur for that portion of the damage amounting to less than 2,000 francs per ton, while compensation would occur only when the amount of damage exceeded 2,000 francs per ton. Indemnification would, therefore, occur more frequently.

121. This position carried little weight. Arrangements once set down are difficult to dislodge, and the shipowners found themselves stuck with the Civil Liability Convention as a starting point for negotiation. This situation was exacerbated by the realization that insurers could now provide shipowners with the coverage required by the 1969 Convention, thus undercutting one of the major arguments used by shipowners at the 1969 Conference.

122. The group was composed of Brazil, Canada, France, Germany, Ghana, Greece, India, Japan, Liberia, Netherlands, Norway, U.K., U.S., and the USSR.

123. The group reported back to the main committee during its twenty-first meeting and, after very brief discussion, its report was accepted by a vote of 33-0-7.

124. The shipowner would be disentitled from indemnification only if his noncompliance with the listed conventions was "wholly or partially" a cause of the incident. This applies to all shipowners whether or not their flag state is a party to conventions. [Art. 5(3).]

The resolution of this issue was by no means the end of the problems for the conference. Many divisive issues remained, and although they were of only minor relevance to environmental protection, they are important in understanding the overall result of the conference. They are also interesting in that the shipowner was again isolated against the policies of the oil and environmental interests. Most contentious was the question of the proper amount of indemnification for the shipowner and the procedure for implementing it. Three early proposals were immediately rejected, and the committee again found itself badly split.[125] One of the main protagonists, France, for rather eccentric reasons had completely reversed its policy since 1969, and it now favored substantial reductions in the burden on the shipowner.[126] Interestingly, France again came in conflict with the United Kingdom, which had itself reversed its pro-shipowner policy since 1969. Not only had the U.K. learned since the last conference that the insurers could actually cover the higher liability limit of 2,000 francs per ton (the TOVALOP limits had been implemented with no problem for the P & I Clubs), but OCIMF had convinced the governments of both the United Kingdom and the Netherlands that their interests lay not with their shipowners but with their oil industries. In both countries, their largest fleets were owned by oil companies and, of course, each imported massive quantities of oil.[127] High indemnification of the shipowner was not among their priorities.

125. The draft provision for a full 1,000 francs per ton indemnification was rejected by a vote of 14-16-7. The British proposal for a very limited indemnification of only 300 francs per ton was also rejected (20-23-4), as was a moderate Dutch proposal for a 500 francs per ton indemnification (vote of 13-15-8). The last suggestion would have left the primary shipowner liability at 1,500 francs per ton. The British delegate pointed out that in one vote the states on one side accounted for 72 percent of the world oil imports while those on the other, only 16 percent (LEG/CONF.2/C.1/SR.23, p. 3).

126. The French delegation had recommended a limitation of liability at 3,000 francs per ton in 1969 but was now intransigent in its demand that this limit be reduced to 1,250 francs per ton. The French position in 1971 was a curious one. One of the major goals of the delegation was to restrict the amount of insurance the shipowner would have to get directly from the London-based market. Whereas in 1969 the delegation was led by an official with environmental sympathies, in 1971 it was dominated by the Ministry of Finance which hoped to undercut the London insurance market and the money flowing into Great Britain in consequence. Its proposal was a complex one and was advanced despite the fact that it would cost the industry-based Fund dearly and, thus, would conflict with France's other interests as a large oil importer with a tanker fleet that is highly integrated with the major oil owners. A very frustrated OCIMF delegation (which contained a French lawyer to deal with what was seen as the French legal and logical eccentricity) argued strenuously against the French delegation.

127. The British delegate commented quite candidly that the insurance market could profitably accommodate "the required coverage and the alternative was a heavier burden on the oil industry, a major interest in Britain in its own right and as a tanker-owner" (LEG/CONF.2/C.1/SR.23, p. 5). In 1969 the Dutch delegation had urged that the cargo-owner and not the shipowner bear the primary burden for oil pollution damage. In 1971, it

With such intransigent interests, it is not surprising that not until the penultimate day of the committee's meetings was final agreement reached. Even so, the decision (which was very similar to the original Dutch proposal of three weeks earlier) was not so much a product of compromise as of attrition.[128] During the twenty-third meeting, figures were selected almost at random and over thirty votes were taken until, well after midnight, an acceptable majority was finally achieved. The result was based on a typical submission by Norway on behalf of all Scandinavian states.[129] This "Scandinavian compromise" provided that the fund would indemnify the shipowner for that 500-franc portion of his liability above 1,500 francs per ton.[130] For that amount, the fund was to have the option of acting as the primary "guarantor" (insurer) in replacement of the regular insurance companies. (For state positions on this and related issues, see Chart 11.)

Explanations of the policies of most states are contained in the text or footnotes to this chapter. Group 1 consists of the environmentally active states plus those oil industry states which tended to take similar positions (U.K., Netherlands, Germany). Group 2 consists of maritime states that did support the final "Scandinavian compromise." Japan's position here is undoubtedly also influenced by its position as a large oil importer, which moderated its shipowning interests. Group 3 consists of France, the USSR and its allies, some core maritime states, and miscellaneous others.

---

now sought to protect oil industry interests. The British and Dutch delegations worked very closely with OCIMF throughout the conference and actually took some of their arguments and positions from the OCIMF delegation. Their opposition to the French/Norwegian suggestion that the Fund should take over the role of "guarantor" was based, in part, on OCIMF's "estimate" that this would result in such an increase in "administrative" costs for the Fund that it would require a 500 percent increase in contributions. (Interviews.) Germany, which was similarly situated to Britain and the Netherlands, followed similar policies on these issues.

128. In fact, France commented that it "could no way be regarded as constituting a compromise." Statement by Mr. Jeannel, LEG/CONF.2/C.1/SR.24, p. 14.

129. Although most states with large shipowner interests (Greece, Liberia, USSR, France, Japan, Norway, and Denmark) supported substantial shipowner relief, Norway and its three Scandinavian allies—Denmark, Sweden, and Finland—provided the compromise that reduced shipowner indemnification to 500 francs per ton. It was a common phenomenon throughout the negotiations for Norway to take a strong shipowner position and then, with the consultation of its regional partners (especially the more environmentally concerned Sweden), provide a compromise submission. Danish policy is closely akin to Norway's, as it too has a powerful shipping lobby and a significant independent fleet (considering the size of the country). Finland is a consistent, but silent, partner in the group.

130. Article 5(1) states that the Fund shall "indemnify the owner and his guarantor for that portion of the aggregate amount of liability under the Liability Convention which: (*a*) is in excess of an amount equivalent to 1,500 francs for each ton of the ship's tonnage or of an amount of 125 million francs, whichever is the less, and (*b*) is not in excess of an amount equivalent to 2,000 francs for each ton of the said tonnage or an amount of 210 million francs, whichever is the less."

With the negotiation of detailed articles for the convention's implementation, a successful "package" finally had been produced.[131] In review, it is interesting to look at the progress of the conference on its central points. The draft convention had contemplated a shipowner liability of only 1,000 francs per ton or 105 million francs as an overall maximum, with the fund to take over the role of insurer ("guarantor") for the balance to 2,000 francs. The final act maintained a basic shipowner liability of 1,500 francs per ton (maximum of 125 million francs), with indemnification only if the shipowner complied with stipulated conditions and with the fund acting as guarantor only at its own discretion. Some of the exceptions to liability had been removed for the fund, but some important ones had been retained. The liability limits of the fund were set at double those of the shipowner under the 1969 Convention, with a provision for a further doubling. The result clearly reflects a compromise among all the interests involved in the long negotiations—shipping, oil, and environmental.

AN ASSESSMENT

This chapter has not attempted a detailed legal analysis of every practical aspect of these instruments, nor of the many nuances in interpretation and application that could be provided by a consideration of a number of borderline cases.[132] We have focused on the more controversial issues that are central to the development of the law. Chart 12 shows how the major legal issues have developed since 1967. The major thrust of the post-*Torrey Canyon* activity clearly was in the field of liability and compensation and, later, in the development of the 1973 Convention on the Prevention of Pollution from Ships. In the compensation field, we have an example of the innovative compromises that can be achieved through the interaction of diverging, often hostile, interests.

TOVALOP and CRISTAL were clearly responses to activities undertaken by governments in opposition to industry. Yet the industry agreements were, for over seven years, the only complete operational mechanism

131. In addition to the basic provisions, there are approximately forty-five articles dealing with the implementation and operation of the Convention. The Fund is to be constituted through a levy on oil companies importing in excess of 150,000 tons of oil per calendar year into a contracting state. Only a small proportion of the $30 million per incident limitation of the Fund is to be collected initially, leaving the balance to be collected (when needed) on a "call" basis similar to that of CRISTAL. (Art. 10–15.) The remaining articles are the traditional "Final Clauses." One of these, article 40, provides that the Convention should enter into force upon the ratification of at least eight states in which are located "persons" who "have received during the preceding calendar year a total quantity of at least 750 million tons of contributing oil." This figure was estimated to be approximately one-half of the anticipated annual oil movement by sea within the next few years.

132. For a more detailed discussion of the IMCO conventions and of TOVALOP/CRISTAL, see numerous articles in the *Journal of Maritime Law and Commerce*.

on an international scale for compensation for oil pollution damage. TOVALOP alone was an inadequate remedy, but with the continuing pressure on governments after the successful conclusion of the 1969 IMCO Conference, CRISTAL was then drafted by the oil industry. The combination attests to the ability of multinational industrial interests to contribute positively to international environmental protection—when motivated to do so.

TOVALOP is in force for over 99 percent of the world's tonnage, and CRISTAL covers about 95 percent of their oil shipments.[133] Between 1971 and 1978 the latter paid out approximately $4,800,000 on twelve claims.[134] This is, as expected, but a small proportion of the claims paid under TOVALOP, but the overall scheme has been an extremely active and valuable one. Furthermore, despite their supposedly "interim" nature and despite the entry-into-force on June 19, 1975 of the Civil Liability Convention, the industry agreements have been amended so as to allow them both to continue in operation even after the Fund Convention becomes operative. This extension will allow compensation to be provided even for those states not party to the intergovernmental agreements. Interestingly, the TOVALOP/CRISTAL agreements were to be revised again in 1978 so as to make them almost exactly parallel to the 1969 and 1971 Conventions. When this is done, the private arrangements will have come full circle—from a reaction to proposed new intergovernmental schemes to an exact reflection of them. In addition, the activities of TOVALOP have been expanded to provide a convenient vehicle for industry intervention in other aspects of maritime environmental control. In its 1975 report, TOVALOP's chairman wrote that "there will still be an active role for the Federation to play, particularly in advising governments on pollution avoidance and contingency plans, providing technical assistance, pinpointing causes of pollution statistically and reflecting overall membership interests."[135]

133. Interviews with CRISTAL officials. Most of the oil companies not registered in CRISTAL are state owned, not privately owned. In fact, in one of the largest oil pollution incidents since the *Torrey Canyon* disaster, the 1974 spillage of 50,000 tons of oil onto Chilean shores from the tanker *Metula*, the cargo was owned by the Chilean government company *Empresa Nacional del Petroleo*, which was not registered in CRISTAL. The ship was on charter from an American oil company which was registered in TOVALOP. The TOVALOP liabilities have been settled with the Chilean government. Had higher damages occurred (a matter of some debate), Chile would not have been eligible for compensation from CRISTAL.

134. Interviews with CRISTAL officials.

135. Statement by A. B. Kurz, Chairman, *Report and Accounts: 1975*, International Tanker-Owner Pollution Federation, p. 4. TOVALOP has, for example, been active in preparing pollution control "contingency plans" in the Caribbean, Straits of Malacca, Bantry Bay, and Bahrain. These are undertaken at the request of oil companies and governments. TOVALOP also participated in an advisory capacity in the salvage of the supertanker *Metula*.

CHART 12

International Provisions for Compensation for Oil Pollution Damage

	General international law: Pre-1969	*Oil industry package**		*Intergovernmental package*	
		TOVALOP	*CRISTAL*	*1969 CLC*	*1971 Fund*
Basis of liability	fault	fault/reversed burden of proof	strict	strict	strict
-exceptions to strict liability			*exceptions*: -war -third party -caused by claimant -unidentified ship -natural phenomenon -government negligence	*exceptions*: -war -third party -caused by claimant -unidentified ship -natural phenomenon -government negligence	*exceptions*: -war -third party -caused by claimant -unidentified ship
Bearer of liability	shipowner	shipowner or bareboat charterer	participating oil companies		
Amount of liability -primary	1,000 francs/ton or value of ship after accident	$100/ton to $10 million limit		2,000 francs ($150)/ton to 210 million franc ($15 million limit)	

–secondary			$30 million		450 million francs ($35 million) with option to double
Means of assuring compensation	voluntary insurance	insurance with no national enforce-ment	guaranteed fund	compulsory insurance with enforcement	guaranteed fund
Type of damage covered	damage reasonably foreseeable	governmental cleanup expenses	direct costs and losses	direct and conse-quential costs and losses	direct and conse-quential costs and losses

*This package was to have been amended in 1978 to parallel exactly the intergovernmental package but early negotiations to achieve this were unsuccessful.

The desired goal is, however, the creation of a uniform *intergovernmental legal* regime and, despite many criticisms of the 1969 and 1971 IMCO conventions, there was widespread support for their implementation. Only the United States stands in public opposition to them.[136] Indeed, many of the criticisms directed at the two conventions deal with oversights by the drafters rather than major substantive omissions, and even those which do focus on the latter also recognize that the conventions constitute a valuable first step. That a larger purpose for the conventions was not achieved is a political not a legal criticism. What *was* produced was a result of "the mix of many events and factors . . . and the adjustment of interests in the light of national policy and the range of acceptability of results, both domestically and internationally."[137]

From this perspective, some very significant accomplishments were achieved. Together, the 1969 and 1971 conventions form a uniform international scheme where compliance is required with only one set of regulations, where all claims from an incident can be consolidated under one jurisdiction, and where any judgments rendered will automatically be enforceable on all other contracting parties.[138] In utilizing strict liability, the Civil Liability Convention is an important development in international law, as is the more nearly absolute liability imposed on the fund. Moreover, the scope of the regime is such that it could not be achieved by a national scheme. For example, the establishment of an accepted international framework of responsibility provides access to vessels on the high seas without the problems of jurisdiction which might ensue from a national scheme. The 1971 Fund also creates an inexhaustible fund, the resources of which are determined by the demands on it. If the Fund were required to pay, for example, a total of $45 million compensation for a number of incidents in one year, contributions required from cargo-owners for the following year would be adjusted to satisfy such payment. Most national funds would not have such flexibility and would require contributions to be made in advance to a dormant fund. The 1971 Fund, while it does guarantee

136. The Civil Liability Convention (CLC) entered into force on June 19, 1975 and the Fund Convention on October 16, 1978. In *Hearings*, Subcommittee on Oceans and the International Environment, Committee on Foreign Relations, United States Senate, 93rd Congress, 1st Session, April 17 and 18, 1973, almost all American commentators favored ratification of both conventions. This situation has changed in the last two years with the introduction of the new "Super-Fund" legislation. This legislation, if enacted, would set shipowner liability limits of $300 per ton while establishing a $200 million "Super-Fund." The legislation would not be compatible with ratification. (See HR.6803, 95th Congress, 1st Session.) On the status of ratifications, see Chart 17.

137. Lowell Doud, Submission to Subcommittee on Oceans and the International Environment, Committee on Foreign Relations, United States Senate, 91st Congress, 1st Session, April 17 and 18, 1973, p. 163.

138. CLC, arts. IX and X.

compensation regardless of the frequency of oil spills that may occur in the future, does not require contributions until after the event. Additional innovative developments are evident in many other areas of the convention: the mandatory insurance requirements are to be applied to states not party to the convention; the "conditions" are to apply to a shipowner seeking indemnification whether or not the flag state is a party to the convention which is the basis of the condition; and compensation from the fund is to be paid to a state party to the convention whether or not the flag state of the polluting ship is a party.[139]

While many environmentalists would demand higher limits of liability than $35 million, there is a provision in the convention to double the limits to $70 million. Even higher limits could be achieved by amending the convention.[140] In any event, the experience of TOVALOP and CRISTAL has shown that by far the greatest costs come from the numerous small spills from shall ships. Although it is certainly necessary to be able to handle the inevitable large spills such as the *Amoco Cadiz*, these have to date constituted but a small proportion of pollution liability costs.

At the same time, important criticisms can be made. At a very general level, it is clear that neither convention substantially changes the basic nature of legal obligations for environmental damage. At the Stockholm Conference in 1972, it was agreed that states would be responsible for actions taken under their "jurisdiction *or control*." No such national responsibility was attached by the earlier IMCO Conventions. Moreover, the Stockholm Principles obligate a state to prevent pollution damage to "areas beyond the limits of national jurisdiction."[141] These areas are excluded from coverage under the IMCO liability conventions. On a more specific level, therefore, although both direct and indirect "pollution damages" are to be compensated under both conventions, the scope of coverage is restricted to damage in the territory or territorial seas.[142] It is also limited to

139. CLC, art. VII(11) and Fund, arts. 5(3) and 4(1)(*a*).

140. On higher liability limits, see the letter of Senator Muskie to Secretary of State William Rogers, November 17, 1971, *Hearings*, Subcommittee on the Oceans and the International Environment, p. 161, fn. 6. Whether the figures should be raised further will likely be decided after some experience is had with the Fund. It is quite possible that the limits will have to be raised especially as the larger new supertankers now plying the oceans begin to deteriorate. One spill from a coastal refinery at Mizushima, Japan, was estimated to have caused more than $75 million damage. In fact, due to the special circumstances of the spill and the extraordinarily thorough cleanup effort, the total cost was put at closer to $200 million. See C. W. Nicol, "The Mizushima Oil Spill—A Tragedy for Japan and a Lesson for Canada," Environmental Protection Service, West Vancouver, Canada, May 1976. Present estimates of the cost of the *Amoco Cadiz* spill range upward of $100 million.

141. See Stockholm Declaration on the Human Environment, principles 21 and 22.

142. See CLC article I(6) for the definition of "pollution damage." "Indirect" damage includes losses consequential to the pollution such as losses to foreshore tourist industries.

"persistent" oils[143] and only when carried by a ship that is "actually carrying oil in bulk as cargo."[144] Furthermore, as mentioned earlier, the claimant must usually identify the particular source of the oil, sometimes a difficult task given the multitude of possible sources. This means that often there will be no compensation for damage from unidentified oil slicks which are themselves the product of normal operational tanker activities of benefit to oil transportation interests in general. Unfortunately, this was the price for the fund's acceptance by the oil industry and the many states sympathetic to it. Finally, it should be pointed out that some provision is necessary for lessening the burden of proof on the innocent victim. Even where the source of the spill can be easily identified, great difficulties arise in demonstrating the extent and exact causation of damage. Some commentators have argued that the fund should have the option of undertaking an independent investigative role.[145] While this could become an onerous task (although CRISTAL has not found it so), the suggestion certainly points to the absence of an independent organization that could, at the international level, provide both independent advice on combating a pollution incident and an independent assessment of the costs of preventative measures and pollution damage to all parties. Presently, only the industry has been involved in such activities and, as the *Metula* incident demonstrated, the advice of special interests may vary from that of more impartial observers.[146] Some independent authority—likely through the administrator of the fund —will be required in order to ensure effective pollution compensation.

143. CLC art. I(5). "Persistent" oils are heavy black oils such as crude oil as distinct from light refined oils. (See Chapter 2 for discussion of "persistent" and "nonpersistent" oils.)

144. CLC. art. I(1). This is more restrictive than TOVALOP, which also covers tankers when in ballast.

145. For example, see Lawson Hunter, "The Proposed International Compensation Fund for Oil Pollution Damage," *Journal of Maritime Law and Commerce* 4 (1972), p. 117, and Wood, "Integrated Approach," p. 61. Admittedly such an investigative role (combined with liability for unidentified spills) would substantially alter the nature of the Fund. Moreover, had environmental groups participated in the actual preparatory work for the convention, such a scheme might at least have been considered.

146. The Fund would be "independent," as it would be an intergovernmental body despite the fact that the oil industries pay the costs. On the *Metula* dispute, see "Row over 'Chocolate Mousse' Oil Spill," *Sunday Times*, February 18, 1975; and Noel Mostert, "The Age of the Oilberg," *Audubon* 77 (May 1975), p. 18. Mostert cites the report of Roy W. Hann, an observer for the U.S. Coast Guard, who criticized TOVALOP observers for a lack of concern for controlling the pollution once it had escaped from the *Metula*. TOVALOP officials strenuously disagreed with this viewpoint. There is great pressure on them to take a neutral stance, they argue, in order to retain the trust of the government on the one hand and the P & I Clubs on the other. (Interviews.) Other officials interviewed about the *Metula* incident did corroborate Dr. Hann's analysis on this specific incident. Considering the variety of interests represented among observers at oil spills and the varying and uncertain characteristics of each spill, such disagreements are unavoidable. In its fast payment of claims (such as the $1.1 million paid out within nine months of the *Argo Merchant* disaster), TOVALOP has fulfilled its primary function very well, however.

One last criticism of the Fund Convention concerns the application of the conditions for indemnification. While the idea of predicating indemnification on compliance with the "requirements" of certain conventions is an imaginative one, it is unnecessarily restrictive to exonerate the fund from indemnification *only* if it proves the ship's noncompliance was "as a result of the actual fault or privity of the owner and not just the master."[147] Furthermore, article 5(3) refers to compliance with "requirements," while many of the most important regulations are only "recommendations." Moreover, it has been argued that the effect of indemnification is, after all, only to reimburse the shipowner's insurer and that this will have little impact on the shipowner's conduct.[148] If so, the "conditions" on indemnification may be ineffective in actually reducing pollution.

As with any legislation, many particular criticisms of these conventions can be made. But it will be only by actual experience with the operation of the Civil Liability and Fund conventions that their success or failure can be judged. In any event, in setting up the administrative structure of the fund now that the convention has entered into force, suggestions will be considered and, if acceptable, implemented. More significant change must await experience and will require amendment. With the long-awaited entry-into-force of the Fund Convention on October 16, 1978, this experience finally is being obtained.

147. Fund, art. 5(3). The requirements of the Loadline Convention and the Convention for the Safety of Life at Sea would usually be imputed to the knowledge of the owner, but the other two stipulated conventions, the 1954 Convention and the Collision "regulations," are more operationally oriented so that violations of them would not likely be imputed to the shipowner's actual fault. This retains the traditional rule that there is no *respondeat superior* (vicarious liability) in admiralty matters.

148. On these criticisms, see Lowell Doud, "Compensation for Oil Pollution Damage: Further Comments on the Civil Liability and Compensation Fund Conventions," *Journal of Maritime Law and Commerce* 4 (1973), p. 525; Hunter, "Proposed Fund"; and Wood, "Integrated Approach."

Chapter VI

Jurisdiction and Enforcement

> We do not doubt for a moment that the rest of the world would find us at fault, and hold us liable, should we fail to ensure adequate protection of [the] environment from pollution or artificial deterioration. [We] will not permit this to happen either in the name of freedom of the seas, or in the interests of economic development. We have viewed with dismay the abuse elsewhere of both these laudable principles . . . [and] we are aware of the difficulties faced in the past by other countries in controlling water pollution and marine destruction within their own jurisdictions.
>
> (Canadian Prime Minister P. E. Trudeau, House of Commons, October 24, 1969[1])

With these challenging words, the Canadian prime minister introduced a dramatic development into international environmental law: the coastal pollution-control zone. For years coastal states had been compelled to rely on the goodwill of distant flag states to implement and enforce pollution controls on their ships. But as the threat of coastal contamination steadily grew, this deference was no longer enough.

For centuries, flag states have retained almost exclusive jurisdiction both to legislate standards that apply to their ships and to enforce these standards. This was the law of the sea, and it ensured the flag states a nearly monopolistic control over pollution prevention. Certainly, IMCO could prescribe "international" standards, but legislative jurisdiction to implement them rested elsewhere. The flag state could "ratify" the IMCO agreements or not. Coastal states, on the other hand, could traditionally legislate only for their internal waters and, subject to "innocent passage," for their territorial seas. They had no legislative authority at all in the high

1. House of Commons *Debates* (Canada), October 24, 1969, p. 39. This speech announced the government's plan to introduce a bill to control environmental standards in Canada's Arctic waters. The Arctic Waters Pollution Prevention Act, tabled in the House on April 8, 1970, established a 100-mile "pollution control zone" within which the Canadian government claimed authority to set environmental ship standards. For a discussion of Canadian policy in this field, see R. Michael M'Gonigle and Mark W. Zacher, "Canadian Foreign Policy and the Control of Marine Pollution," in Barbara Johnson and Mark W. Zacher, eds., *Canadian Foreign Policy and the Law of the Sea* (Vancouver: University of British Columbia Press, 1977), pp. 100–157.

seas beyond.[2] The flag state monopoly was secure, for it also extended to the lion's share of the enforcement jurisdiction: the powers of inspection, investigation, and prosecution. Indeed, even more than the distribution of legislative jurisdiction, these powers of promoting compliance have emerged in recent times as the most crucial and contentious areas in the international politics of marine pollution control.

IMCO is ostensibly a "technical" and "regulatory" body and, as such, should not be concerned with allocating general legislative and enforcement authorities. Nevertheless, as a result of the gaps and contradictions in international law and the frequent pressures to change it, IMCO's negotiations often have spilled over into these areas. Indeed, as the one continuing regulatory body with a long history of environmental "management," the experiences of IMCO provide an excellent opportunity to evaluate the competing pressures to which a flexible and evolving jurisdictional regime is subject. These experiences are especially enlightening when looked at in conjunction with the three United Nations Conferences on the Law of the Sea with which IMCO has had close ties (in 1958, 1960, and 1973–1978).[3]

LEGISLATIVE JURISDICTION

Early Conferences, 1954–1960

Coastal state concern for the environmental impact of foreign shipping is a recent phenomenon. Prior to the 1954 Conference, little international attention had been paid to the pollution issue, and what proposals did surface seldom, if ever, called for changes in the customary "freedom of the seas." High seas oil pollution often, however, resulted in pollution of a state's coastline, and it was this that the 1954 Conference was convened to remedy. But at that time there were no proposals for altering traditional jurisdictions. A fifty-mile coastal "prohibition zone" was implemented by the conference, but it implied no transfer of authority from the flag to the coastal state. The high seas/territorial seas distinction was retained, so that it was the flag state exclusively that controlled standards and enforcement to within a distance of three to twelve miles from the shores of another state.

2. Myres McDougal and William Burke have characterized this issue as a process of claim and counterclaim to exclusive authority between "exclusive" national actors and the "inclusive" international community. See the *Public Order of the Oceans* (New Haven: Yale University Press, 1962), chap. 1. To the extent that flag states retain the sole power to ratify an international convention, the "inclusiveness" of the "community control" is less than perfect.

3. The chapter will focus on both the allocation of legislative jurisdiction between coastal and flag states and their powers and policies for promoting compliance. This latter topic will be particularly concerned with problems of self-reporting, inspection, surveillance, and prosecution.

This was made quite explicit at the time in order to avoid, in the words of the Soviet delegate, any "international disputes in connection with enforcement."[4] As a result, article XI was incorporated into the convention:

> Nothing in the present Convention shall be construed as derogating from the powers of any Contracting Government to take measures within its jurisdiction in respect of any matter to which the Convention relates or as extending the jurisdiction of any Contracting Government.

Jurisdiction was not defined but it was assumed, of course, to refer to the territorial seas and internal waters.

This result was not a surprising one at the time, and no one expressed dissent. In 1958, at the first U.N. Conference on the Law of the Sea, the status quo was now formally, if vaguely, confirmed in treaty form. Four conventions were produced in 1958 and 1960, and nothing was changed.[5] Indeed, the only convention of any significance for these issues was the Convention on the Territorial Sea and Contiguous Zone (1958), and it basically incorporated the customary international law (article 17). But a hint of things to come could be detected in the creation of a limited new protective coastal jurisdiction, the "contiguous zone." This zone was to go to a maximum distance of only twelve miles from the coast, and it allowed a coastal state to take action to prevent the infringement of a limited number of regulations including "sanitary regulations within its territory or territorial sea" (article 24). The significance of this article is not so much in the very few specific powers that it conferred (although one could perhaps read *sanitary* to include pollution prevention) but in the fact that this was the first explicit recognition of a *functional* zone conferring any actual jurisdiction on the coastal state beyond its territorial sea.[6]

Signs of Change, 1967–1971

The sight was a devastating one, as the United Kingdom helplessly floundered to contain the spreading black tide of the *Torrey Canyon*. No one was in control, but the message was clear: the benevolent authority of the

4. Statement by the delegate from the USSR, General Committee, 1954 Conference for the Prevention of Pollution of the Sea by Oil, May 7, p. 4.

5. The Convention on the High Seas recognizes that the freedoms of the high seas must not restrict the exercise of these freedoms by others (art. 2). The Convention also urges states to draft regulations to prevent oil pollution (art. 24) and prevent the dumping of radioactive waste (art. 25). Likewise, the Convention on the Continental Shelf (arts. 5 and 7) contains some vague protective obligations.

6. For an excellent discussion of the functional significance of the contiguous zone, see McDougal and Burke, *Public Order*, pp. 565–730. This zone, in conferring coastal state jurisdiction, is significantly different from that created by the 1954 Convention.

flag state had foundered, and coastal state deference to it had been badly battered. That subservience was the passive foundation on which the legal superstructure had long ago been erected. Now the ground was shifting.

IMCO was not the proper forum in which to renegotiate the *modus vivendi* between ship and shore interests, but it was a good place to start. The *Torrey Canyon* contributed to the momentum that was building toward yet another law of the sea conference, and the IMCO meetings of 1969, 1971, and 1973 played an escalating role in the conference's preparations.[7]

The development in 1969 of the Intervention Convention was for IMCO a first foray into the realm of delineating jurisdictional competences. This instrument is, in its entirety, a "jurisdictional" convention, for it delineates the geographical and substantive scope of the rights and duties of coastal and flag *states* in their mutual relations. For this reason, it is known as the "Public Law Convention" and is clearly distinguishable from mere standard setting. As one coastal state delegation commented, the convention was nothing but "a particular manifestation of the more general residual authority envisaged for the coastal state in areas adjacent to its territorial sea."[8]

But intervention on the high seas was a preexisting coastal state right (encompassed by self-defense or self-help), and although they denied it, most maritime states did recognize that this was so. Indeed, it was really only because of the longstanding existence of the right that IMCO was allowed to deal with it. The agency would never have been permitted to "create" such a power, which would have been a major amendment to the maritime legal regime. In fact, the fate of a proposal by Belgium, Denmark, and the Netherlands to extend the application of the convention into the territorial sea demonstrated the limits of the organization.[9] Such an extension would have substantially restricted coastal powers in an area where they had hitherto had almost unlimited sovereign rights. Such a change would have been, in essence, a jurisdictional change, although some states supported it on the grounds that it would not really constitute a limitation on coastal sovereignty but would simply be a "contractual" arrangement applicable only to the intervention question. This is an oft-repeated argument in favor of jurisdictional amendments at IMCO and, from a short-term, strictly legal perspective, it has merit. However, the argument ignores the important fact that international law develops from the extrapolation of narrow precedents, and in 1969, with a new law of the

7. The 1962 Amendments to the 1954 Convention had expanded the coastal prohibition zone from 50 to 100 miles but had not dealt with jurisdictional issues at all.

8. U.N. doc. A/AC.138/SC.III/L.26, August 31, 1972, p. 20.

9. LEG/CONF/C.1/WP.2.

sea conference imminent, this was especially important.[10] Such an extension would seem to have been in the narrow interests of the maritime states, but many opposed it. The wider long-term implications of involving IMCO in the revision of what was properly the law of the sea were clear to them. After all, an expansion of IMCO's role could cut both ways in the future—adversely affecting maritime as well as coastal interests. Such a development was not relished. Only eight maritime states supported the initiative, while twenty-four coastal and maritime states opposed it.[11] IMCO's role was kept closely within its accepted limits.

A similar fate met expansionist amendments submitted on the draft Civil Liability Convention. Canada and a small number of other states with significant offshore fisheries sought to extend the geographical scope of the convention so as to provide them with compensation for damage beyond their territorial seas.[12] Maritime states, of course, opposed such a proposal, arguing that no state had a legal interest in any waters beyond the territorial seas and that IMCO would have been well beyond its powers had it tried to create one. The debate died quickly, but it foreshadowed the later demands at UNCLOS III for an extended "coastal zone." And it did point to an "important vacuum in international law in that there is an absence of present remedy to impose liability for pollution of the high seas where such pollution does not affect territorial waters because no action lies on behalf of the international community."[13]

Another initiative was put forward to allow a coastal state to require that all ships in its territorial seas—including those on innocent passage—carry insurance.[14] Again an important hiatus in environmental protection had been identified, but again the desire not to interfere with the larger legal

10. The Seabed Committee negotiations were beginning at this time to expand beyond seabed issues into other areas of the law of the sea. In December 1970, the General Assembly formally expanded its agenda to all law of the sea matters and called for the convening of a formal conference in 1973.

11. These eight supporters were Belgium, Denmark, West Germany, Greece, Ireland, Italy, Netherlands, and the United Kingdom. The opponents included developed "coastal" states; developing nations that strenuously opposed any restriction of coastal state sovereignty; the Soviet bloc that, along with the United States, opposed any reduction in coastal state control in the territorial sea; and many maritime states that were unconvinced of the beneficial long-term impact of the proposed amendment.

12. Canada later criticized the failure of the conference to accept this geographical extension and then cast the sole negative vote against the final convention. Soon after, it enacted the Arctic Waters Pollution Prevention Act with its 100-mile pollution control zone and amended its Territorial Seas and Fisheries Zones Act, extending its fisheries zones well beyond the territorial seas. See M'Gonigle and Zacher, "Canadian Foreign Policy," pp. 113–122.

13. Aaron L. Danzig, "Marine Pollution—A Framework for International Control," *Ocean Management* 1 (1973), p. 356.

14. LEG/CONF/C.2/WP.46.

regime of the oceans prevailed. Despite the fact that this proposal had the support of every Mediterranean state except Greece and Israel and virtually all coastal states on major trading routes, the proposal was defeated.[15] Again, particular needs were sacrificed on behalf of a wider legal system. And those who quietly hoped for a precedent-setting change were disappointed. Innocent passage was preserved intact.

At the 1971 Fund Conference, the maneuvering was more explicit. The law of the sea conference was now definitely planned for 1973, and preliminary negotiations had begun. Delegations were more aware than ever of the precedent-setting implications of their decisions.

The expansion of the coastal zone was the first priority of many. Australia, ever mindful of its vulnerable Great Barrier Reef, proposed that the right to compensation be extended to a distance of 50 miles from the coastline.[16] On a similar tack, Argentina and India proposed the replacement of the term *territorial sea* with more open-ended phrases such as *maritime zone* and areas under *national jurisdiction*.[17] The latter proposals were quite consistent with the territorially expansionist law of the sea policies of these two developing maritime states. Had they been accepted they would have symbolized important international recognition of a coastal state's proprietary interest in such broad areas, and therefore many coastal states supported them. In particular, Canada was eager to obtain international acceptance of its Arctic Waters Pollution Prevention Act with its 100-mile zone, and its delegation (which included a representative from the law of the sea section of its External Affairs ministry) sponsored an amendment extending the Australian proposal to 100 miles.[18] Despite its purely regulatory and nonjurisdictional nature, the 100-mile coastal prohibition zone in the 1954 Convention was now relied on by the delegation as an indication of the existence of a coastal state interest.[19] But again maritime states were

15. The alignment of states on this issue is most instructive. *For:* Australia, Cameroon, Canada, China, Egypt, France, Indonesia, Ireland, Italy, Ivory Coast, Libya, Malagasy, Peru, Philippines, Portugal, Singapore, Spain, Syria, Yugoslavia. *Against*: Belgium, Denmark, Finland, Ghana, Greece, Israel, Japan, Liberia, Netherlands, Norway, Poland, Sweden, Switzerland, USSR, UkSSR, U.K., U.S. *Abstain*: Brazil, Germany, India, Korea, Romania, Thailand, Venezuela. Clearly, the major shipowners opposed the proposal *en masse* (with a few exceptions for some Mediterranean shipping states that had initiated the amendment in the first place). The USSR and U.S. strongly opposed any tampering with innocent passage. So too did Israel, which saw it as yet another excuse for Arab interference with its shipping. Coastal and developing states supported the proposal wholeheartedly, although some developing states with maritime interests (Brazil, India) abstained.

16. LEG/CONF.2/C.1/WP.32.

17. LEG/CONF.2/C.1/WP.56 (Argentina) and LEG/CONF.2/C.1/WP.5/CONF.1 (India).

18. LEG/CONF.2/C.1/WP.53. The nature of Canadian policy in this area is explained more fully in the following chapter.

19. LEG/CONF.2/C.1/SR.19, p. 9.

intransigent, and the proposed changes were rejected.[20] IMCO stayed within its mandate.

The 1973 Conference

Unlike previous IMCO meetings, the success of the entire 1973 Conference hinged on the resolution of crucial jurisdictional issues. The IMCO Conference took place after the conclusion of the preparatory meetings for UNCLOS III, but before its formal convening, and as a result it was certain to have a real, tangible impact on the larger United Nations conference.[21] Its decisions would not be of merely symbolic significance. Indeed, they would likely be decisive in later determining which and how states would control and enforce the standards for shipping off all the world's coastlines. Here was an issue of great importance—one of the ''most basic and most difficult,'' a ''key to agreement'' for the law of the sea conference.[22]

Behind this diplomatic activity was the growing eagerness of coastal states to achieve a change in the traditional powers for legislating ship standards. Put simply, they wanted control and, despite IMCO's designation as a technical or regulatory body, great pressure existed at the 1973 Conference to initiate the change. Some states in the Seabed Committee had already advocated granting almost unrestricted power to coastal states to set higher standards in their coastal zones. At the same time, many maritime states were feeling great unease at the continued, if vague, recognition in article 11 of the 1954 Convention of the power of a coastal state to ''take measures *within its jurisdiction* in respect of any matter to which the Con-

20. The states opposing any change were Belgium, Brazil, Bulgaria, Germany, Greece, Italy, Japan, Liberia, Netherlands, Norway, Poland, Romania, Sweden, USSR, and the U.K. Interestingly, the United States, despite its opposition to an expanded coastal zone, recognized the gap in the coverage provided by the Civil Liability Convention and supported the Australian and Canadian proposals. Perhaps the American representatives agreed with the French delegation that providing a right of *compensation* now would reduce the demand for full coastal state *jurisdiction* later. Indeed, American faces were still smarting from the slap dealt them by the Canadian Arctic extension, and the delegation was definitely afraid that ''if damage on the high seas were not included, States either unilaterally or in combination might establish a regime that might not be in harmony either with the 1969 Convention or with the present Convention, one of the aims of which was precisely to establish a uniform regime'' (LEG/CONF.2/C.1/SR.19, p. 3). Recent American legislation, the Clean Water Act (1978), has in fact extended American compensation jurisdiction to the edge of the economic zone.

21. This IMCO meeting also followed the Stockholm Conference on the Human Environment by a year. In Stockholm, a number of recommendations and principles on the marine environment had been approved, and these provided a normative basis for later, more detailed, proposals.

22. U.N. doc. A/AC.138/SC.III/L.26, August 31, 1972, p. 17 (Canada). This referred to the determination of the ''appropriate jurisdictional authority to prescribe necessary standards [and] the determination of the appropriate authority to enforce'' them.

vention relates'' (emphasis added). The 1973 Convention was to replace the 1954 Convention, and the search for a substitute for article 11 developed into the most discordant debate of the month-long gathering. In fact, the debate continued well into the first few sessions of the later law of the sea conference.

The issue was known as the ''special measures'' controversy, for it concerned the extent to which a coastal state should be allowed to set special standards for its ''coastal zone.'' This concern over the *content* of coastal jurisdiction was compounded by another basic disagreement as to the proper geographical *extent* of the coastal jurisdiction.

Concerning the *extent* of jurisdiction, it was apparent that many states felt that the traditional division of legal authority between a narrow territorial sea and the high seas was dangerously anachronistic. Could IMCO change this? Considering that article 11 of the 1954 Convention simply used the inconclusive term *jurisdiction* to define the area of coastal authority, any debate on a substitute was, in the words of one IMCO official, ''absolutely unnecessary.'' If the words were left unchanged, the customary law of the sea would apply, and any decision to change it could simply be left to the law of the sea conference—just as it should be. But each group of states had its own axe to grind. Many maritime states wanted to state expressly now that the territorial sea was the area of coastal state authority without leaving a decision to the uncertainties of UNCLOS III.[23] At the same time, the United States certainly opposed any new vague phrase (such as ''waters under the jurisdiction'') that could possibly imply recognition of a 200-mile ''patrimonial sea'' or ''economic zone'' as the area of coastal authority. This fear was exacerbated by the arguments from the other side, in particular from the coastal ''territorialists'' who wanted to use that phrase for just that purpose—with the extent of coastal state jurisdiction to be decided by state practice such as their own unilateral proclamations.[24] Moreover, although it was clear that the continued restriction of coastal state jurisdiction to the territorial sea was not acceptable even to the great middle group of participants, many states continued to cling to it simply as a way of extracting concessions on other matters from those for whom the usage or nonusage of the term was a priority consideration.[25] The result was a protracted, if somewhat artificial, debate. The final compromise was

23. Those states speaking in favor of retaining the territorial sea division were Liberia, France, U.S., U.K., USSR, and Nigeria (although Nigeria defined it as thirty miles in width).

24. The states which argued for the acceptance of a 200-mile patrimonial sea were largely Latin American: Argentina, Chile, Ecuador, Mexico, Peru, and Uruguay as well as Tanzania.

25. But many of these states were willing to delay consideration of this issue until UNCLOS III: Australia, Canada, Iceland, India, Indonesia, Kenya, New Zealand, Philippines, Sweden, Trinidad and Tobago, and Venezuela.

appropriately vague. Under article 4, the flag state was to prohibit violations "under the law of the Administration" of the ship, while the coastal state would also prohibit such violations "within the jurisdiction" of the state. Additionally, it was stipulated that the latter clause was, by article 9(3), to "be construed in the light of international law in force at the time of application or interpretation of the present Convention." This latter formulation left to UNCLOS III the task of deciding the limits of particular coastal state jurisdictions which would then be applied retroactively to the 1973 Convention. Again, the result at IMCO was "no change."

The true central focus of the conference was the *content* of coastal jurisdiction and not its geographical scope. Early drafts of the convention had simply incorporated article 11 from the 1954 Convention, but this met with disapproval from both ends of the political spectrum. On the one hand, many maritime states argued that the article, though it did not actually confer any power on the coastal state, failed to restrict it. The traditionally loose formulation of the coastal power had long been acceptable to shipping states, but now the likelihood of that power being widely used and of its being expanded beyond the territorial sea was increasingly threatening and many hoped to see it restricted. In the words of the Liberian delegate, the provision was "an invitation to take unilateral action . . . totally at variance with the objectives of the Conference."[26] On the other hand, many coastal states argued that the article was not as *positive* a statement as it should be. Canada, in particular, sought international approval of its unilateral Arctic legislation, and it wanted the article to contain a clear statement confirming this broader coastal authority.

At the February preparatory meeting for the conference, Canada introduced a sweeping proposal: coastal states should have legislative authority for environmental protection in waters under their jurisdiction. However, as usual with these preparatory meetings, participation by nonmaritime states was negligible and the proposal received little support. To most maritime states it represented a fundamental challenge to international shipping. To the coastal states, it presented an opportunity finally to achieve environmental control of their coastal waters, and before the October conference itself, it was obvious that the proposal had extensive backing. The issue had been long discussed—most recently at the summer 1973 session of the U.N. Seabed Committee. If the October IMCO Conference were to succeed at all, some compromise would be necessary. No sooner had the meeting begun than a group of five maritime states submitted their suggestion.[27] They agreed to recognize *some* coastal powers but

26. Statement by Mr. Wiswall (Liberia), MP/CONF/SR.12, p. 12.

27. United Kingdom, Netherlands, Norway, Sweden, and Greece. MP/CONF/C.1/WP.36.

hoped to avoid the main navigational impediments that such powers could cause. The amendment allowed for the setting of stricter *discharge* standards by the coastal state in any areas "within its jurisdiction" but prohibited its setting stricter *design*, *equipment*, or *manning* standards in any area—including the territorial sea and internal waters—except where the environment was "exceptionally vulnerable." If such an article were really inevitable, this was a good compromise for the maritime states. Moreover, Canada was satisfied by it. Canada's main concern was to protect its "exceptionally vulnerable" Arctic, and though that country had been the main protagonist of the more sweeping coastal position, it was content despite the article's very restrictive implications for other larger areas of the oceans. With its support, the issue proceeded quickly, and after a counterproposal the same day by coastal states,[28] an agreed formula was achieved and put to a vote in the committee.[29] By a margin of 29-10-9 a new article (article 8) was accepted and incorporated into the draft text to be submitted to the plenary.

> (1) Nothing in the present Convention shall be construed as derogating from the powers of any Contracting State to take more stringent measures, where specific circumstances so warrant, within its jurisdiction, in respect of *discharge* standards.
>
> (2) A Contracting State shall not, within its jurisdiction, in respect to ships to which the Convention applies other than its own ships, impose additional requirements with regard to *ship design* and *equipment* in respect of pollution control. The requirements of this paragraph do not apply to waters the particular characteristics of which, in accordance with accepted scientific criteria, render the environment *exceptionally vulnerable*. [Italics added.]

The acceptance of this text was a major victory for Canada and, to a lesser extent, for its coastal allies. It was also a stunning departure from tradition for IMCO, particularly considering the earlier opposition to both an extended coastal state jurisdiction and IMCO's entry into this field. Indeed, despite insincere assurances that the article would not create a precedent as to the "nature and extent" of coastal jurisdiction[30] but would apply only contractually among the parties to the convention, article 8 would obviously be a powerful precedent for the Law of the Sea Conference. Unfortunately for its supporters, the debate was not yet closed.

The American delegation was resolutely opposed to article 8. Despite its new-found environmental commitment, the United States (and particularly

28. MP/CONF/C.1/WP.37, submitted by Australia, Brazil, Canada, Ghana, Iceland, Indonesia, Iran, Ireland, New Zealand, Philippines, Spain, Uruguay, and Trinidad and Tobago.

29. MP/CONF/C.1/WP.43.

30. Statement by Mr. Lee (Canada), MP/CONF/SR.11, p. 8.

its Department of Defense) was opposed to any amendment with as great a potential for interference with freedom of navigation. At the same time, this article prohibited higher standards within areas under a coastal state's jurisdiction—including possibly its territorial seas *and internal waters*. This was a sweeping intrusion on state sovereignty, and it undermined the main American instrument for controlling ship standards, the Ports and Waterways Safety Act. As such, the agreement was totally unacceptable, and the delegation immediately mounted an intensive behind-the-scenes campaign to defeat it.

The American diplomatic technique was, in the circumstances, a logical one. Recognizing the very tenuous balance of support for the compromise, the strategy was to "play both ends against the middle." The article had, after all, been acceptable only because, while it would expand coastal state jurisdiction, it would at the same time prevent future unilateral action on design and construction standards. With these inherent contradictions, the U.S. delegation, on the one hand, sought to convince the more extreme coastal states that the article was too *restrictive*, limiting them in the exercise of their full coastal jurisdiction. On the other hand, to the maritime powers, it was explained that article 8 was a major concession that would unnecessarily prejudice their negotiating position at UNCLOS III and, ultimately, their freedom of navigation.[31]

From the beginning the coastal state territorialists[32] and many maritime states[33] had opposed the new article. It was, therefore, on those states that had supported the article as the "best compromise available" that the American delegation concentrated its efforts (see Chart 13). They were amazingly successful—six of the original cosponsors of the article reversed their votes in plenary.[34] Moreover, the wavering USSR now acquiesced to

31. Ironically, while the delegation was actively engaged in this Januslike tactic in Committee I, it was also telling delegations in Committee II that unless segregated ballast tanks were accepted in the convention, the United States would implement them unilaterally under its Ports and Waterways Safety Act.

32. The leaders here were Ecuador and Tanzania.

33. These states included Belgium, France, Germany, Italy, Japan, and Switzerland.

34. The United Kingdom and the Netherlands had cosponsored the original article 8 but withdrew their support under concerted American pressure. National representatives interviewed at IMCO reported that the change in the British position came about as a result of a request by the British prime minister's office. When asked about the dramatic shift in his delegation's position after leading the compromise, it was reported that the British representative replied that he had not been "the captain of the proposal, only its pilot. And, as everyone knows, the pilot is the first to abandon a sinking ship." (Interview.) Three of the states that switched were states with broad coastal ambitions which now seemed convinced that the article had potentially restrictive implications for their larger UNCLOS ambitions. These states were Brazil, Iran, and Uruguay. It is interesting to note Brazil generally described itself as a coastal state on this issue and Argentina itself as a maritime state, even though Brazil has a substantially larger shipping fleet. Both states have great interest in coastal territorial expansion as well as in developing their shipping fleets. It is not easy to predict which interest will predominate for what country at what time.

CHART 13
State Positions on "Special Measures" Compromise

	Opposed in Committee and Plenary	*Supported in Committee but opposed in Plenary*	*Supported in Committee and Plenary*
	Belgium	Netherlands	Australia*
	France	Sweden	Canada*
	Germany (F.R.)	U.K.	Iceland*
Developed	Italy		New Zealand*
	Japan		Spain*
	Switzerland		Cyprus
	U.S.		Denmark
			Greece
			Norway
		USSR (abstained)	Poland
		Bulgaria	
		ByeloSSR	
Soviet bloc		Germany (D.R.)	
		Hungary	
		Romania	
		UkSSR	
	Argentina	Brazil* (abstained)	Ghana*
	Cambodia	Iran*	Indonesia*
	Cuba	Ireland*	Peru*
	Ecuador	Uruguay*	Philippines*
	Korea	India	Trinidad and Tobago*
	Mexico (abstained)	Kenya	Chile
	Monaco	Tunisia	Egypt
Developing	Singapore		Jordan
	Tanzania		Liberia
	Venezuela		Nigeria
			Panama
			Saudi Arabia
			Sri Lanka
			Thailand

NOTE: Positions of states are determined by their statements in Committee and Plenary and their vote in Plenary.
*Cosponsors of compromise proposal.

CHART 14
Compliance Measures Accepted, 1954–1973

1954	*1962*	*1973*
I Reporting Own Compliance		
Obligation to report to IMCO on laws and administrative actions implementing Convention. Article 13(*a*)	Article 13(3)	Article 11(1)(*a–c*)
Obligation to send IMCO information on adequacy of reception facilities (IMCO Assembly Res. 154, 1968). Article 8	Deleted	Reinstated as Article 11(1)(*d*)
Obligation to require vessels to record certain oil movements in oil record book (altered in 1962, 1969, and 1973). Article 9 and Annex B	Article 9 Annex B	Annex I Reg. 20, App. III
Obligation to send IMCO and concerned states information on actions *re* violations reported to it. Article 10(2)	Article 10(2) Article 12(*b*) and delete *concerned states*	Article 4(3), 6(4) Article 11(1)(*e–f*)
	Obligation to require new vessels over 20,000 dwt to report to flag states discharges exceeding regulations. Article 3(*c*)	Article 8 Protocol I
	Obligation to send IMCO and concerned states information on discharges from its vessels over 20,000 dwt which exceed regulations and which are reported to it. Article 3(*c*)	Article 8(4), amended *if it considers it appropriate*

CHART 14 (Continued)

1954	*1962*	*1973*
		Obligation to send IMCO information on investigations of vessel casualties. Article 12(2)
II **Inspection and Surveillance**		
Right to inspect documents required by Convention (until 1971, only oil record book). Article 9	Article 9	Annex I, Reg. 20(6)
Obligation not to delay vessels as a result of inspection of documents. Article 9	Article 9	Article 7(1); Annex I, Reg. 20(6)
		Obligation to inspect the ship equipment required by Convention at regular intervals and to issue certificates of compliance. Annex I, pp. 4–7
		Obligation to investigate casualties of own vessels. Article 12(1)
		Obligation to watch coastal area for possible discharge violations and facilitate gathering of information. Articles 6(1), and 8; Annex I, Regs. 9(3), 10(6)
		Right to inspect vessels in own ports for violations of discharge regulations. Article 6(2)
		Right to inspect vessels in own port for violations of

CHART 14 (Continued)

1954	*1962*	*1973*
		discharge regulations if accusations received from other parties (IMCO Assembly Res. 151, 1968). Article 6(5)
		Right to inspect vessels in own ports for violation of construction or equipment regulations if "clear grounds" exist for believing they do not correspond with certificate specifications. Article 5(1-2)
		Obligation to pay compensation for undue delays caused by inspection. Article 7(2)
III **Actions against Suspected or Proven Violations**		
Right to inform a flag state of a violation by one of its vessels (1973: expanded to obligation). Article 10(2)	Article 10(2)	Obligation to report a violation to flag state (previously a right). Articles 6(2–3), 8(3)
Obligation to investigate accusations of violations by own vessels and to initiate proceedings if the investigation provides evidence of guilt. Article 10(2)	Article 10(2)	Articles 4(1), 6(4)
Obligation to impose penalties and to make them the same "irrespective of where the violation occurs." Articles 3(3) and 6.	Articles 3(3), 6	Articles 4(1), (4)

CHART 14 (Continued)

1954	*1962*	*1973*
	Obligation to send IMCO information on inadequate reception facilities reported by its vessels (although construction made nonobligatory in 1962). Article 8(3)	Annex I, Regs. 10(7)(*b*)(*vi*), 12(5)
	Obligation to impose penalties "adequate in severity" to deter violations. Article 6(2)	Article 4(4)
		Obligation to detain a vessel violating ship standard regulations until it can proceed without threatening marine environment. Article 5(2)
		Obligation to undertake proceedings against a violation within own jurisdiction—or forward information to flag state. Article 5(2)
		Obligation to apply all regulations to vessels of nonparties as well as parties to the Convention. Article 5(4)

NOTE: The 1978 regulations, relating largely to periodic inspections by classification societies, are outlined in the section on the TSPP Conference.

Chart 15

Compliance Measures Rejected, 1954–1973

	1962	*1968/69*	*1973*
Reporting Own Compliance			
Inspection and Surveillance		Right to board vessels on high seas to determine if discharge violations were committed (France)	Right (unqualified) to inspect vessels for possible violations of construction and equipment regulations (Canada-Spain)
		Obligation to inspect vessels in ports for possible discharge violations (France) (1973: Canada-Spain)	- same -
Actions against Violations	Obligation to inform IMCO of perceived violations (U.K.)		Obligation to deny entry to vessels known not to conform with ship standards (U.S.)
			Obligation to delay vessels not complying with ship standards unless going to nearest suitable repair port or unless posing a threat to port or marine environment [modified obligation accepted in art. 5(2)] (U.S.)
		Obligation to prosecute vessel on basis of an accusation by another	Right to undertake preemptive proceedings for all ship-standard viola-

	1962	1968/69	1973
		party (i.e., assign probative force to a report by another state) (France)	tions by own vessels—even when apprehended in waters under the jurisdiction of a foreign state (Netherlands)
		Right to participate in a foreign state's prosecution of a vessel which resulted from its report (France)	
		Right to prosecute a vessel for an oil-record-book violation while in its port (a limited type of port-state enforcement) (France)	Right to undertake proceedings against a vessel for violations outside own jurisdiction while it is in its port (port-state enforcement) (Canada-U.S.)
	Obligation for *vessels* to prove innocence when accused of violation (reversed burden of proof) (1962 and 1967: U.K.)	- same -	"Visible sheen" as *prima facie* proof (reversed burden of proof) (U.S.)
	Obligation to impose penalties according to an international standard (1962: Malagasy; 1967: Italy)	- same -	
		Obligation to impose penalties set by an international tribunal (France)	
	Obligation to impose "severe" penalties (Yugoslavia)		Obligation to confine penalties to fines unless otherwise agreed to by flag state (Netherlands)
	Obligation to impose penalties on master and/or crew where guilty (as opposed to shipowner) (Yugoslavia)		

NOTE: Only the most important proposals which were rejected in 1973 are listed. Many minor suggested amendments are excluded.

the American demands. The Soviet Union could have accepted the proposed article. It did restrict future unilateral moves by coastal states in general while recognizing the jurisdictional interests of Arctic coastal states. This was important to them, for the Soviet Union after all had openly supported Canada's Arctic initiative as being compatible with its own sovereignty claims in the Arctic.[35] But the IMCO Conference took place against the backdrop of renewed Arab-Israeli hostility (the Yom Kippur war), and it was widely believed by delegates that the abstention of the Soviet Union in the final plenary vote was a gesture, under pressure, to détente with the United States. In the end the article still received majority support (the vote was 26–22–14) but fell far short of the two-thirds majority required in plenary for inclusion in the convention.[36] Now, with no replacement for the 1954 Convention's article 11, the content of coastal state jurisdictional powers was completely left to UNCLOS III.[37]

ENFORCEMENT

Beyond the legislative competence to set ship standards lies the power to enforce them. Ensuring compliance with accepted pollution-control regulations has unquestionably been the most difficult problem in promoting the protection of the maritime environment. It involves not only the specific rights and obligations of enforcement but the very character of the technical regulations themselves.[38]

Numerous enforcement provisions have been considered in IMCO over the last twenty years, but they all can be classified according to one of three functions: reporting of compliance, inspection and surveillance, and actions against violations. Chart 14 summarizes the provisions incorporated in the conventions concerning these three enforcement functions; Chart 15 details the major proposals that have been rejected.

35. In order to gain the Soviet support, the Canadian delegation had threatened to ask IMCO to declare the Arctic a "special area." This the Soviet Union strongly opposed, as it could have had the effect of undermining its own sovereignty claim by "internationalizing" the area. As a result, the Soviets encouraged the Canadian position on special measures but "on last minute instructions from Moscow" they declined to support Canada in the final vote. This shrewd maneuver upset Canadian plans, and it was angrily condemned as a "double cross." By taking a number of Soviet bloc states with it, the move was fatal to the new article. Delegates amusedly reported a mix-up in the Soviet signaling system so that, after all this, the Polish representative, much to his embarrassment, mistakenly voted *for* the article. Interviews.

36. A number of states—Australia, Brazil, Canada, Ireland, New Zealand, Peru, Philippines, and Uruguay—expressly reserved their positions on the scope of their powers.

37. The discussion of UNCLOS III is found in the concluding section of this chapter.

38. For example, regardless of how enforcement authority is allocated, if the vessels of contracting states are prohibited from discharging oil in a concentration above 100 ppm but are not required to install technologies which would measure the oil content of an effluent, the chances for compliance are rather low.

The Early Conferences, 1954–1962

Enforcement of agreements has always been a sad chapter in the history of international politics, and the story of environmental regulation is no exception. Especially for the early years, when there was little genuine commitment to establishing workable pollution-control standards, it is not surprising to find a dearth of reliable compliance mechanisms. The 1954 Conference was, after all, the first international environmental conference in nearly thirty years, and the attention of the delegates was taken almost completely with the complexities of the discharge regulations themselves and not with their enforcement. Moreover, the participating states had very little experience with enforcing international regulations on the high seas. A number of maritime states had established informal understandings with shipping companies that their tankers would not discharge oil within fifty miles from shore, but these were not legally enforceable except as they were incorporated into national legislation and even then only within the three-mile territorial area.[39] At the same time, most coastal states had not experienced enough coastal pollution to be particularly concerned about compliance with regulations. Combined with the traditional deference to flag-state control on the high seas (and to the principle of freedom of the seas), this lack of experience by maritime states and interest by coastal states produced an enforcement system notable only for its inadequacy.

The 1954 Conference did accept a number of British proposals on the subject, but these were so innocuous as to provoke only minimal debate.[40] The linchpin of the enforcement system was "self-reporting," with all its attendant inadequacies. States were to report to IMCO all laws and actions implementing the convention, including the adequacy of their own reception facilities and results of investigations and proceedings concerning reported violations. In addition, there was the important requirement that tankers record their cargo movements, including the disposal of oil residues, in an "oil record book." This was to assist the detection of violations but, dependent as these obligations were on voluntary and conscientious self-incrimination, they had no chance of success without some additional enforcement mechanism.

With self-reporting the basic enforcement mechanism, the stipulations for coastal state inspection and surveillance were meager and almost self-defeating. Contracting parties were allowed to inspect only the ship's oil record book and were obliged not to delay the vessel as a result. The value

39. These maritime states were Belgium, France, Japan, the Netherlands, Norway, Sweden, and U.K., and the U.S.

40. 1954 Conf. Doc., Enforcement Provisions Suggested by the United Kingdom Delegation, General Committee, May 7, 1954.

of this right was almost negligible, as a record book was so easily falsified by a recalcitrant crew. But even this modest concession to coastal interests evoked the angry opposition of the USSR which, as ever, opposed any infringement of a vessel's autonomy while in foreign ports.[41] With such constrained powers of inspection, a state's ability to detect violations rested largely on its surveillance of its coastal waters and its ability to identify any oily discharges in excess of 100 ppm. This power was almost meaningless. A surveillance program of any degree of thoroughness would have been prohibitively costly for most states and, in the end, quite ineffective anyway. Establishing the oil content of an effluent was technically almost impossible on the open oceans. The coastal states were powerless.

Finally, even where a violation was actually detected outside the territorial sea, the coastal state was allowed only to *inform* the flag state of the perceived infraction.[42] It was then the flag state's obligation to investigate the accusations and prosecute where its investigations substantiated the charges. The potential for abuse in such a system is obvious, especially considering the extreme difficulty of establishing the oil content in discharges. The result was a convention with which some vessels could not even comply and which, outside of territorial waters, all could violate without serious fear of detection or prosecution.

That this was the case became clear when, prior to the 1962 Conference, IMCO undertook a survey of the twelve states for whom the convention had been in force for four years.[43] As was to be expected, the survey revealed deficiencies in both the content and application of the regulations. Most parties had been lax in applying the provisions within their territorial waters, and the regulations themselves did not permit effective application beyond. Of the almost six hundred offenses cited within territorial waters, 83 percent were reported by Germany and the United Kingdom.[44] Fifty percent of these did result in successful prosecutions, so it was clear that, with effort, offenses could be detected and prosecuted. But the experience of all states outside territorial waters provided only the most dismal prospects. In total, only fifteen offenses had even been detected there, and *none* had been successfully prosecuted.[45]

41. 1954 Conf. Doc., General Committee, May 10, 1954, p. 13. It is noteworthy that the USSR has resurrected this issue only recently in IMCO's Legal Committee with its proposal for a special legal regime to protect flag-state jurisdiction in foreign ports.

42. The traditional rights of a state to legislate and enforce regulations within its territorial sea was recognized in the previously discussed article 11.

43. 1962 CONF/2.

44. Finland and to a lesser extent Canada, Denmark, and Norway exhibited a modest attempt at applying the convention, but Belgium, Ireland, and Sweden reported no offenses, and France and the Netherlands only one apiece.

45. 1962 CONF/2, pp. 7-10.

The basis for this disappointing performance was not difficult to understand. Reliance on the "oil record book" and on flag-state prosecutions had proved to be inadequate,[46] and coastal-state surveillance did not make up for it. The German and British delegations affirmed the difficulty posed by a reliance on the 100-ppm standard, above all for proving discharge violations in open water. The fact that the great majority of their convictions occurred for offenses committed in ports lent considerable support to this point. This was compounded by the impossibility of surveillance at night or in foggy weather and, as the Canadian delegation complained, by the shortage of personnel necessary to patrol a large coastal area.

The existing compliance system was clearly inadequate, and its improvement was now a major issue. Some minor changes were made, first of all, in the self-reporting provisions[47] and those relating to penalties—in particular, the imposition of penalties for discharge violations recorded in the oil record book.[48] But, the efficacy of these new provisions still depended on an incredible degree of candor by ships' crews. They were, alone, not likely to produce any dramatic advancement, and in other areas where obvious improvements could be made, most shipping states were not willing to strengthen the system if it imposed any new burdens on their industries.

Under the 1954 Convention, the inspection system was particularly constrained, and to give it some potential for effectiveness, two suggestions were put forward. The United Kingdom, still in 1962 an important proponent of improved environmental controls, advocated that states be given the right of in-port inspection for any vessel on which incriminating information had been received.[49] France, concerned about the declining state of the Mediterranean, went so far as to suggest that such powers be given with regard to *any* vessel at all in a state's ports.[50] Both proposals engendered vehement opposition. The Liberian delegation attacked them as likely to cause costly delays in port (particularly, of course, for flag-of-convenience ships). The Soviet Union, ever wary of any interference with their vessels,

46. Ibid., pp. 2–4.

47. The only new reporting provisions considered by the conference were ones whereby a state would require its new vessels over 20,000 dwt to report discharges exceeding the regulations, which it would then forward to IMCO. The implementation of these obligations would have required a great deal of pressure by flag states on their shipowners and, like similar earlier provisions, the chances of implementation were low. As it turned out, despite the entry-into-force of the 1962 Amendments, they have remained "paper" provisions (see Chapter 4).

48. Previously the oil record book was only *evidence* of a violation. Other new provisions obliged states to impose penalties "adequate in severity" to deter violations and to report inadequate reception facilities in ports which their vessels visited (despite the fact that the legal obligation to build such facilities was deleted by the conference participants).

49. 1962 CONF/C.2/8.

50. 1962 CONF/2, pp. 15–16.

saw them as tantamount to bestowing extraterritorial rights on the inspecting states.[51] Although the British proposal originally had substantial support, these attacks eroded any chance for its success or for that of the far more sweeping French proposal.[52] The inspection system was left unchanged.

The Soviet Union's qualms about its jurisdictional autonomy were widely shared by other states when it came to improving the prosecution of violators. This was one area in which coastal states possessed woefully inadequate powers under the convention and in which flag states had been very deficient in fulfilling obligations. But the desire of all states to protect their legal autonomy prevented any progress being made. For example, a British proposal that would have shifted the onus onto the master to show that he did *not* commit an alleged violation (a "reversed burden of proof") was rejected as being incompatible with the legal systems of the continental European countries.[53] Although it would have obviated much of the difficulty in obtaining evidence of a violation—and been a momentous step toward reliable enforcement—it could not, according to many continental European states, be adapted to Roman law.[54] Another British proposal received the same short shrift. This amendment would have required parties to inform IMCO as well as flag states of perceived violations. It was rejected not only because of the pressure it would have put on the flag state to investigate the allegation in good faith but, again, because of the suspicion and conflict it aroused about IMCO's interference in a flag state's legal system on behalf of other states.[55] Finally, attempts to provide effective penalties by setting them according to an international standard or by imposing them directly on the master and crew were again rejected as infringing on the autonomy of states' legal systems.[56] This unswerving protectiveness made nonsense of the discharge standards themselves—including the new total prohibition on "polluting" discharges for ships over 20,000 dwt. The system was a sham.[57]

51. 1962 CONF/C.1/SR.7, p. 9; C.2/SR.2, p. 6; and C.2/SR.9, pp. 2–5.

52. The British proposal was initially accepted in committee by a vote of 15-3-3 (1962 CONF/C.2/SR.9, p. 7), but following a vote of 11-8-9 in the plenary, it was withdrawn (1962 CONF/C.1/SR.7, p. 11).

53. 1962 CONF/2, pp. 12 and 26.

54. This argument was put forward by France, the Netherlands, Denmark, and Norway, and the proposal was also opposed by Belgium, Canada, Sweden, and the U.S. 1962 CONF/C.3/SR.4, pp. 5–9. Given the technical difficulties of working with a "ppm" standard of proof, it was becoming obvious that an impasse had been reached. Without a radical change in the methods of proof, no effective enforcement would ever occur.

55. No one supported Britain on this proposal, and the USSR, France, Liberia, Denmark, Italy, and Greece opposed it. 1962 CONF/C.2/SR.4, pp. 5–9.

56. 1962 CONF/C.2/3, C.2/7, and C.2/SR.8, pp. 14–16.

57. Indeed, the deletion of the requirement of reception facilities (see Chapter 4), with which such ships would have needed to comply, was perfect proof of this hypocrisy.

Clearly then, the compliance provisions accepted in 1962 were insignificant, and the most important proposals were defeated. But the debates did focus states' attention on certain questions which were to become very important in future international negotiations on the oil pollution problem. The British proposal for reversing the burden of proof drew attention to the crucial issue of designing accords so that it would be possible to obtain evidence that could be used to challenge a ship's compliance with the discharge rule. The French and British proposals on inspecting ships in ports highlighted the need for removing some of the thick wall of national sovereignty which surrounded ships engaged in international trade in order to give coastal states some ability to obtain the information required for prosecutions. The British proposal for forwarding accusations of violations to IMCO as well as to the flag states suggested the need for someone to oversee a flag state's compliance with the regulations. Finally, the various proposals for more stringent penalties dramatized the lack of deterrence in the present system and the need for sanctions which would have a real deterrent effect.

IMCO Negotiations, 1967–1969

The 1967 *Torrey Canyon* disaster, although certainly not an instance of "operational" pollution, greatly dramatized the pollution problem and inevitably affected IMCO's work in all its aspects. Following the extraordinary session of the IMCO Council in May 1967, both the Subcommittee on Oil Pollution (SCOP) and the Legal Committee held a number of meetings to discuss proposals for promoting compliance with the 1954 Convention. In addition, the heightened feeling of urgency exerted pressure on the negotiations already underway in the SCOP to formulate new discharge regulations based on the LOT system—a system which was eventually to have a decisive impact on the ability to enforce compliance with pollution standards. In the end, it was only with respect to the actual discharge standards themselves that any real change took place. Although the pace of all negotiations quickened after the *Torrey Canyon*, no change in actual enforcement practices occurred until the 1973 Conference, six years later.

Despite the devastating impact of the *Torrey Canyon* on the United Kingdom, the British delegation was becoming increasingly mindful of its shipping interests, and in the 1967–1969 IMCO negotiations it contented

---

Furthermore, without any monitoring equipment, a tanker's captain could argue that the residues were discharged evenly throughout a long voyage at less than 100 ppm. While this was exceedingly unlikely in fact, it did provide a possible legal defense when a tanker arrived in port without residues. Interview with Gordon Victory, former official, Marine Division, U.K. Department of Trade.

itself with resurrecting proposals which had already been rejected. The other maritime states, although very generous with their statements of concern, were still reluctant to amend their inadequate fifteen-year-old compliance system.[58] Some relatively unimportant proposals were approved at the sessions, but there was no attempt to amend the enforcement provisions of the 1954 Convention. Within these meetings, only France submitted any radically new proposals. In this period, that country reacted quite strongly to the impact of the *Torrey Canyon*, which had badly polluted its Normandy coastline. It was only later that its commercial interests began to reassert their authority and bring the government back to the maritime fold.[59]

The French proposals were singularly unsuccessful. Some nominal improvements in the self-reporting provisions were adopted in 1968 as *nonbinding* Assembly resolutions, but their impact, like that of their predecessors, was inconsequential.[60] It was rather to advance the efficiency of the inspection provisions that the French amendments were directed. The delegation proposed that all states be *required* to inspect *all* vessels in their ports for possible discharge violations and that all states be allowed to board vessels in their adjacent coastal zones to detect violations.[61] These suggestions went right to the heart of the enforcement problem. Rigid port inspection would finally have cut through the fraudulent façade of the existing system, while coastal zone inspection would have revolutionized the whole law of the sea. The amendments provoked a host of objections and denunciations: they would impose long and unwarranted delays on innocent ships; they were outlandishly expensive and unnecessary; they were technically difficult and, in the case of coastal zone inspection, positively dangerous; and in so upsetting the balance of coastal and flag state rights (especially in the "coastal zone"), they were outside IMCO's authority. The proposals garnered absolutely no support, although a more modest British proposal to allow a state to inspect a vessel in port after receiving evidence of a violation fared much better. Despite having been rejected in 1962, the proposal was now passed as an Assembly resolution. This resolution was, of course, nonbinding.

58. The IMCO documents on the SCOP and Legal Committee meetings provide little indication of the debates, but the authors were able to secure access to short reports of these meetings from two governments. They indicated some of the major points of contention and the positions of several states.

59. See discussions in Chapters 4, 5, and 7. French policy until 1970 was largely controlled by its environmental ministry and showed a distinct coastal orientation.

60. A/ES.VI/Res.147 and 154. These resolutions *requested* that states require vessels to report excessive discharges to coastal states in the vicinity and that these states send information on the location and character of their own reception facilities. Apart from the fact that the provisions were not passed as formal amendments, they were likely to have very little effect as long as there was no systematic review of a state's fulfillment of them.

61. OP. III/SP.1; C/ES.III/3/Add.4, p. 4; and LEG/WG(I). III/WP.15, pp. 1–2.

In this period, all proposals to improve the prosecution of suspected violations were rejected. Despite proof that the 100-ppm standard was functionally unenforceable on the open oceans (scientific evidence showed that aerial photographs could not differentiate discharges of oil exceeding 100 ppm from oil discharges below that figure or even from emissions of other substances), a renewed proposal for a reversed burden of proof was rejected.[62] These scientific findings did, however, have a beneficial impact on states' acceptance of the 1969 Amendments that set new discharge standards, the violation of which could be more easily detected. But, as had happened before, most of the proposals for improved control over prosecutions were rejected as challenging the autonomy of national legal systems. This was certainly the reaction to the submission that reporting states should have a right to participate in flag state proceedings originating from their reports,[63] as well as to a proposal that would have assigned "probative" force to the report (that is, *required* the initiation of legal proceedings by the flag state).[64] Finally, France sought unsuccessfully to allow contracting states to undertake proceedings against ships which had commited violations recorded in the oil record book—regardless of where they occurred. Interestingly, a revised version of this proposal would emerge as the major enforcement initiative at the 1973 IMCO Conference and the U.N. Law of the Sea Conference, but in 1968 it was seen as an unprecedented infringement of the flag states' high seas jurisdiction. After fifteen years, the enforcement system remained virtually unchanged.

With the profound political intransigence that was manifested in these years, it was only a change in the discharge regulations themselves that could possibly alter the situation. That change was the introduction in the 1969 Amendments adopting the load-on-top system. With its new discharge limits (60 liters per mile to a maximum of 1/15,000 of the ship's cargo-carrying capacity), load-on-top provided a qualitative change in the nature of the compliance system. With it, vessel crews would have a much better knowledge of the volume of oil they were releasing than with the old

62. OP III/4, p. 4; and OP III/WP.1, pp. 4-5. The research study concluded that any conviction would demand an admission of guilt by a ship's master—an occurrence which was rather unlikely. OP III/4, pp. 3-4; and interviews with U.K. government officials. For a similar evaluation, see the statement of John Kirby, a Shell executive, in "The Clean Seas Guide," *Conference on Oil Pollution of the Sea, 7-9 October 1968 at Rome* (Winchester, England: Wykeham Press), p. 209.

63. This would certainly have impinged on the autonomy of a flag-state's judicial system. Belgium unsuccessfully suggested a compromise procedure to allow the reporting state to follow the proceedings. See OP III/WP.1, pp. 4-5; LEG/WG(I); I/2, pp. 8-9; and LEG/WG(I); I/WP.11, p. 2.

64. The maritime states demanded the option to make their own evaluations of their vessels' potential guilt before undertaking prosecutions against them. See C/ES.III/3/Add.4, p. 4.; LEG/WG(I); I/WP.11, p. 6; and LEG/WG(I).II/2, p. 4.

"ppm" standard and the 1/15,000 stipulation would make it possible to determine a tanker's compliance with the regulations while it was at its loading port. The right to inspect vessels in such ports was not to be discussed for a few years yet, but the possibility of greatly strengthening the compliance system was imminent.[65] This was the first significant improvement in the enforcement system, and it is fitting that the change was but an incidental by-product of an industry initiative (LOT) designed primarily to avoid a potentially costly and inefficient alternative (see Chapter 4).

Throughout this period, work was also going forward on the issues of coastal state intervention and compensation. No significant enforcement issues were involved in these matters except with regard to pollution liability insurance. And here a very imaginative innovation was adopted. Under the Civil Liability Convention it was agreed that all contracting states would require full insurance for *all* ships entering their ports and not just for ships from contracting states. This was intended not so much as an enforcement provision against contracting parties but as an obstacle to those noncontracting states that felt they could achieve a competitive advantage by remaining outside the convention and foregoing full insurance. When motivated by a desire to advance their own commercial interests, the major maritime states were able to produce a jurisprudentially significant[66] provision that would have a substantial beneficial impact for coastal states while also providing an inducement to wider acceptance of the convention.[67]

The 1973 Conference

By 1973, there was no longer any denying the inadequacy of the existing enforcement system. As one study noted, "The enforcement record under the 1954 agreement is . . . dismal, with very few cases being prosecuted, and even fewer significant penalties being assessed for violations."[68] An American delegate to the 1973 Conference put the matter more succinctly

65. Less than a year after the passage of the 1969 Amendments, Sweden submitted a proposal that states be obligated to inspect all tankers at their loading ports. OP 9/12, pp. 5–6, and OP 9/2/6.

66. Its jurisprudential interest arises from the application of the convention to states *not* party to it. Some jurists argue that it has been the absence of this type of compulsory international legislation that has prevented international law from being true "law." See, for example, H. L. A. Hart, *The Concept of Law* (Oxford: Clarendon Press, 1961), p. 231.

67. 1969 Civil Liability Convention, art. VII (II). H. Steyn, a Mobil Oil executive, testified that the effect of this article was that "no nation with a tanker fleet of any importance will be able to afford *not* to ratify the convention once the major oil consuming nations have done so." *Hearings*, Subcommittee on Air and Water Pollution of the Committee on Public Works, U.S. Senate, 91st Congress, 2nd Session, July 21 and 22, 1970, p. 87.

68. William T. Burke, Richard Legatski, and William W. Woodhead, *National and International Law Enforcement in the Oceans* (Seattle: University of Washington Press, 1975), p. 48.

when he remarked that compliance was "spotty, at best."[69] So ineffective were the 1962 Amendments to the 1954 Convention that the U.K., Sweden, and Japan introduced the LOT system in domestic legislation, even though it would unquestionably violate the amendments. Their conclusion was that since the 100-ppm stipulation was unenforceable, no problems would result by introducing load-on-top even before the 1969 Amendments became law.[70] Until the 1969 Amendments finally came into force in January 1978, there was virtually no international legal control of operational polluting discharges outside of ports. This is a remarkable state of affairs and a severe indictment of this international legislative process.

A 1975 IMCO report substantiates this assessment and points to a glaring lack of good faith by the IMCO membership. For example, despite the existence of a reporting requirement in the 1954 Convention, no information on prosecutions was supplied to IMCO until after a 1974 request from the Marine Environment Protection Committee (the only exception being the responses to a questionnaire prior to the 1962 Conference). Even then, very few states responded to the MEPC circular, so what information has been available often has come from national reports and not from submissions to IMCO.[71] For all intents and purposes, self-reporting did not exist.

Reliance on coastal state inspection and surveillance had also proved itself unworkable. National reports revealed a huge disparity in the seriousness with which different states attempted to enforce the convention. As usual, the vast majority of violations that were detected occurred in ports or inland waters (where coastal states have always had their own power to control pollution). Even the United States Coast Guard, which possessed the world's most extensive coastal enforcement capabilities, conducted patrols in port twice daily but only twice a week in the territorial sea and contiguous zone and only occasionally ("random flights") beyond that point.[72]

Such detections as did occur on the high seas resulted in few successful prosecutions by the flag states to which they were referred. The Canadian submission to the MEPC indicated that only occasionally did its reports of

69. Minutes of October 12 meeting of Committee I of the 1973 Conference given to the authors. There are no official records of Committee I available to the public. The tape recorder transcribing the sessions malfunctioned so that even the IMCO Secretariat does not possess a record. Two participants in the Committee I meetings have given the authors copies of their own minutes.

70. The implications are that the complete discharge prohibition for new vessels over 20,000 dwt was unenforceable, as were the coastal zone prohibitions for all other ships.

71. MEPC/Circ.17. Only five states and Hong Kong replied even to this MEPC request for information. Several additional countries have supplied information since mid-1975.

72. *Marine Environmental Protection Programs: An Analysis of Missions Performance* (Washington, D.C.: U.S. Coast Guard, Dept. of Transportation, August 1975), pp. 135–140.

violations lead to convictions, and it often never even heard from the flag states.[73] In one case, Liberia, not a single response was forthcoming for eleven complaints made. The French[74] and British[75] reports confirmed the universality of this experience. In addition, an American study concluded that even where discharges were identified, few penalties were imposed because of the "inability to determine source/cause or to provide adequate evidence for penalty assessment."[76] Moreover, the average level of penalties has been so low that it has certainly not constituted a major deterrent to future violations. Despite the rise in the maximum limits of these penalties under many national legislative schemes, courts still hesitated to apply them rigorously in individual cases.[77]

Notwithstanding these obvious deficiencies in enforcement, the oil and shipping industries claimed in the early 1970s that the pollution problem had been much reduced as a result of the voluntary compliance of 80 percent of world tanker tonnage with the LOT system. Such an argument was contradicted by the continuing pollution of many coastal areas, and this prompted American and other governments' officials to challenge the efficiency of its implementation. Worried by these allegations, at least four of the "seven majors" quietly carried out surveys of tankers at their Middle Eastern loading ports in 1971 and 1972. To their surprise, they found that, indeed, only a third of the tankers were using the LOT system well; another third were using it very poorly; and another third were not using it at all. If there was anything encouraging in this, it was that the oil-company-owned tankers' record was 50 percent better than that of the independently owned—or so the oil industry reported.[78] But the results were disappointing. To most individual shipowners engaged in this massive and highly competitive business, the state of the marine environment—and their small contribution to it—were just not important.[79]

73. MEPC/Circ.17.

74. "Cas de Pollution par les Hydrocarbures de 1968 à 1975" (Ministere de la Marine Marchande) (typescript). This report was sent to IMCO in 1976. The record of Liberia, in at least acknowledging reports, improved greatly over the time period, although it successfully prosecuted only one of twenty cases which it claimed to have investigated.

75. "Report on the Exercise and Performance of the Functions of the Department of Trade and Industry under the Prevention of Oil Pollution Act 1971 during the Year Ending 31 December 1973" (typescript). Issued by Marine Division, U.K. Department of Trade and Industry.

76. *Marine Environmental Protection Programs*, p. 95 (see fn. 72).

77. In the U.K. the maximum fine has been raised from £1,000 to £50,000; in Canada, from $5,000 to $100,000. But in the U.K. the highest actual fines are about £1,500 and in Canada, about $15,000. The maximum fine in the United States is $10,000 according to its national report, although the average is one-tenth of that.

78. The existence of these studies was made known to the authors in interviews with some industry officials. The studies were not themselves shown, however. The worst offenders, it was reported, were those tankers on single-voyage—as opposed to longer term—charters.

79. This is a situation similar to Garrett Hardin's thesis of "the tragedy of the commons."

The industry studies have remained secret, but they did prompt OCIMF and the ICS in 1972 to issue the *Clean Seas Guide* setting out instructions on the use of the LOT system. Indeed, OCIMF now suggested that the oil companies would be willing to assist parties to the 1969 Amendments (not then in force) to implement the regulations by carrying out *their own* inspection program at the oil-loading terminals.[80] Thus, prior to the 1973 Conference, both government and industry recognized the inadequacies of both the 1954 Convention and its load-on-top replacement. None of the controls currently available was working.

During 1972 and 1973, numerous proposals for improving compliance were put forward in the negotiations for a new draft convention. The United States, with the support of a small number of coastal developed countries such as Australia, Canada, and New Zealand, was at the vanguard of these early consultations. The developing countries were almost completely absent from them—only four attended the preparatory conference in February 1973. But they were certainly not going to miss the main conference itself in October, a fact of which the many maritime states were well aware. IMCO was becoming less of a club for the "regulars" and, with the still worsening pollution situation, some conciliatory gestures would be necessary. But even with this broader participation and the new reformist zeal of the U.S. and its allies, the maritime states were eager to keep IMCO within the narrow constraints of the law of the sea and to defeat any sweeping revisions in the enforcement regime.

The bankruptcy of self-reporting was now evident even to the most naive of observers, but the new convention dutifully incorporated the traditional requirements. At the same time, some cautious improvements were made in the more important coastal state rights of inspection. Earlier, France and Britain had tried unsuccessfully to give some teeth to this power, and the issue was raised again in 1973. Compliance with the new construction and equipment standards and the 1969 discharge regulations now permitted much simpler verification. However, the penchant for incrementalism was well established at IMCO, and the new inspection regulations were no exception. Distinctions were drawn among the various types of inspection. To provide evidence of conformity with the new equipment and construction standards, all flag states were now obligated to inspect their vessels at regular intervals and to issue certificates of compliance which the port state was forced to accept unless there were "clear grounds" for believing they were invalid. This was, of course, unsatisfactory to the "environmentalist" states (Australia, Canada, and the U.S.) which demanded an unfettered

See Garrett Hardin and John Baden, *Managing the Commons* (San Francisco: W. H. Freeman, 1977), p. 16.

80. OP X/2/4, p. 4.

right of coastal state inspection.[81] Too long had they been the helpless victims of accidents caused by reliance on the flag state and on those certificates which are so easily obtained from a lax classification society or with the aid of a bribe.[82] But the developed and developing maritime states and the Soviet bloc were adamant. One "concession" was enough. Indeed, these many states secured the acceptance of a "balancing" obligation for the coastal state to compensate vessels for any "undue delays" caused by inspecting vessels for construction or design-standard violations.[83] But a beginning had been made, especially in an area which had important ramifications for accidental as well as operational pollution.

Inspection rights for determining discharge violations were treated separately, and with them more progress was made. A few states such as Canada and the United States hoped to make these inspections an obligation rather than a right but, of course, the majority of maritime states opposed such an extreme provision for fear of the delays it would cause and in realistic recognition of the fact that no contracting party would ever seriously undertake such an extensive obligation.[84] Predictably, the Soviet bloc continued to object to any coastal state inspection powers at all.[85] But the conference did agree to a general *right* of inspection for port authorities —on their own initiative or at the request of another contracting party which had observed a contravention by the vessel. This was a major innovation. For the first time, there was created the possibility of an effective and reasonably economical detection system in ports. Its actual success is, however, more problematical. Governments—and particularly those of oil

81. See the Canadian-Spanish proposal, MP/CONF/C.I/WP.26. This may have been a meaningless debate, since despite the "clear grounds" stipulation in the Safety of Life at Sea Convention (SOLAS) government inspectors in ports have been able to inspect vessels for compliance with SOLAS regulations whenever they wished. In fact, some governments make a practice of doing random checks on vessels in their ports, although they are certainly more reluctant to conduct them on foreign vessels than on their own. One government official interpreted the "clear grounds" clause as meaning "Do not conduct witch hunts."

82. Interviews with government officials. An American official said that the inspection business was rife with payoffs. Governments do have some control of the quality of the inspection in that they recognize only certain classification societies, but some supposedly reputable societies or inspectors will "bend" their standards in order to secure business—or a payoff.

83. This was proposed by Canada and Spain in MP/CONF/C.1/WP.26. Canada backed this proposal only to mollify the concerns of other states regarding its own radical port-state enforcement proposal. It barely got through Comittee I by a vote of 19-13-10, and many developing states opposed it. And, of course, the port-state enforcement proposal was defeated in plenary.

84. Interviews. The formal proposal was submitted by Canada and Spain. MP/CONF/C.1/WP.26. The European supporters were Sweden, Norway, West Germany, and France. They were probably quite aware that few states would comply with it.

85. The Soviets also attempted to have the right of inspection confined to the "slop tanks" but were defeated 12-20-9. MP/CONF/C.1/WP.47, p. 5.

exporters—will have to devote substantial resources to the inspection system if the right is ever to be come a reality.

Having achieved some improvement in the provisions for inspection, the conference turned its attention to the prosecution of violations. A number of important provisions were accepted despite the opposition of many maritime states. For example, coastal states retained explicit authority to prosecute construction and equipment violations as well as discharge violations. The Netherlands had proposed that flag states have a preferential right to control these prosecutions, but it was clear that such a provision would have limited existing coastal rights in their internal waters and territorial seas. This was quite out of step with the main direction of the debates (which was to increase coastal enforcement), and although it had some obvious supporters, it was rejected.[86] Other powers and even obligations were imposed on the coastal states, so an improved set of provisions was drafted for dealing with violations of the regulations.[87]

Of all the proposals concerning the power to prosecute vessels contravening the regulations, the most innovative and far-reaching was that for "port-state enforcement."[88] Prosecution of violations was, it will be remembered, characterized by the territorial seas/high seas dichotomy. On the high seas, flag state control was exclusive. The inadequacy of this arrangement had been repeatedly demonstrated to IMCO, and coastal states were increasingly unwilling to submit to it. If granting full power to a "coastal state" to prosecute (and arrest) any violator in its coastal zone (of perhaps 200 miles) was a bit too radical for most participants, perhaps granting a similar power to a "port state" would not be. Port-state enforcement suggested just that: a state would be allowed to initiate proceedings against any vessel *in its ports* for any violation which was committed *anywhere*. To a dispassionate observer, this would seem to be a reasonable proposal which satisfied the maritime states' demand that vessels not be stopped during voyages. But in the context of traditional law of the sea, it was still an exceedingly contentious issue.

Canada and the United States had both supported port-state enforcement at the February 1973 preparatory conference, but that meeting was

86. MP/CONF/WP.16. Some maritime states (France, Germany, Greece, Japan, and the U.K.) were worried about discriminatory and costly proceedings and fines against their vessels in foreign ports.

87. Coastal states were *obligated* to prevent a ship from sailing if it "presented an unreasonable threat to the marine environment" unless it was proceeding to a repair yard [art. 5(2)]. They were also *obligated* to report violations outside of their jurisdiction to flag states and to either initiate proceedings or send reports to flag states with respect to violations within their jurisdiction [arts. 4(2), 6(2–3), and 8(3)].

88. The original proposal can be found in MP/CONF/8, p. 2. Amended versions can be found in MP/CONF/C.1/WP.25 and WP.34.

dominated by the maritime state "regulars" and the proposal was rejected.[89] In October, however, developing states constituted a majority of the participants (thirty-nine out of seventy-one), and the Canadians and Americans expected that the proposal would obtain broad support from them. Furthermore, since the first rejection of the port-state enforcement proposal, the two countries had substantially amended it to overcome the objections expressed by its opponents. Originally, the sole restrictions on the port state were that it would prosecute a violation which occurred in the territorial sea of another state only with that state's permission and that it could not prosecute where another state had already done so.[90] The new submission stipulated many new conditions: penalties were restricted to fines; the vessel was to be released upon the posting of a bond; a party intending to prosecute a vessel had to inform the flag state; and, most importantly, the port state was obligated to terminate proceedings if the flag state initiated them within a period of sixty days.[91] The amendment was flatly rejected (see Chart 16).

Familiar patterns were evident in the roll call vote on the amendment. The threat of expanded coastal state power beyond the territorial sea (with its commensurate erosion of exclusive flag state jurisdiction) once again raised the old specters of foreign interference with shipping and increased economic costs from prosecutions and delays in port. These concerns differed among various groups of states. The Soviet bloc had long sought to maintain inviolate control of its ships and crews abroad, while European tankerowners were concerned about discriminatory treatment by the oil-exporting Arab states. Liberia, of course, feared discrimination by everyone against flag-of-convenience vessels. For both the Soviet bloc and maritime states, these specific concerns were compounded by their hesitation to concede any "bargaining chips" on an issue which might later be offered as part of a "package deal" at the Law of the Sea Conference.[92]

It was also this concern about its potential impact on UNCLOS III that resulted in the unexpected opposition of many developing states to the proposal. Not surprisingly, a small group of developing maritime states opposed it for reasons similar to those of their European counterparts. In fact, their concern was even greater, as they feared discriminatory treatment from competitor developing countries as well as from the developed countries. But they were joined by many developing *coastal* states which saw port-state enforcement as a premature sacrifice of their more extensive

89. PCMP/4/30. It was a new proposal, then, which seems to have originated in Ottawa. However, the U.S. was its major advocate.

90. MP/CONF/8, p. 2.

91. MP/CONF/C.1/WP.25 and WP.34.

92. This concern was exacerbated by the fact that the Canadian proposal originally used the phrase "port-state *jurisdiction*." The term "port-state *enforcement*" was coined only later.

CHART 16
State Positions on Port-State Enforcement

	Support	*Oppose*	*Abstain*
Western Europe and Others	Australia Canada Denmark Ireland Japan New Zealand Iceland South Africa Sweden U.S.	Finland France Germany (F.R.) Greece Italy Monaco Norway Spain U.K.	Belgium Netherlands
Soviet Bloc		Bulgaria ByeloSSR Germany (D.R.) Poland UkSSR USSR	
Developing	Chile Indonesia Iran Philippines Tanzania Trinidad-Tobago	Argentina Cuba Ecuador Egypt Iraq Jordan Kuwait Liberia Mexico	Brazil Cambodia Ghana Morocco Nigeria Peru Singapore Uruguay

Defeated: 16-25-10

coastal claims to environmental jurisdiction in the entire "economic zone." For many of these countries, a coastal zone jurisdiction was far preferable to port-state enforcement. Many more ships would pass through their 200-mile economic zone than would ever enter their ports. This was certainly the situation of the African states through whose zones countless tankers passed during their voyages between Europe and the Persian Gulf. Moreover, a coastal zone jurisdiction was a tangible geographical extension, whereas port-state enforcement was more of a procedural power for the coastal state. The prospect for gaining acceptance of the more substantial jurisdiction was certainly not to be jeopardized by establishing a restrictive alternative at IMCO.

From another perspective, it was also the impact on UNCLOS III of an IMCO acceptance of port-state enforcement that led some developed states such as Japan and the United States—as well as the oil industry spokesmen—to support it. They foresaw that its approval would reduce the need for coastal state jurisdiction which, like the Europeans, they opposed because of its potential for interference with navigation for either environmental or political reasons. Unlike the Europeans, however, they believed that the *acceptance* at IMCO of such a measure would reduce more extreme coastal state demands at UNCLOS III.[93] Canada, the creator and a major proponent of the scheme, was ambivalent about the implications of the proposal's acceptance for the U.N. Conference. Like the United States, Japan, and most of the other backers of the proposal, Canada's ability to control vessel-source pollution along its coast would be enhanced by port-state enforcement, as most shipping passing near its coast eventually entered its ports. Unlike these states and like many developing countries, however, Canada did favor unfettered coastal state jurisdiction as its ultimate goal, but it had not initially considered that the acceptance of port-state enforcement at IMCO would prejudice the prospects for realizing the ultimate goal at UNCLOS.[94] In discussion with the other delegations at the conference, the Canadians' perspective was changed so that, by the time of the vote, they too began to withdraw their support for the proposal. Indeed, members of the Canadian delegation indicated to others that, while Canada as its originator would still vote for it, it would be unwise for them to do so.[95] Not all of Canada's coastal state allies agreed with this judgment, and they were quite divided in the final voting. In this confusion, the measure was rejected.

This defeat of the port-state enforcement proposal reconfirmed IMCO's traditional reluctance to wander too close to the boundaries set for it by the prevailing laws of the sea. It also meant that the traditional coastal/flag state separation of prosecuting authority still remained unchanged. Many other

93. At the same time the U.S. saw real merit in the proposal, whereas most Europeans did not.

94. Canada did not see port-state enforcement as sufficient, as it desired the right to legislate and enforce national standards in a coastal zone as it had done in Arctic Waters Pollution Prevention Act (1970). See the prior discussion on its coastal state, "special measures" proposal.

95. There were significant differences within the Canadian delegation on this policy change. In particular, members from Canada's Ministry of Transport saw the submission as a boon to IMCO and to enforcement generally, and they insisted that IMCO's decisions could be kept distinct from what occurred at UNCLOS III.

small improvements in the enforcement system were made but, as ever, no significant changes were acceptable.

The fate of the American proposal to reverse the "burden of proof" was typical of the lesser amendments submitted. The detection of discharge violations had been facilitated with the change in 1969 from the old 100-ppm standard to one based on quantitative measurement of 60 liters per mile for all vessels (to a maximum of 1/15,000 of the cargo-carrying capacity for tankers), but proof of most violations still required gathering evidence immediately after the discharge. Consequently, the U.S. proposed that a "visible sheen" of oil in the wake of the vessel would be *prima facie* proof of a violation unless the master could prove that it did not, in fact, contravene the regulations.[96] Despite the *caveat* that such a test would be applied only to the extent that it accorded with "the fundamental law of the Contracting State," it encountered the same opposition from the European "civil code" countries as had met the earlier British proposal for a reversed burden of proof. Moreover, it was suggested that no monitoring or control system then existed to allow the master to prove his innocence and that a sheen surrounding the vessel did not necessarily mean it came from that vessel. These technical objections the American delegation was able to surmount by making its application contingent on the development of such monitoring devices, but the basic opposition to incurring any legal or economic inconvenience in order to facilitate prosecutions remained, and the proposal was withdrawn.

In retrospect, it can be said that the 1973 Conference strengthened the enforcement system by accepting several significant measures (especially inspection rights), but its contribution fell far short of the revolution some called for. Exclusive flag-state enforcement jurisdiction on the high seas remained unchallenged, and no *shift* in authority from the flag to the coastal state occurred. The United Nations Conference on the Law of the Sea was scheduled to begin within a few months, and the task of working out that issue fell entirely to it.

One important facet of compliance that had seldom been considered in the past was the promotion of ratifications. In 1973, a number of provisions were included which specifically dealt with this topic. One such stipulation which will have a profound effect on the implementation of the convention was that any tanker built soon after the conference's completion (and *before* ratification by its flag state) would have to meet both the 1971 tank-size limitation regulations and the new requirement for segregated ballast tanks for it to enter the ports of contracting states.[97] This meant that flag states

96. PCMP/4/17; MP/CONF/C.2/WP.3 and WP.26.
97. Annex I, regs. 1(6), 13, and 24.

would not have the usual incentive to delay ratification in order to give their shipowners a competitive advantage by avoiding the costs of meeting the construction regulations. Since *all* shipowners would want to ensure that their tankers could enter the ports of most states—especially those of the major oil-importing countries—they would all have to order new tankers built to the specifications of the convention. As a result, even though shipping firms in the major maritime states disliked the tank-size and segregated-ballast regulations, most strongly supported this new article as a way of protecting them from a competitive disadvantage when their states eventually ratified the convention.

At the Civil Liability Conference in 1969, states took the unprecedented step of applying part of the convention (the requirement of insurance) to noncontracting states. This was an important jurisprudential leap, and the United States delegation at the 1973 Conference now argued that an even broader provision should be incorporated into the new convention. It proposed that all contracting parties should be obligated to apply it *in toto* to other states after it had been in force for five years.[98] This would both expand the substantive scope of the convention's coverage and encourage states not to delay ratification. But the provision was just too sweeping, and it engendered strongly hostile reactions. Both the Soviet bloc and the Western European states objected to the application of an entire treaty to nonparties which were outside the convention's jurisdiction. In addition, the developing maritime countries objected that the proposal could easily impede the development of their shipping industries. As a result, a more limited provision was accepted which required only that the convention be applied "as may be necessary" to ensure no special treatment be given to nonparties.[99]

The 1973 Conference created an excellent "technical" convention (see Chapter 4). But the compliance system underlying it was only an adaptation of the long-ineffective traditional regime. Both the Western European maritime states (and Japan and Liberia) and the Soviet bloc had a common interest in defeating any measure which would significantly increase the likelihood of prosecution and cause delay for their vessels. The competitive businesses of Western Europe opposed any increased economic costs being placed on them. As it was certainly their governments which would be the first to ratify the convention, these were legitimate fears, for

98. The U.S. proposal, which had been defeated by a vote of 3-18 at the preparatory conference, was included as an option in the draft convention submitted to the October conference (MP/CONF/4, p. 8). The vote in Committee I in October was 6-1-18.

99. Article 5(4) states, "With respect to the ships of nonparties to the Convention, Parties shall apply the requirements of the present Convention *as may be necessary* to ensure that no more favourable treatment is given to such ships." (Emphasis added.)

any increased costs would translate into either reduced profits or a loss of competitive advantage. The motivation for the Soviet bloc was political rather than economic: they opposed any foreign interference with or prosecution of their vessels. Although there were certainly differences among these groups, their generally shared interests substantially limited what the conference could do.[100]

The developing states—except those few with immediate maritime aspirations—were largely motivated by a desire to increase the *rights* of coastal states in advance of UNCLOS III.[101] At the same time they were reluctant to accept *obligations* which would impose costs on their governments so that, while supporting an unlimited *right* of inspection to determine discharge violations, they opposed *obligations* to carry out such inspections and to detain vessels not conforming with ship-standard regulations in their ports.

The protagonists throughout the four-week conference were the small but familiar group of wealthy, Western environmentalist states without large fleets engaged in international trade. The backbone of this grouping was Australia, Canada, New Zealand, and the United States (although they were sometimes assisted by others such as Iceland, Ireland, and a few continental European countries). In general, costs were not a concern of these four highly developed states. Their concern was to get compliance with the technical regulations, and as a result virtually every proposal to improve enforcement emanated from this group.

In the final analysis, the outcome of these negotiations was determined by whether the developing countries sided with the Western maritime and Communist states or with the "environmentalists." If measures protected or increased coastal state *rights*, they joined with the environmentalist clique. Generally too they went along with their proposed coastal state *obligations* if they were not too costly or could be ignored. But any threat to their larger jurisdictional or economic interests pushed them into the maritime states' voting bloc. The developing states, for the first time, held and exercised the balance of power at IMCO.

100. Being under some domestic pressure from environmental interests the European states were willing to accept a modest increase in prosecutions in order to curtail pollution. As a result they accepted an unlimited right of inspection to detect discharge violations, while the Soviet group opposed all measures which would curtail flag-state control. At the same time, the Soviet group was excessively protective of its sovereignty in the territorial sea and inland waters, even where it meant some sacrifice of flag-state control. As a result they did not support the Dutch proposal to give flag states a broad preferential right to prosecute ship standard violations, and they opposed restricting national penalties to fines. This changed at UNCLOS III, where they accepted minor incursions in traditional coastal-state powers in their own territorial seas.

101. Their opposition to port-state enforcement, however, reflected the fear that approval of this right would compromise the later acceptance of complete enforcement jurisdiction (including the power of arrest) in the "economic zone."

The result of this pattern of coincidence and conflict of interests was a compliance system with some additional, but modest, enforcement opportunities. The right to inspect for construction and equipment violations was limited in most cases to examining sometimes unreliable certificates, but the right to inspect for discharge violations was markedly improved. Even here though, the value of inspection would be dependent both on the development of monitoring devices (the records from which would be acceptable in courts) and on the willingness of states to undertake thorough surveillance and inspection programs. Considering that only slight improvements were made in the provisions relating to actual prosecutions, it was not likely that the changes would have a deep or widespread impact. After twenty years of indifference by flag states, no major reallocation of enforcement authority toward the coastal states occurred. And with this coincidence of interests evident at the IMCO Conference, what were the hopes for UNCLOS III?

IMCO Negotiations and the TSPP Conference: 1974–1978

The 1973 Conference on Marine Pollution was only a skirmish before the real battle at UNCLOS III. There, basic maritime standard-setting and enforcement powers were to be renegotiated, and attention soon shifted to this larger forum (see section following). In the meantime, IMCO's Maritime Safety Committee (MSC) and its new Marine Environmental Protection Committee (MEPC) focused the organization's attention on the enforcement problem as never before. Their deliberations and those of the unexpected 1978 Tanker Safety and Pollution Prevention (TSPP) Conference produced a number of potentially valuable additions to the enforcement system.

Increased concern at IMCO for enforcement of safety and pollution regulations was initially spurred by the OECD report on flags of convenience, issued in 1974.[102] One source of the competitive advantage of these fleets over their European counterparts was thought to be their lax enforcement of international regulations, a laxity which also unfairly tainted even the most conscientious shipowner. On closer investigation, however, it was discovered that the flags of convenience, though certainly greater offenders than the traditional maritime states, were not the only villains. All national fleets had their share of "substandard ships." To control these, wherever they were, was the real target, and the MSC and Assembly soon drafted and approved guidelines for identifying them.[103]

102. A summary of the data can be found in *Maritime Transport 1974* (Paris: OECD, 1975), pp. 88–105.

103. MSC XXXIV/18, annex XV; MSC XXXV/21, annex X.

The new focus of concern led to important extensions in monitoring compliance with IMCO conventions. Many delegates felt that, as an organization, IMCO had a strong history in formulating rules and an improving record in achieving their ratification but a still frightful one in ensuring their implementation. To rectify this, IMCO was given new powers. First, the MSC in 1975 recommended that reports of violations of the SOLAS Convention be transmitted not only to the flag state but to IMCO as well and that the flag state later report on its own investigations and prosecutions. The flag state was also to notify IMCO of investigations into all its ship casualties. Information on the classification societies or governmental agencies which had issued certificates to these ships also would be forwarded. These proposals were accepted by the Assembly.[104] Second, the MSC in 1977 requested the Secretariat to keep a list of casualties which it would review at each meeting.[105] Incidents were to be removed from the list only on receipt of a satisfactory report from the flag state.[106] Third, encouraged by this bold initiative, the MEPC followed suit, establishing a similar arrangement for casualties causing marine pollution.[107] It also called on states to forward information on all discharge violations to IMCO as well as flag states and asked that the Secretariat distribute the information to all members. The MEPC, in addition, agreed to review the reports at one meeting each year "with a view to monitoring the extent to which the incidents are followed up."[108] A similar procedure was agreed to for reports on inadequate reception procedures.[109]

It is too early to say how well these improvements will work. They are not legally binding, so it will be up to the states themselves to comply with the resolutions. At the same time, the framework is now in place for IMCO to monitor states' promotion of compliance with conventions, to highlight the weaknesses in their enforcement programs, and to exert pressure on them to improve these programs. The organization can, in addition, more easily identify how the technical regulations might be improved.

At the 1978 TSPP Conference, enforcement issues were not prominent. In one area, though, some important progress was made—the inspection and certification of tankers for compliance with construction and equipment standards under the MARPOL and SOLAS conventions. In accordance with the Carter initiatives, the United States put forward a number of new proposals. In short, it proposed (1) that more frequent or "intermediate" surveys be conducted for tankers over ten years old; (2) that

104. Res. A.321 and 322 (IX).
105. MEPC VIII/18, p. 2.
106. Res. A.390 (X); MSC/MEPC/6.
107. MEPC VII/19, annex II, sec. II; MEPC VIII/WP.7, pars. 68–69; MEPC VIII/12/1.
108. Res. A.391 and 392 (X); MEPC VIII/WP.7, par. 65.
109. MEPC VIII/WP.7, pars. 11–13.

nationally appointed surveyors (generally from classification societies) be able to detain a vessel until deficiencies are corrected; (3) that surveyors be obligated to inspect or survey a vessel at the request of port officials; (4) that flag states be obligated to send IMCO information on the identity and responsibilities of its surveyors; and (5) that the surveyors be required to carry out ad hoc inspections of vessels between the periodic surveys.[110]

Some debate occurred over these proposals, although most passed without difficulty. The greatest opposition was to the requirement for ad hoc inspections, a provision that was potentially disruptive for the classification societies which act as inspection and certification agents for shipowners and most states. The societies are competitive, each vying with the others for business, and ad hoc inspections could alienate their clientele. In addition, as the International Association of Classification Societies (IACS) and certain states pointed out, such a requirement for ad hoc inspections could also, in effect, require surveyors from the same society or from different societies to check up on the work of others—thus undermining morale and causing harmful competition. To remove these "problems" IACS proposed more frequent, scheduled surveys instead, a proposal that would also incidentally increase their business. Some states that opposed random and possibly discriminatory interference supported this alternative (Japan and the USSR), but most maritime interests (and certainly the International Chamber of Shipping) rejected such a costly and time-consuming new proposal. The result was a compromise—authorized surveyors would conduct ad hoc surveys except on ships where the flag state required "mandatory annual surveys."[111] To mitigate the possibility of friction arising with a society's client shipowners, IMCO was instructed to draw up criteria for selecting ships and carrying out the ad hoc inspections.[112]

The beneficial effects of this one small change could be substantial. The quality of work by surveyors has been notoriously uneven, and the possibility of a sudden "spot check" between scheduled inspections should improve this. These spot checks will also reduce the attractiveness of substandard ships, especially if they are to be held in port while the repairs are made. The TSPP Conference may improve the quality of inspections in other ways as well. This is basically a result of the introduction of crude-oil-

110. The U.S. proposals and the reactions of other participants are spelled out in the reports of the intersessional group which met in May, June, and July (TSPP I/9, II/2 and III/8) and in the report of the working group on inspection and certification of the October MSC/MEPC meeting (MSC/MEPC/WP.11). The proposals, with some minor changes, were integrated into SOLAS Protocol, chap. I, part B, and MARPOL Protocol, annex I, regs. 4, 8.

111. SOLAS Protocol, reg. 6; MARPOL Protocol, reg. 4.

112. TSPP/CONF/12, res. 10.

washing, a system which, unlike load-on-top (LOT), requires inspection in the oil *discharge* port and not the oil *loading* port. One of the greatest weaknesses of LOT has been its dependence on the oil-exporting states, which have always lacked interest in inspecting returning tankers. The transfer of this responsibility to the developed oil importers should be a major step forward.

Some additional improvements by the TSPP Conference were the provisions for encouraging early ratification of the conventions. As mentioned in Chapter 4, the United States was adamant in its demand to be able to implement the new protocols before they legally entered into force. By setting fixed dates for the application of the regulations to new ships and by urging early application of the regulations to existing ships, the conference paved the way for early American action and the almost-as-early acceptance by other states. If the United States carries through on its commitment to implement the new agreement as soon as possible, this simple provision may prove the biggest aid to compliance yet devised.

IMCO's work in this recent period may substantially improve the enforcement of its convention regulations.[113] But many of its recent resolutions are without legally binding effect and, hence, are unenforceable. With the past history of flag state indifference, one cannot be optimistic that a sudden qualitative improvement will now occur. Moreover, no matter how much tinkering IMCO does with its powers to oversee the compliance of flag states or with the operation of the classification societies, many of the basic rules for setting and enforcing maritime environmental standards are beyond IMCO's powers. IMCO must work within the strictures set down by U.N. Law of the Sea Conferences.

UNCLOS III: THE REVISION OF LEGISLATIVE AND ENFORCEMENT JURISDICTIONS

While IMCO was preparing for its 1973 Conference, negotiations were under way for the convening of the Third United Nations Law of the Sea Conference. The decision to hold this conference had been taken in December 1970 in the U.N. General Assembly, and preparatory negotiations for it took place during the years 1971–1973. Three conference committees were charged with various aspects of the law of the sea. In 1958 and 1960 environmental issues had been a tangential concern at best. Now, in 1974, one committee, Committee III, dealt largely with the issue of marine pollution as well as with the lesser issues of marine scientific research and

113. By August 1977 the MSC had received forty reports of violations and eleven responses from flag states.

The voyage of the American tanker *Manhattan* through Arctic waters in 1969 sparked Canadian legislation for a one-hundred-mile pollution-control zone. This act was a dramatic, unilateral manifestation of coastal power that challenged traditional maritime freedoms and contributed to the already growing demands for a new conference on the law of the sea. Amid continuing controversy on this issue, the conference opened four years later. Here, a Canadian Eskimo protests the ship's intrusion into his own "personal sovereignty" while the ice-bound tanker waits before plowing over his igloo. (Photo: CP.)

the transfer of technology. Committee II, which dealt with fishing and traditional navigational issues, also dealt in part with jurisdiction over the marine environment.

The work of Committee III built upon and extended that of IMCO. Indeed, it produced forty-six articles on the preservation of the marine environment,[114] and the two issues of greatest consequence—seen as the

114. Informal Composite Negotiating Text, U.N. doc. A/CONF.62/WP.19 (July 15, 1977). The authors have not undertaken a detailed political analysis of the UNCLOS negotia-

The Third United Nations Conference on the Law of the Sea resuming its fourth session in New York in March 1976. This conference has continued for over five years since it first started meeting in 1973. The results will have important consequences for IMCO's regulatory activities in setting and enforcing maritime environmental standards. (Photo: United Nations.)

"key to agreement"—concerned the jurisdiction to set and the jurisdiction to enforce environmental standards.[115] These were the very issues which had preoccupied IMCO in 1973, and they were now the major topics of debate in the UNCLOS environmental negotiations. Moreover, although the numbers were greater, the same pattern of regional and interest groupings emerged as had come to dominate IMCO in recent years.

Five substantive sessions of the conference had been held by 1977: one in Caracas in 1974, one in Geneva in 1975, and three in New York in 1976

tions. The following commentary seeks to explain the general political and legal trends as they affect IMCO's future work in the area.

115. U.N. doc. A/AC.138/SC.III/L.26, p. 17 (August 31, 1972). This study will focus on these issues. It will not discuss other important jurisdictional issues such as the status of international straits—where concern was not focused on environmental matters.

and 1977.[116] After an initial meeting in Caracas where delegations put forward their policy preferences, it was decided in Geneva to request the chairman of each of the three committees to draw up draft articles representing the trend in the negotiations in their respective committees. In order to achieve a consensus on these articles, early voting was to be avoided, and successive versions which reflected the continuing developments were to be drafted instead. This was done, and an informal "Single Negotiating Text" (SNT) was issued at the end of the Geneva session.[117] A "Revised Single Negotiating Text" (RSNT) was issued at the spring 1976 session in New York, and a commentary on the RSNT at the second New York session.[118] At the 1977 session, the drafts were consolidated into an "Informal Composite Negotiating Text" (ICNT).

Part 12 of the ICNT was drawn up by the chairman of Commitee III, Ambassador Yankov of Bulgaria. It focuses on vessel-source pollution, although it also contains several provisions imposing obligations to prevent land-based, seabed, and air pollution and pollution from dumping. However, these latter are treated in very general terms.[119] It is only in treating vessel-source pollution that very specific rights and obligations are set out and, not surprisingly, these retain the flag state as the basic standard-setting authority for international shipping. There is, for the first time, a general obligation on states actually to establish international ship standards through "the competent international organization" (that is, IMCO), but it still falls to the flag state to apply these standards through national laws.[120] There is now, at least, a legal obligation on these states to establish such laws, but no penalties are provided for their failure to do so.

Coastal state *legislative jurisdiction* is expanded in certain limited areas, but it is actually *restricted* in much broader ones. In a state's internal waters, it remains what it has always been—full sovereign control.[121] In the territorial sea, however, the coastal state's traditional authority has now been severely curtailed. The existence of the authority to set standards for vessels in the area is still recognized—including those relating to "the preservation of the marine environment . . . and the prevention and control of pollution"[122]—but this is still subject to the "right of innocent passage." It is in the definition of *innocent passage* in part 2[123] that the first restriction occurs, for a

116. An initial session dealing with procedural issues was held in December 1973.

117. Single Negotiating Text, U.N. doc. A/CONF.62/WP.8 (May 6, 1975).

118. RSNT, U.N. doc. A/CONF.62/WP.8/Rev.1 (May 6, 1975), and the Commentary on RSNT, part III, U.N. doc. A/CONF.62/L.18 (September 16, 1975).

119. See articles 193–210 and 236–238 of the ICNT for these general obligations.

120. Art. 212(1)(2).

121. Art. 25(2), part II.

122. Art. 21(*f*), part II.

123. Part II of the ICNT is that part produced by Committee II of the conference. As Committee II dealt with the delineation of national jurisdictions, there was some overlap with

polluting passage is considered *not* to be innocent only where the ship engages in an act of *willful and serious* pollution.[124] Few pollution incidents would be *both* willful and serious (even the *Torrey Canyon* disaster was not willful), so almost all passage will be innocent passage. Although a state may legislate for these ships, it must not legislate so as to restrict this freedom. But here a second and even more restrictive article now applies. Even when legislating so as *not* to affect innocent passage, the coastal state's laws and regulations "shall not apply to the design, construction, manning or equipment of foreign ships unless they are giving effect to generally accepted rules or standards."[125]

The potential restrictiveness of this regime on a coastal state's jurisdiction in the territorial sea is enormous. Traditional maritime law as reflected by article 17 of the Geneva Convention on the Territorial Sea and Contiguous Zone states that foreign ships exercising their right of innocent passage have to comply with coastal state laws, which must themselves conform to other rules of international law.[126] In the words of Canadian officials, the "constructive ambiguity" of this Geneva Convention provision creates a workable situation whereby both coastal state sovereignty and the maritime state right of innocent passage are protected. As one commentator argued in defense of this traditional formulation, the enactment of pollution legislation by some twenty-five states has produced no "patchwork quilt" of varying regulations: "maritime commerce continues." For a coastal state to use its customary power to implement regulations which were seriously at variance with international ones would, it was argued, be "self-defeating" as "many of us are [dependent] on the smooth flow of international shipping to ensure our economic well-being."[127] The replacement of the traditionally vague coastal power over its territorial sea by the explicit restrictions of part 2 creates, in the words of the Canadian delegate, "a new order of absolute sovereignty for flag states within the territorial sea."[128]

The most important issue concerning legislative competence in the years leading up to UNCLOS III was the right of states to set environmental standards outside the territorial sea—particularly in the projected 200-mile

---

the work of Committee III. Indeed, part XII, dealing with the marine environments, is seen as an elaboration within the jurisdictional framework of part II and is therefore subservient to it.

124. Art. 19(*h*), part II.

125. Art. 21(2), part II.

126. Such a provision is also found in the 1954 Convention (article XI) but was deleted from the 1973 IMCO Convention.

127. Statement to Committee III Plenary, September 14, 1976, unofficial Canadian government text, p. 5. Referring specifically to Canada's domestic legislation, the Canadian delegate to Committee III commented that Canada had "endeavoured where possible to incorporate international norms in our national laws."

128. Ibid.

"economic zone." This was certainly the question behind the 1973 IMCO debate on "special measures." By the time of the Caracas session in the summer of 1973 it was evident, however, that very few states were now strongly committed to this particular expansion of coastal state authority. With the virtual acceptance of the idea of the economic zone itself, a particular power to set higher environmental standards there was itself no longer so important to the diverse coastal state bloc. Moreover, the Soviet bloc, the developed and developing maritime states, and the landlocked and geographically disadvantaged countries were dead set against such an expansion, which might undermine the traditional freedom of navigation on the high seas. In these circumstances a few "environmentalist" developed states such as Canada and Australia and a few "territorialist" developing nations were able to make only very few inroads into the traditional supremacy of the flag state in the area.

The ICNT reflects, therefore, the predominant maritime interest in uniform international standards applied by the flag state. Despite the recognition in article 56(*b*) of part 5 of the ICNT of a coastal state's "jurisdiction with regard to the preservation of the marine environment" in its economic zone, the exercise of that jurisdiction is largely restricted in part 12 to enforcement or to standard-setting for "ice-covered" areas. Initially, in Geneva, the SNT simply provided for the right of coastal states to legislate special measures in "areas of the economic zone, where particularly severe climatic conditions create obstructions or exceptional hazards to navigation."[129] This was the "Arctic exception," but it was so permissive and its wording so vague that many maritime powers expressed concern lest it be applied too freely. In New York, therefore, a specific "Arctic exception" was provided as well as another more restrictive article for other areas of environmental vulnerability. The "Arctic exception," article 235, is set out in a separate section of part 12 ("Section 8: Ice-Covered Areas") and it reads,

> Coastal States have the right to establish and enforce non-discriminatory laws and regulations for the prevention, reduction and control of marine pollution from vessels in ice-covered areas within the limits of the economic zone, where particularly severe climatic conditions and the presence of ice covering such areas for most of the year create obstructions of exceptional hazards to navigation, and pollution of the marine environment could cause major harm to or irreversible disturbance of the ecological balance. Such laws and regulations shall have due regard to navigation and the protection of the marine environment based on the best available scientific evidence.

129. Art. 20(5), SNT.

This article clearly was an attempt to appease Canada and its intense commitment to its Arctic Waters Pollution Prevention Act—a commitment it had pursued even at the risk of violating international law. At very little cost to the maritime states (which did not often navigate in these ice-covered areas anyway), Canada's major concern was assuaged and its relative moderation assured in ensuing debates. This done, the general provision for other vulnerable areas has been weakened with each session of the conference. As mentioned, the broad provision of the Geneva SNT was found to be too permissive and was amended in New York. The revised provision, article 21(5) of the RSNT, restricted coastal special measures to areas where such measures were required "for recognized technical reasons" and then only where "the competent international organization . . . determine[s that] the conditions in that area correspond to the requirements set out." In other words, the coastal state power was made subject to an international veto. With Canada quite inactive on an issue which no longer concerned it, Australia took over the leadership at the 1976 summer session. Hoping to gain sufficient jurisdiction to set special coastal discharge standards that would protect the Great Barrier Reef, Australia argued for increased coastal state flexibility, and in this it was supported by other states concerned about passing tanker traffic—Egypt, Malaysia, India, China—as well as by the more "territorialist" states such as Kenya and Ecuador. In opposition to this approach, the Netherlands and Germany argued that the role of the "competent international organization" should be strengthened, a position which met with the approval of Turkey, Japan, Bulgaria, Argentina, the United Kingdom, and Liberia. In the end, the negotiating group for Committee III accepted a revised text of article 21(5) which gave the international organization a much expanded role while at the same time restricting any possible coastal state legislative authority solely to discharge standards. The provision in the ICNT, article 212(5), simply reaffirms the practice accepted in 1973 at IMCO of establishing "special areas" through that organization.

In conclusion, with regard to the coastal state power to set ship standards, UNCLOS III will likely produce a convention that recognizes the special Arctic exception but confers no greater jurisdiction in the economic zone. Indeed, while leaving intact the power to control standards in internal waters, the existing sovereign rights in the territorial sea will undoubtedly be restricted. If any compromise of flag state jurisdiction is to occur, it is in the area of enforcement.

Articles 214 and 217 of the ICNT oblige states generally to enforce the laws and regulations adopted with regard to land-based national and international seabed pollution and pollution from dumping. Ship-source pollution is covered by articles 218 to 221. In the Geneva SNT, a simple flag state obligation to comply with international standards was provided. Where requested by another state, the flag state was to investigate an alleged violation and prosecute if "satisfied" that there was sufficient evidence.[130] This basic obligation was augmented at New York in the RSNT by the following requirements:[131] Vessels would not be allowed to leave their home ports when they did not comply with "applicable" international standards of design, construction, equipment, and manning; all ships would be certified and inspected at regular intervals; and the flag state would investigate all violations of international rules whether or not requested to do so. These obligations were tightened in the negotiating group of the second New York session. If properly implemented they could substantially improve the efficacy of flag state enforcement. Unfortunately there is still no provision for effective international review of flag state enforcement nor any sanctions should such enforcement prove inadequate.

Coastal state enforcement powers in the ICNT are not broad, but they are certainly much expanded over traditional powers and over those contained in the original SNT. In the ICNT, the coastal state's power to investigate, arrest, and prosecute a ship for violations of any applicable national or international standards in the territorial sea is affirmed.[132] It is, however, to be exercised "without prejudice to the right of innocent passage," a phrase again raising the uncertainties of part 2 of the ICNT.[133]

For the first time in international law, the new law of the sea texts grant coastal states a legal environmental interest in the waters beyond their 12-mile territorial seas, in the 200-mile economic zone. This is a potentially important first step for coastal state environmental protection but only in the long term.[134] In the immediate future, the powers granted to the coastal state in this zone are apparently important but in fact quite insignificant. Flag-state authority continues to prevail. Article 221(3) permits a coastal state to require information from a passing ship concerning its registry and ports of call should that ship violate applicable standards in the economic zone. Where this information is lacking or inadequate and there has been a "substantial discharge *and* significant pollution," the coastal state may

130. Art. 26, SNT.

131. Art. 27(2–4), RSNT. This became article 218(2)–(4) of the ICNT.

132. Art. 221(2), RSNT.

133. Some manifestations of the potential interest in this extended environmental jurisdiction are already evident. The recent passage in the United States of the Clean Waters Act includes a provision (sec. 311) extending jurisdiction over environmental damage out to the edge of the 200-mile zone. In all probability, this exceeds the UNCLOS rules and has been a source of much consternation within the many American government departments that have tried unsuccessfully to have it repealed.

134. The implication is that a coastal state will be able to stop and arrest a passing ship only if the violation results in a "serious and willful pollution."

physically inspect the ship.[135] This power is really quite limited in practice, since the likelihood of detecting a discharge and being able to order a vessel to stop while still in the zone is slight. Moreover, coastal state prosecution of such a vessel is permissible only where the violation is "flagrant or gross" and results in a "discharge causing major damage or threat of major damage."[136] It is also noteworthy that the article lacks any *preventative* element which would allow an inspection where pollution is "likely."

Even this coastal state power of prosecution was, however, not included in the Geneva SNT and was added to the RSNT over the protests of the maritime states (particularly the United Kingdom, USSR, and Greece). Coastal state interference with the flow of commerce was clearly their greatest concern, and as a result, when the new coastal power was agreed to, provisions were also made for "flag state preemption" of both coastal- and port-state enforcement (discussed further on). Article 221(7) of the ICNT allows a ship to proceed on its course despite a flagrant violation where the flag state has given a "specific undertaking" of responsibility for the compliance and liability of the affected vessel. More generally, however, article 229 provides for the mandatory suspension of *any* proceedings by a coastal state for *any* violation beyond the territorial sea if the flag state itself undertakes prosecution within six months. This preemption does not apply where the proceedings relate to "a case of major damage" or where the flag state "has repeatedly disregarded its obligations." Despite this not very satisfactory caveat, unless the *bona fides* of the flag state can really be assured, the provision could effectively destroy the impact of the coastal- and port-state prosecutory power.

The major enforcement innovation of the Law of the Sea Conference is the adoption of port-state enforcement. Drafted by the Bulgarian chairman of Committee III, the articles on this new method of enforcement were clearly patterned after the British proposal in Geneva in May 1975. Port-state enforcement had, since the 1973 IMCO Conference, been seen as a compromise between flag- and coastal-state enforcement, and as such, the British proposal was solidly endorsed by maritime interests.[137] Coastal states with significant passing traffic were not enthusiastic about it as an

135. Art. 221(5). Article 221(4) requires a flag state to "ensure" that their vessels comply with requests under 21(3).

136. Art. 221(6).

137. U.N. doc. A/CONF.62/C.3/L.24 (May 1975). It was cosponsored by the United Kingdom, Belgium, Denmark, West Germany, Greece, the Netherlands, and from the Soviet bloc, Bulgaria, East Germany, and Poland. In debate (U.N. doc. A/CONF.62/SR.19 and 20), it was also supported by Finland, Japan, Liberia, Norway, and the USSR. The delegate from the United Kingdom commented during these debates that "there were practical difficulties in stopping and boarding large ships in a busy sea lane, and any evidence so obtained would be equally available in the next port of call" (SR.19, p. 4).

alternative to coastal-state enforcement, and others at least hoped that the British proposal would be strengthened before it was incorporated into the Single Negotiating Text.[138] However, the original provisions reflected the conservatism of the British proposal and, undoubtedly, of the Bulgarian chairman, although the subsequent session resulted in an improved text for the port states.

In the SNT, a port state was required to investigate and refer to the flag state for prosecution any violations of international standards by ships in its port. The port state could, however, undertake prosecutions for discharge violations occurring in the port state's coastal zone or, if requested, in the zone of another coastal state.[139] Flag state preemption was provided but without the caveats later incorporated in the RSNT. This regime was strengthened in the RSNT, which provided that a port state could not only investigate any violations anywhere but could also prosecute *discharge* violations anywhere (and not just those occurring in the coastal zone).[140] This "universal" port-state enforcement had been strenuously opposed by the United Kingdom and was a major topic of debate at the summer session of 1976. Furthermore, with the expansion of the right of prosecuting discharge violations from those occurring in the territorial sea to those on the high seas as well, article 231(1) now provided that a port state could prosecute for *any* violations (including structural violations) in the economic zone. This is all subject, of course, to the flag-state preemption. On the other hand, by article 220, the port state can prevent a vessel from sailing if its violation of international standards poses a threat to the marine environment.

In conclusion, the ICNT provides for the retention of the flag state as the basic standard-setting jurisdiction. Coastal states are allowed to set special standards either in ice-infested waters or, with the approval of IMCO, in other areas of exceptional vulnerability. They can freely regulate standards in their internal waters but can regulate so as to affect innocent passage in the territorial sea only according to the restrictive criteria set out in part 2. Concerning the enforcement of standards, the flag state is again retained as the basic authority outside the territorial sea, especially with its power to "preempt" either coastal- or port-state prosecutions. Coastal states have only a very weak enforcement authority in the economic zone ("substantial violations"), but can still prosecute violations in their territorial sea and internal waters. Port states have new, wide-ranging powers to prosecute *any*

138. Tanzania, India, Senegal, Spain, Indonesia, and Nigeria all suffered from the effects of passing ships, and therefore they argued in favor of coastal state enforcement, as did Canada despite its limited passing traffic. New Zealand, Iran, and the United States hoped for a stronger port-state regime than that proposed by the U.K.

139. See art. 27(1), (3), and 28(2), SNT.

140. Art. 28(1), RSNT; later 219(1), ICNT.

discharge violations and any structural violations in the economic zone. Both coastal and port states may impose only monetary penalties,[141] and they are themselves liable for any unlawful or excessive detention or prosecution.[142]

In the tradition of the IMCO conventions that preceded it, the new environmental law of the sea is a product of incremental amendment. It is not a startling or radical development and, indeed, to some, "the ICNT envisions 'business as usual' in the tanker trade."[143] Certainly, from an ideal perspective it is woefully inadequate. Yet in fact the new rules are an improvement, and they will be acceptable to the wide spectrum of national policies. They may also be improved before the conference concludes.[144] In any event, in a few years they will become a new context within which IMCO will work. As ever, the flag state will retain the basic legislative authority, but coastal and port states will now have a legitimate environmental interest in a 200-mile coastal economic zone and a wider power to exercise some real enforcement authority. Unless flag states accept and implement the new higher international standards, committed coastal interests could still act to challenge this regime.

141. Art. 221, ICNT.
142. Art. 233, ICNT.
143. Letter to authors from an American UNCLOS representative.
144. The first 1978 UNCLOS session opened under considerable pressure on environmental issues as a result of the *Amoco Cadiz* incident only weeks earlier. Of particular concern were coastal state standard-setting powers in the territorial sea; coastal state enforcement powers in the economic zone; and coastal state powers of intervention. France led the battle on these issues, particularly on the possibility of establishing reciprocal regional controls over tankers on innocent passage. No formal amendments to the ICNT were accepted, however, and it is unlikely that major changes will occur.

PART THREE

Political Processes of Environmental Control

Chapter VII

The Environmental Law of the Oceans: Determinants of Change

The immense, the infinite, bounded by the heavens, parent of all things . . . perpetually supplied . . . neither seized nor enclosed. . . .[1]

Grotius's centuries-old argument in favor of freedom of the seas (*mare liberum*) has long been our conventional wisdom. Unchanged for four hundred years, the sailors' business was business as usual. Pollution was an injury inflicted on a generous but helpless environment. Even twenty-five years ago, no international law protected the oceans.

But pollution eventually became a "social cost" and then, quickly, a political issue. Since the first international controls in 1954, a remarkable transition has occurred: a vast, new environmental dimension has entered the realm of international law and politics. The Intergovernmental Maritime Consultative Organization has been central to the development of that dimension. It has produced myriad technical regulations on accidental pollution, rules for preventing intentional pollution, and a sweepingly detailed intergovernmental scheme for assigning liability when prevention fails—as it often does. And with the Law of the Sea Conference, IMCO has stumbled toward providing a regime that might actually enforce the rules it has made. Some slow progress has occurred.

Yet the law as we have seen it develop leaves much to be desired. Even for that portion which is actually in force, loopholes abound, obligations are incomplete. The law is, however, only what it was made to be: "The content and evolution of law become, in essence, functions of the various underlying social and political patterns of interest and influence in the society."[2] Traditionally, legal studies in this area have avoided considering

1. Hugo Grotius, *Mare Liberum* (New York: Oxford University Press, 1966) (first published in 1608), p. 37.

2. Oran Young, "International Law and Social Science: The Contribution of Myres S. McDougal," *American Journal of International Law* 66 (January 1972), p. 62. This simple but profound insight is the central lesson of the work of Myres McDougal. In fact, the McDougalian approach focuses on much the same concerns as are considered in this book:

such "underlying" concerns. As one commentator has noted, "the scholarly emphasis remains on the *outcome* (*output*) aspect of the whole process of law development or law changes. The *input aspect and the actual process aspect* of bargaining and decision-making are generally neglected."[3]

Equally relevant to domestic and international law, these observations are particularly applicable to environmental regulation, which must necessarily involve both legal regimes. IMCO's environmental rules must be seen, therefore, not only in relation to the necessities of the problem itself but also as a product of the policies and influences of many governmental and nongovernmental actors.

A related and important concern is the role of the international organization itself: "Inevitably, the movement to protect the biosphere and defend the earth is concerned with institutions as well as with behavior, because through institutions human effort is guided and social goals attained."[4] But here too one must retain a flexible perspective. As many authors have argued in recent years, international organizations are not just "highly institutionalized structures," since they also encompass many complex patterns of informal interaction.[5] To understand changes in the international regime of environmental control, therefore, one must look to the nature of the pollution problem itself, to the policies and influences of interested states and nongovernmental actors, and finally to the operation of the organization in which the regime is shaped.[6]

---

participants, perspectives, arenas, sources of influence, strategies, outcomes and effects. For an elucidation of this typology, see McDougal et al., "The World Constitutive Process of Authoritative Decision," in C. Black and R. Falk, eds., *The Future of the International Legal Order* (Princeton, N.J.: Princeton University Press), 1969, p. 73. For a review of the McDougalian approach and a complete bibliography of his works, see W. Michael Reisman and Bruce H. Weston (eds.), *Towards World Order and Human Dignity* (New York: Free Press, 1976).

3. A. Sheikh, *International Law and National Behavior* (New York: John Wiley, 1974), p. 310; italics in original. Daniel Serwer's "International Cooperation for Pollution Control" (*UNITAR Research Reports*, no. 9, 1972) is a good example of the type of technical/political analysis that is necessary.

4. Lynton Caldwell, *In Defense of Earth: International Protection of the Biosphere* (Bloomington, Ind.: Indiana University Press, 1972), p. 6.

5. As John Ruggie has commented, one must shape one's analysis in terms of "the structural characteristics of the emerging system and not of the dominant image" (that is, the old hierarchical, state-centric, and institutionally monolithic image) in order to provide "'solutions' to the 'problem' of fundamental institutional transformation." "The Structure of International Organization: Contingency, Complexity and Post-Modern Form," *Peace Research Society Papers* 18 (1972), p. 91.

6. For other, similar definitions of what is involved in "regime change," see John Gerard Ruggie, "International Responses to Technology: Concepts and Trends," *International Organization* 29 (summer 1975), pp. 570–574. See also Robert O. Keohane and Joseph S. Nye, *Power and Interdependence* (Boston: Little Brown, 1977). Keohane and Nye's entire work is a study of international regime change, and their classification of the factors affecting it is very close to those identified here.

THE PROBLEM

Three aspects of the ocean oil-pollution problem itself have had a visible impact on the development of international controls. First, as with almost all environmental issues, observers have been highly dependent on scientific analysis for their identification of the problem and for their appreciation of its consequences. Second, oil pollution is (again like most environmental matters) simply a minor undesirable consequence of larger, more "important" economic activities. Third, the control of oil pollution depends on technological innovation, a commodity in limited supply.

Oil pollution has been varied in its impact. In localized areas it has had visibly damaging effects, although in most areas it has had a slow, incremental development, the sources and effects of which often have gone unnoticed by the public. At that level its identification has been dependent on systematic scientific analysis, which has prompted some states to support more stringent controls but in general has had limited effect. Moreover, the scientific dispute over long-term consequences of persistent oil has long mitigated against any significant politicization of the issue, relegating it instead to specialized bureaucratic rather than political actors.[7] At this level, private interest groups usually prevail.

It is not surprising that when conflicting scientific analyses emerge (as they often do, given the need to "interpret" the data), governments tend to back those conclusions which support their existing policy preferences. As a result, scientific information itself frequently becomes a basis for *bargaining* between opposing interests rather than being treated as material deserving of independent *analysis*. With little independent scientific advice emanating from IMCO itself, this has greatly affected the decision-making process.[8] Yet this situation is not surprising: "All rule creation involves bargaining because rules benefit some interests more than others."[9] There are, of course, honest differences of scientific opinion on environmental pollution, but these pale in comparison with the instances of scientific or technological

7. Environmental issues dependent upon scientific information need to have "the relevant data . . . politicized by mobilizing social interests of sufficient intensity to generate the process of public policy-making." Zolenek J. Slouka, "International Environmental Controls in the Scientific Age," in Lawrence Hargrove, ed., *Law, Institutions and the Global Environment* (Dobbs Ferry, N.Y.: Oceana, 1972), p. 208. This problem is exacerbated by the public's general lack of interest in scientific arguments.

8. Through GESAMP (see Chapter 3) some independent studies have been done but in response to requests from the member states for specific information. But there is no body attached to IMCO that undertakes independent, investigative research.

9. Robert Cox and Harold Jacobson, *The Anatomy of Influence: Decision Making in International Organizations* (New Haven, Conn.: Yale University Press, 1973), p. 39. This phenomenon is a common one, as the experience of numerous international fisheries-management bodies would show.

arguments being used to shore up policies motivated by other concerns. One noteworthy example of this subservience of science to politics is the complete reversal of roles between the United Kingdom and the United States in the scientific debates between 1954 and 1973.[10] The debates at the 1978 TSPP Conference on the merits of retrofitting segregated ballast tanks is another. Such examples abound.[11]

Another important effect of differences in scientific opinion has been to create a fundamental conflict on the basic question of who has what burden of proof. Environmental problems are often long-term subtle developments, so environmentalists have argued that "to postpone action on environmental issues until all the evidence was in . . . could result in irretrievable failure."[12] Yet, understandably, states and commercial interests are reluctant to accept costly "solutions" to a problem they believe might not exist. The result has been to impose a gradualism on IMCO's environmental activities.[13] Visible manifestations of coastal pollution did provide the impetus for the first oil pollution conference, but without obvious proof of the seriousness of oil's polluting effects, most conference participants were—*and have remained*—unperturbed. This has been a common attitude to environmental controls.

10. In 1954, when the United Kingdom was the leading environmental protagonist, the United States attacked British studies that stressed the persistence of *black* oils in the marine environment. In 1973, the policies of these states were reversed, and an American study by Max Blumer pointing out the serious polluting effects of *white* oils was now challenged by the British government and its Warren Springs scientists. The arbitrary judgments underlying the "scientific" studies were made very clear in the American criticism of a British study that now minimized the environmental consequences of oil pollution: "From the foregoing it appears that the authors of the report evaluated oil pollution as harmful only when its effects were: (*a*) obvious, (*b*) widespread, (*c*) visible, (*d*) critical, (*e*) direct. The harmful effects of oil pollution need to be evaluated not only as indicated above, but also with respect to harmful effects which are: (*a*) subtle, (*b*) localized, (*c*) invisible, (*d*) indirect, (*e*) long term." For more information on this particular debate, see Charles S. Pearson, *International Marine Environment Policy: The Economic Dimension* (Baltimore: Johns Hopkins University Press, 1975), pp. 84–93. All the studies done by states for the 1973 Conference drew conclusions favorable to their country's stated policies. Perhaps the only technical-scientific issue which did transcend policy orientations was double bottoms, with environmentalists like Canada and Australia *opposing* them on technical grounds.

11. This phenomenon permeates IMCO's work but is found in many institutions. See Matthew Holden, Jr., *Pollution Control as a Bargaining Process: An Essay on Regulatory Decision-Making* (Ithaca, N.Y.: Cornell University Water Resources Center, 1966).

12. Caldwell, *In Defense of Earth*, p. 226. Richard Falk cites popular biologist Paul Ehrlich's warning that "The trouble with almost all environmental problems is that by the time we have enough evidence to convince people, you're dead." *This Endangered Planet: Prospects and Proposals For Human Survival* (New York: Random House Vintage, 1971), p. 188.

13. In this vein, Meadows and Randers wrote of the "mismatch between the time-span of environmental problems and the time horizons of institutions designed to deal with these problems." Dennis L. Meadows and George Randers, "The Time Dimension," in David A. Kay and Eugene Skolnikoff, eds., *World Eco-Crisis: International Organizations in Response* (Madison: University of Wisconsin Press, 1972), p. 48.

To catalyze political action, IMCO has instead been extremely dependent on catastrophe. It was only with the disastrous *Torrey Canyon* accident that the oil pollution problem ceased being just a minor localized issue or a remote statistical trend and became instead a tangible political reality. Only then did IMCO begin hard negotiations on accidental and, interestingly, on operational pollution as well. The incident was, in the words of IMCO's former secretary-general, a "godsend" for the organization and, he might have added, for other environmental organizations throughout the world as well.[14] Since 1967, other spills have periodically maintained the momentum. The grounding of the *Argo Merchant* in December 1976 off the eastern United States was the most notable.

This dependency on disaster is not a characteristic unique to IMCO's environmental responsibilities[15] nor just to IMCO itself. Indeed, writing of the threat of nuclear catastrophe, Richard Falk has observed,

> [O]nly a traumatic course of events provides the learning experience needed to bring about modifications in the organization and values underlying human experience. Educational efforts, appeals to history and reason, warnings about the eventual breakdown of the system and even crises at the brink are not able to overcome the rigidity of vested interests, habits of affiliation, and bureaucratic practice. . . . World order goals remain abstract for most people.[16]

But in IMCO the problem is acute. The organization is dependent on its members to initiate action, and it is not even *all* catastrophes which produce results but only those off the coasts (or, at least, threatening the perceived interests) of developed, politically responsive countries with active roles in the agency. Many large spills have had minimal impact on IMCO's activities.[17] Moreover, the generally low-level, bureaucratic treatment of the problem is so entrenched that even incidents such as the *Torrey Canyon* off the British coast do not guarantee an effective, long-term national or institutional response. At the time of an incident, what is normally an incremental problem becomes an urgent concern capable of generating high political interest even, in this instance, from the British prime minister

14. Interviews.

15. The first Safety of Life at Sea Convention (1914) was drafted only after the sinking of the *Titanic*, and American efforts to get IMCO to revise the international rules relating to ship fire protection were successful only after the loss of the *Lakonia* (1963) and the *Yarmouth Castle* (1965). After these incidents—which killed many Americans—the United States through its roving ambassador, Averell Harriman, threatened IMCO with unilateral action unless something were done.

16. Richard Falk, *A Study of Future Worlds* (New York: Free Press, 1975), p. 45.

17. Huge spills off Chile (*Metula*) and Spain (*Urquiola*) have had almost no noticeable effect in IMCO in comparison with spills that mobilized "initiating" actors such as the United Kingdom and the United States.

himself. But without some additional internal pressure, the interest soon fades, and ensuing IMCO negotiations are returned to the bureaucratic level where narrower interests prevail. Moreover, while the catastrophe does have a general mobilizing effect on states' treatment of pollution, it is often unrelated to the most serious pollution sources and does not focus political attention on them.

A second and even more important fact about the nature of the oil pollution problem is that it is but a minor, undesirable side effect of the global economic system. Economic development, technological expansion, and rising standards of living all have had an obvious and central role in the development of the environmental crisis.[18] More specifically, the world's great and growing dependence on oil and its recently developed capabilities for constructing giant supertankers are themselves direct products of the burgeoning world economy. This is the ultimate cause of the oil pollution problem, but short of the economy's uncontrollable contraction or even collapse, almost no amount of pollution will change it. Instead, pollution control must itself fit into the prevailing economic and political systems. It has been and will likely continue to be only a peripheral activity needed to contain the unwanted effects of economic activity.

As a result, alternatives for pollution control are limited by what the larger system will permit. At the international level, this system is one of great competition and interdependence—large oil and shipping companies operating vessels of many nationalities to and from most countries in the world through waters of all jurisdictions. Pollution results from the indispensible activities of large international industries, and it requires an industry-based technological solution to resolve it. This is the particular setting for the oil pollution issue, and it shapes the policies and influence of all negotiators, both states and industry alike.

Confronted with this highly competitive and interdependent global system, states seek to ensure only the acceptance of regulations in which they and their industries will be treated no less favorably than any others which are affected. Any additional costs are unwanted but most important is the desire of all participants to avoid being subjected to any "competitive disadvantage" or "unfair burdens." This attitude is the major obstacle to rapid progress. Indeed, given the vast variety of domestic and foreign actors in the shipping field, buck-passing (or rather, bill-passing) has been refined to an art in international negotiations.[19] The meaningless obligations

18. These processes are the first source of regime change that Keohane and Nye consider in *Power and Interdependence*. For a discussion of their "economic process model," see pp. 35-42, 129-132. What is of importance is the extent to which "economic processes" are brought under conscious political control when their operation creates specific undesirable consequences such as oil pollution.

19. The general scale of preferences as to who should assume the costs has been (1) other

that have inevitably emerged from the proposals to require the installation of reception facilities is a telling example of the obstinacy of this attitude.

The benefits from this situation accrue to industry. International business has always opposed restrictions, and responsibility for oil pollution is simply the latest in a long history.[20] What is significant in this situation, however, are the opportunities for influence which it also confers. Although relied on by many states, the oil, shipping, and insurance corporations are concentrated in a few. These latter states are, of course, protective of their own competitive interests. In order to impose new environmental controls on them other states must themselves either reach an international agreement, act alone, create an alternative, or acquiesce. Considering the complexity and diversity of the industries' operations across numerous boundaries, unilateral action that would create a variety of national standards is anathema, and any such action by all but the most powerful and self-sufficient states would undoubtedly fail. For maritime interests, "uniformity" has been the battle cry in a struggle to retain free access to all the world's oceans and to prevent the imposition of uneven costs. This argument has much validity given the nature of international shipping, and it has had a pervasive impact on the alternatives available to any state. Unilateralism is seldom a feasible option. For a state dissatisfied with international negotiations, another choice is to create an alternative, to go to the extreme of setting up its own commercial enterprise. This is a costly and little-used strategy.[21] International agreement or acquiescence are therefore, in all but the rarest cases, the only options available.

Not only must the solution to the oil pollution problem be acceptable to the prevailing international economic structure, but it must also be implemented through technological developments based in the industries themselves. Not surprisingly, as the industries control much of the information

states or industries therein, (2) an industry that is comparatively weaker than an alternative industry in one's own state, (3) any industry, (4) local governments or port authorities, and (5) the central government.

20. The attempt to impose third-party environmental costs on shipowners in the 1960s and 1970s is strikingly similar to the struggle, fifty years earlier, to impose "second-party" costs on them. Until 1920, the shipowners bore no liability for damage to the goods they carried for a cargo-owner. It was only unilateral legislation by the United States, Canada, and Australia that prompted the shipowners to agree to the Hague Rules (1924) imposing liability for such damage on them.

21. For example, in rejecting demands for much-increased limits of liability, the British insurance industry did not expect disgruntled states to attempt to circumvent it. Its expectations were proven correct in one case, when the Canadian government passed its Arctic Waters Pollution Prevention Act (1970). Although calling for higher limits in the act, the government—on the advice of the London insurers—eventually restricted the liability in the act's regulations to the agreed international level. The Americans, however, did set up their own alternative enterprise, the Water Quality Insurance Syndicate, to provide insurance for their new Water Quality Improvement Act (1970). But even then the limits of liability were well within those acceptable to the London insurers.

and the technology necessary for pollution control, they retain a tremendous influence on the eventual outcomes.[22] Unless states are willing to require new technologies in conventions prior to their actual development (a very rare occurrence), they are limited by what industry will provide. The entire progress of technical standards since 1954 reflects the constraints imposed by a dependence on technologies which have been developed and made public by the shipping and oil industries. The 1954 and 1962 discharge regulations for nontankers were, in effect, emasculated because the necessary technologies were supposedly unavailable. Meanwhile, the industry kept its own ''load-on-top'' system for tankers under wraps until it—and not governments or IMCO—decided to unveil it. This was also to an extent the case with crude-oil-washing, a system which had been considered as early as 1967 but was rejected as ''uneconomical.'' Only when its use became profitable after the OPEC price rise was the system touted for its environmental advantages. Even then the oil industry supported it as a mandatory requirement only as a way to rebut the more expensive proposal for the retrofitting of segregated ballast tanks.[23] To this day, no effective monitoring system for white oils has been developed or, at least, unveiled. Technical problems always arise in developing new controls, but the greatest impediment is the unwillingness of industry to act except under political pressure and, even then, only to the degree that is ''necessary.''

The oil industry is not alone. One of the most dramatic examples of the near-veto power held by industry was the 1969 Civil Liability Conference. For weeks, the insurers adamantly rejected the proposed new liability regime as ''uninsurable.'' Their real concern was to ensure that they maintained control over any expansion in coverage, prompting one oil industry official to allege that the insurers ''intentionally misled'' the British delegation and the conference.[24] The P & I insurers could certainly have apportioned the increased liability among themselves, and as it turned out, the reinsurers were also quite well able to cover the proposed risk (at least twice over!). At the time, however, the facts were just not available to government negotiators to counter the arguments of the industry. Conference

22. Perhaps the basic point of Robert Engler's recent book, *The Brotherhood of Oil* (Chicago: University of Chicago Press, 1977), is the power the oil industry draws from its manipulation of information. Knowledge of oil reserves, the state of the technology, and so on is concentrated in the industry and not in the government. Engler describes the U.S. administration as ''economically illiterate, factually empty-handed, and readily cowed by industry's ideological rhetoric and political influence'' (p. 36). See also pp. 6–7, 158–159, 184, 189. See also John M. Blair, *The Control of Oil* (New York: Pantheon, 1976).

23. The American government was well aware of this fact, which was one reason for its reluctance to be pressured into accepting it. As Brock Adams, the Secretary of Transportation, commented at an OCIMF symposium held in Washington just prior to the February 1978 Conference, COW was ''a technique that was developed for its commercial advantages, not its environmental benefits'' (*World Trade*, January 18, 1978, p. 31).

24. Interviews. For a more detailed explanation of this, see Appendix 2.

delegates did attack them as "specious" but stabbed blindly in the dark for an alternative. In the end, industry has provided the much-needed changes, but they have done so reluctantly and only after great pressure.

In conclusion, the nature of the problem itself has inhibited rapid change. Because of the incremental character of important facets of oil pollution and the scientific and technological base required for its regulation, legislative change has been the prerogative of specialized bureaucratic actors and particular economic interests. Only catastrophes have been able to generate the political momentum to break this pattern—and even then politicization has soon receded. In addition, the international and largely nongovernmental character of the shipping business has placed even further restraints on IMCO's members, who are themselves so highly dependent on the very industries they are supposed to regulate. Unilateralism has been rare. Governments have had to operate within the strictures imposed by the information and technologies developed and offered by industry. Combined with other factors, the very nature of the international oil pollution problem has tended to make IMCO's environmental negotiations slow moving and reactive.

CHARACTERISTICS OF STATES

The ultimate decision makers within IMCO, as in other international organizations, are the member states themselves. As we have seen, the context for their work is affected to a great extent by the industries they are trying to regulate, but states do have the final regulatory authority should they wish to exercise it. Representatives of industry and of nongovernmental organizations do participate directly in the negotiations, but at this level their impact is felt through their ability to influence the decisions of governmental representatives. Likewise, in the "intergovernmental" consultations, IMCO Secretariat officials have distinctly limited roles restricted largely to facilitating interstate decisions. The decisions of the states are themselves a product of a variety of factors, and no one characteristic determines a state's policies or influence on even such a limited problem as oil pollution control.[25] Instead, the pattern of decisions reflects (in the words of Secretary-General Srivastava) such a "shuffle of combinations"[26] that it is impossible to predict which characteristics will lead to the adoption of particular policies. At the same time, important patterns can be identified which run throughout the history of the negotiations.

25. *Policies* refer to an actor's preferences, while *influence* can be viewed as the ability either to alter another actor's policies or, simply, to achieve desired outcomes. Our understanding of both was derived from an analysis of papers submitted to and statements made in international negotiations, recorded votes, and interviews.

26. Interviews.

Two such important patterns stand out at IMCO: one is based on specific maritime state/nonmaritime (coastal) state groupings, and the other is based on membership in the larger political groupings of "West European and other" (WEO), Soviet bloc, and developing states. These two patterns, in part, reflect the distinction Cox and Jacobson have made between the "general" and "specific" political environments. Characteristics relevant to the "specific" environment are those that are connected directly or are *intrinsic* to the specific issue, in this case the control of vessel-source oil pollution. Characteristics relevant to the "general" environment bear no necessary relation, or are *extrinsic*, to the specific issue of marine pollution. The value of pursuing this analytical tack is that it affords one an understanding of how and why decision making on particular international issues is linked to the broader structure of international politics. Moreover, it points to how states draw on resources apparently unconnected with an issue in order to affect negotiating outcomes. As Cox and Jacobson have noted in reference to the "general" environment, "Using the threefold alignment classification should help to show the extent to which decisions of different kinds in international organizations reflect the major political cleavages affecting the world."[27] Combined, the "specific" and "general" alignments provide an excellent broad-brush picture of the sources of environmental maritime law.

West European and Other (WEO)

IMCO often has been dismissed as an intransigent "shipowner's club." It would, however, be more precise to view it as a "*Western* shipowner's club," so nearly perfect is the overlap. Indeed, at IMCO's inception, over 75 percent of world tanker registry was in the developed, WEO group. Today, excluding Liberia, the figure is approximately 60 percent. However, since virtually all of Liberia's flag-of-convenience tonnage is beneficially owned by firms in developed Western states, the latter figure is misleadingly low. In addition to this concentration of maritime states in the developed bloc, it is noteworthy that IMCO's major *non*maritime actors are also from this grouping. In fact, in both its active membership and focus of activities, IMCO is very much dominated by the OECD community of developed, capitalist economies. Within this constellation of developed states, three basic divisions can be identified: the developed maritime states; the developed nonmaritime (coastal) states; and, because of its unique situation, the United States.

Developed Maritime. Three major commercial groups are affected by IMCO's environmental regulations: shipowners, ship users, and ship

27. Cox and Jacobson, *Anatomy of Influence*, p. 32.

insurers. One other interest, that of the shipbuilder, is also affected, but in such a limited way that it will not be dealt with separately. Only on the issue of retrofitting segregated ballast tanks have shipbuilders exerted an influence, and even then their impact was slight.[28] The perspective of the shipowner can be understood by concentrating on *tanker* ownership. It is tanker pollution that has been by far the more significant environmental issue at IMCO, but in any event the distribution of tanker and nontanker ownership is closely correlated (see Table 4 and Appendix 1, Table A).

The usual criterion for a state's interest as a "shipowner" is the volume of its registered tonnage.[29] At the same time, *actual ownership* by a state's nationals of vessels registered in foreign countries (especially flag-of-convenience states) is an important concern for some, particularly Greece and the United States. But its effects on their marine pollution policies are quite varied and indirect. For Greece and other maritime states, the existence of the convenience states exaggerates their maritime orientation, making them more receptive to the economic arguments of their own domestically registered shipowners. After all, governmental support for higher and more costly standards would be an indication of insensitivity to the economic interests of their domestic fleet, and this could encourage them to transfer their vessels to convenience flags. Unlike the United States, these countries have a large stake in their registered fleets which they do not want to lose.

The great amount of American vessel ownership has had a very different effect on American environmental policy at IMCO than it has had on that of other shipowning states. Indeed, to the extent that it has affected policy, ship ownership has seemed to encourage American environmentalism. This is a result of the special relationship of the United States to the flags of convenience. They are, in fact, largely an American creation, and the United States is heavily dependent on them. As a result, to the extent the environmental issue is a visible black mark associated with their use, the government is eager to remove it.

There are striking differences in the distribution of shipping capabilities (as indicated by registry in Table 4). One-half of the world's tanker ton-

28. The impact of the shipbuilding interest is quite limited for a number of reasons. Few issues concerning ship standards are of significance to it—tank-size limitation, segregated ballast tanks, and double bottoms being the three exceptions. No liability (Chapter 5) or jurisdictional/enforcement issues (Chapter 6) affect the shipbuilder directly. Moreover, shipbuilders are concentrated in a few countries—Japan, Sweden, and other European states—which already have large shipowner and ship-user interests. The unique impact of the construction interests is, therefore, extremely difficult to discern. See, however, W. R. Stanley and J. M. Goicoechea, "New Dimensions in World Shipbuilding, Shipbuilding Prices and Consumer Cost," *Report of the Sixteenth Annual Tanker Conference, May 10–12, 1971* (Washington, D.C.: American Petroleum Institute, 1971), p. 216.

29. This is the criterion used in the IMCO Convention and found, in the advisory opinion of the International Court of Justice, to be the legally acceptable one. For more on this case, see Chapter 3.

nage is concentrated in half a dozen states, whereas the great "majority of IMCO members do not possess substantial shipping resources. . . . For most of these countries, international efforts at regulation or coordination of maritime affairs is not of overriding national importance."[30] In fact, despite a 400 percent increase in tanker tonnage between 1955 and 1975, the concentration of shipping in the developed maritime states has remained remarkably stable. Great shifts have occurred among these states (for example, the decline of the U.S.[31] and the U.K. and the rise of Japan and Greece), but only Liberia (and it is really an American creation) has been able to make a significant incursion into the monopoly of the "Western European and Other" states.[32]

Fifteen states in Table 4 have consistently held a substantial shipowning interest (more than 1.00 percent of total world tanker tonnage). From this group of states and a few others aligned to them by political or economic ties comes the basic body of "maritime" states. Their policies are not necessarily to be seen as anti-environmental. What is significant, however, is their special sensitivity to the implications for their maritime industries of any environmental proposals and their strong preference for practical, least-cost options. In this light, four groups are discernible as a result of their decisions on the issues analyzed in the last three chapters. Two groups take positions consistently favoring the shipowner, the second of these two being states *not* from the top fifteen shipowners. A third group takes more varied positions, but still generally reflects its high vessel ownership. A fourth group has adopted positions seemingly unrelated to its shipping interests.

The first group can be termed the "core" maritime states for, with Liberia, they control by far the largest proportion of the world's tanker fleet and have provided the backbone of the maritime group at IMCO. These states—Norway, the United Kingdom, Japan, and Greece—have, in the last twenty-five years, controlled between 48 percent and 65 percent of world tanker tonnage. Norway and the U.K. have always been among the

30. Harvey B. Silverstein, *Superships and Nation-States: The Transnational Politics of the Intergovernmental Maritime Consultative Organization* (Boulder, Colo.: Westview, 1978), p. 29.

31. It is interesting to note that the drastic fall in the American share of registered world tanker tonnage in the last two decades has been mirrored by an equally radical shift in its international marine pollution policies.

32. While the decline in the total tanker fleet of the developed countries has been from 76.29 percent of the world total in 1955 to 59.58 percent in 1975, flag-of-convenience states have increased their proportion from 16.97 percent to 31.74 percent in the same period. This mirror-image effect is almost totally attributable to the rise of Liberia. Over that same period, the developing countries, *not* including the flag-of-convenience states, have shown no noticeable increase in their share of the world total. The Soviet bloc has shown a slight increase in this period, from 0.80 percent to 2.47 percent of the world total.

top four shipowners in that period, while Japan and Greece have steadily increased their tanker tonnage (and were, in any event, significant owners of dry-cargo vessels throughout). Except for Greece's failure to be elected to the 1977–1979 Council, all these states have also been represented continuously on both the IMCO Council and Maritime Safety Committee. Except for the aberration of the United Kingdom (particularly in the 1950s and early 1960s), all have been consistently conservative environmental actors.

The great responsiveness of these states to their maritime interests can be understood in terms not only of high *sensitivity* but of high *vulnerability* as well (to use Keohane and Nye's terminology). That is, not only are these states sensitive to "costly effects imposed from outside," but they are vulnerable in that "effective adjustments" away from their reliance on this industry cannot be made easily.[33] For each, shipping confers essential economic benefits which cannot easily be forsaken, although the nature of their respective dependencies varies. The United Kingdom has a long and clearly held maritime tradition both as a shipowner and insurer, and the island relies heavily on foreign trade to supply its sensitive and troubled economy.[34] Japan, the initiator of the "supertanker" boom, *must* run a large tanker fleet in order to maintain its extended lifeline and service its pounding industrial machine. For Greece, its otherwise depressed economy profits greatly from an industry which supplies much-needed foreign exchange.[35] Norway, a small country of some four million inhabitants, has also relied heavily on this one industry. The country prides itself on operating the best independently owned tankers with access to the "responsible" charter markets. With such a large fleet and a small domestic market for its ships (it is a "cross-trader"), Norway has been very sensitive to profit levels

33. Keohane and Nye, *Power and Interdependence*, pp. 11–19. The aberration of the U.K. in the 1950s and early 1960s can be explained by the intense domestic pressure it was under at that time and its consequent desire to seek an international agreement to share in the higher standards which were inevitable domestically.

34. Recent figures show that Britain's direct foreign exchange earnings from its fleet were £789 million in 1974, with investment in shipping in that year amounting to 18.6 percent of all British manufacturing investment. A. D. Couper, "Shipping Policies of the EEC," *Maritime Policy and Management* 4 (1977), p. 136. Its importance can be appreciated from the following news report appearing during the February 1978 conference. It was reported that the British economy (which had seemed to be making a recovery) was undergoing a downturn. The *Guardian* reported that for the first time in five months there was in January "a visible trade deficit of £324 millions, offset by a surplus on invisibles, such as shipping and insurance, of £145 millions" ("Import Rise Breaks Run of Success," February 15, 1978, p. 1).

35. Greece, with its renowned fast-money operators like Onassis and Niarchos, concentrates on providing shipping where cost differentials are large. It is, therefore, extremely protective of the competitiveness of its fleet.

and threats to its competitive advantage. Of all of these states, however, Norway's vulnerability will decrease most in future years, as its relatively small economy is increasingly buoyed by its vast new reservoirs of North Sea oil.[36] These varied national situations occasionally have led to some sharp differences within the group on particular issues. In general, however, these four states and Liberia constitute the bulwark of the international maritime position.[37]

The policies of the remaining "maritime" states have been much more differentiated and, often, *without* regard to the size of their tonnage. Indeed, the second "adjunct" group of conservative maritime states all have small and, in two cases, negligible fleets. The identifying feature of these states—Denmark, Finland, Belgium, and Switzerland—is their political linkage to the first group. Denmark and Finland are heavily influenced by their Scandinavian alliance with Norway, although Denmark has a fair-sized fleet, especially considering its small population.[38] In comparison, Belgium and Switzerland have been influenced not by the interests of their small shipping tonnage but by their transnational allegiance to the industry-based Comité Maritime International (CMI) and, in Belgium's case, by its shipping interests' close ties with the U.K. This transnational allegiance is heavily dependent on the impact of specific individuals or institutions concerned with particular issues and, with a change in any of these, policies may change.[39]

The third group of maritime states includes the middle-tonnage European states—France,[40] Germany, Italy, the Netherlands, and Sweden. These states (and especially the Netherlands, which has important maritime links to the United Kingdom and a significant oil industry investment) usually have followed the "maritime" line but, unlike the first group, their smaller maritime commitment has given great flexibility to other interests and to the initiative of their individual delegates. This pattern was perhaps most obvious at the 1969 Civil Liability Conference, but it has manifested

36. Moreover, with this extensive coastal exploitation going on, the wealthier citizenry of the country will become increasingly conscious of—and vociferous about—the quality of the environment.

37. For many of these states, there is also a substantially higher tonnage per capita. See Couper, "Shipping Policies," pp. 130–131.

38. Other Scandinavian delegates did comment a number of times that Denmark had a strong shipowner lobby and a very conservative bureaucracy.

39. This transnational allegiance has been manifested most clearly in the area of liability and compensation, where the CMI was heavily involved. Belgium also has other transnational links with British shipping and its own interests as a large European port. Switzerland, however, has been actively involved only in issues concerning the Legal Committee and the CMI.

40. French policy has been becoming more firmly maritime in recent years, as a conscious decision has been made to expand the national fleet. Protective domestic legislation mitigates even the impact of this, however.

itself on numerous other occasions as well.[41] With their diversified economies they have necessarily been more responsive to a variety of influences without being as constrained by their shipowner interests.[42]

Apart from the flag-of-convenience states and those few developing countries with smaller fleets but a generally maritime orientation, these three groups constitute the basic maritime bloc. Of the four remaining states among the world's top fifteen shipowners, two—the USSR and Liberia—are not of the WEO group while two others—Spain and the United States—do not fit the maritime pattern. The United States is, as we have said, a special case and will be discussed later. Spain, however, is also something of a unique phenomenon, as it does have a sizable international fleet and yet has consistently supported environmental initiatives. Its policies can be explained by the fact that its ships are largely used to carry domestic imports and so have less concern for competitive advantage. Even more important is that Spain is one of the few countries with overriding environmental interests in this issue. The country has a truly enormous investment in its coastal tourist industry (on both the Atlantic and Mediterranean), and it has not really been linked to nor participated in the European maritime club.

Two other commercial interests are of great importance in affecting the policies of this maritime group: oil and insurance. Both in its role as shipowner and as the cargo-owner, the oil industry, in particular, has had a great impact on the attitudes of many states toward higher international environmental standards. Both the oil industry and independent shipowners oppose the higher cost of improved environmental ship standards, but there are important differences.

Shipping is a cutthroat business; competition is fierce. For the "independents," shipping is their total business, and any costs that could threaten competitive advantage or even just cut into the margin of profit are to be opposed. For the oil companies, these are not major concerns. All of their tankers are on long-term "bareboat" or "time charters," and in a depressed market they are the last to be laid up.[43] As they are the charterers of that 30 percent of the world tanker fleet which they own, their fleet is well used and not so affected by competitive advantage. Additionally, while they

41. At that conference, this group supported strict shipowner liability (except Sweden which, as a result of its Scandinavian alliance, supported cargo liability), while the first group opposed any change in the traditional fault regime. The support of Germany, France, and the Netherlands was particularly important in achieving a change in the law.

42. These characterizations of individual state policies are only approximate. Not only does policy often vary with the particular conference or issue but also with the individual delegate. In addition, the perceptions of observers vary widely.

43. About 30 percent of the market is composed of oil company vessels on bareboat charter, 55 percent of oil company and independent vessels on time charters, and 15 percent of independents on single-voyage or "spot" charters.

do oppose great cost increases or any proposals that could interfere with the efficiency or viability of their operations, shipping is but one small aspect of a much larger "fully integrated structure"[44] and environmental costs can be more easily assimilated. Finally, and very importantly, it is the oil industry which is identified with the oil—they charter the ships, import the oil, and sell the product directly to the public. Not only are they willing to spend vast sums directly on public image advertising,[45] but they will more readily divert some resources toward actual pollution prevention. LOT, COW, TOVALOP, and CRISTAL all emanated from the oil industry and were "sold" to the independents.[46]

This difference has affected the policies of states with a high proportion of oil-company-owned tankers. The size of oil-company ownership is difficult to estimate given the widespread practice of incorporating shadow shipping ventures that seem to be independent. However (by Table E, Appendix 1), it is clear that in at least some countries oil-company-owned tonnage is greater than that owned by the independents (real or otherwise). Interestingly, this pattern of ownership is found in the flexible "middle-tonnage" group of European maritime states just discussed,[47] as well as in the United Kingdom.[48] The effect of this structure of domestic ownership has been to make these states more highly sensitive to the specific goals of the oil industry and less to the cost concerns of the shipowners per se. Generally, it reinforces the moderation of their environmental policy. In contrast, the most conservative maritime powers at IMCO—Norway, Greece, Japan, and to a lesser extent Liberia—seem to have very small percentages of their tanker fleets owned by the oil companies.

44. J. K. Galbraith, *Economics and the Public Purpose* (Boston: Signet, 1973), p. 119. Engler discusses how this structure also allows costs to be distributed throughout the structure in order to minimize them (*Brotherhood of Oil*, pp. 21–22).

45. Exxon, Texaco, Gulf, Mobil, and Standard (Indiana) spent $425 million on public image (largely "informational") advertising in the United States between 1970 and 1972. Engler, *Brotherhood of Oil*, pp. 175–176.

46. Even these were ultimately intended both as cost-saving and public image devices. However, as Noel Mostert has observed, "The independent shipowners, regardless of whether they are under flags of convenience or not, have through their repeated attempts to prevent or delay or misrepresent stronger regulations for their ships, long since disqualified themselves for public sympathy. The tanker industry would be better off without them. The oil companies have indicated that they will be taking a larger share of the tanker market in the future. This, I believe, is an excellent trend, if it gets the independents out of the business. The oil companies, with an anxious eye to their public image, are more careful about their ships." "Supertankers and the Law of the Sea," *Sierra Club Bulletin* 61 (June 1976).

47. Sweden is the one exception to this pattern in the middle group.

48. The United Kingdom is truly the apotheosis of the maritime state and is responsible to a large number of other interests—including the insurers and the London-based International Chamber of Shipping (largely an organization of "independents")—as well as the oil companies. However, OCIMF did affect a switch in the British (and Dutch) positions in 1971 after pointing out the concentration of oil-company owned shipping in their domestic fleets. Their policies on segregated ballast in 1973 were products of this orientation. In 1978, the U.K. acted virtually as spokesman for the oil industry.

One logical source of environmental concern within the WEO group at IMCO would be from the large *oil importers*. It is, after all, these states which risk accidents such as the *Torrey Canyon* and *Argo Merchant* off their shores (both sank near their ports of destination). Unfortunately this has not generally been the case.[49] For one thing (as Table C in Appendix 1 reveals), the large oil-importing states are also the large shipowning states, so the potential environmental threat posed to an oil importer is more than balanced by the immediate concerns of its shipowning interests. In any event, if actual pollution has had a limited impact on the policies of these states, the threat of pollution has had even less. Only for a few developed nonmaritime states (such as Canada and the United States) has this concern had a major effect.

Traditionally, therefore, *cargo-owning* interests have been little concerned with the environment except insofar as it touched on their *shipowning* interests. Certainly, the oil industry was involved in the negotiations at IMCO in the 1960s but as ship-, not cargo-owners. So minor were the costs to cargo-owners that, until 1969, they were content to leave the lobbying at IMCO to the shipowners' association, the International Chamber of Shipping.[50] In recent times, however, this has changed, although even today it is only the occasional issue that has involved the cargo-owner per se. This change occurred first in connection with the negotiations over the 1971 Fund Convention when the issue was the apportionment of costs between the oil and shipping interests. It was these negotiations that gave rise to the oil industry lobby, OCIMF, but the issue was exceptional; only three others—tank-size limitation, reception facilities, and retrofitting segregated ballast tanks—have had cost concerns of specific interest to cargo-owners.[51] More commonly, however, it is these interests that are most sensitive to the rise in public pressure, and as a result they have participated eagerly at IMCO in recent years. While it is as shipowners that they find their greatest interest at IMCO, it is as cargo-owners that they are vulnerable at home. This situation is now an important source of change in IMCO negotiations.

The complex workings of the *marine insurance industry* have, in contrast,

49. For example, the policies advanced on the Intervention Convention certainly did not reflect a state's susceptibility to potential pollution as a result of its high oil imports. In fact, policies were quite the contrary (see Chart 8 in Chapter 5 and Table C in Appendix 1).

50. Only at the 1978 Conference has the effect of new regulations on the cost of oil been an issue. After all, the environmental costs are but a fraction of the shipping costs, and the shipping costs are a small portion of the final cost of the product to the consumer. It is the protection of the industry itself that has always been the concern, and it has been the shipowner and not the cargo-owner who has been threatened.

51. With the tank-size limitation, France and Japan were concerned about its prospects for increasing their oil-importing capabilities from ULCCs. Concerning reception facilities for tankers, it is the oil exporter that would logically have to supply these, and these states have not attended IMCO meetings. On retrofitting segregated ballast tanks, see Chapter 4 for discussion of the 1978 TSPP Conference.

always been of great consequence to pollution control. In its longstanding function of underwriting "risk," the industry has naturally discriminated (if only to a very limited extent) against the more unreliable operators and vessels, acting as a commercial inducement to higher shipping standards. This has had obvious indirect benefits particularly for accident prevention. A detailed examination of the operation of this complex industry is well beyond the scope of this study, except insofar as it has affected one important issue discussed earlier, liability and compensation. (For this, a special analysis of some aspects of insurance is provided in Appendix 2.)

The industry has affected state policies in three principal ways. First, maritime states that opposed increased costs to their shipowners from pollution liability were able to rely on the insurers' arguments that the proposed high limits were uninsurable. This, of course, entrenched their opposition, as they feared not being able to get coverage or, at least, having to pay an exorbitant price for it. Second, those states that demanded higher liability were confronted with claims of the insurance experts that they were visibly incapable of rebutting. Without the commitment to force their limits on an unwilling industry or to create their own insurance scheme,[52] an alternative was necessary—the International Fund was, to a great extent, a result of this dilemma.

The third effect of the insurance industry has been on the policies of the two states with a direct interest in its welfare. Norway, as the center for three Protection and Indemnity (P & I) Clubs, is even more sensitive than usual to the interests of its shipowners when the issue also involves the insurers. But in 1969 and 1971, it was the United Kingdom with its nine P & I Clubs (the "London Group") and its huge underwriting establishment (Lloyd's) that was the omniscient spokesman for the industry. With the huge foreign exchange earnings the business provides, it is not an industry to be taken lightly.[53]

Given the import of these industrial interests, it is not surprising that serious coastal pollution has rarely prompted these states to support higher and more costly ship standards. This was evident as far back as 1954, when a number of European countries who had reported serious pollution along their coastlines opposed the British proposals. The pattern was repeated again when the U.K. fought higher standards for liability and compensation despite the impact of the *Torrey Canyon*. And today it is obvious in the

52. Some did create alternatives. The oil industry had earlier set up the International Tanker Indemnity Association (ITIA) to provide the additional coverage for TOVALOP. The United States later set up the Water Quality Insurance Syndicate to provide insurance under its Water Quality Improvement Act (1970). As the shipowning interests (especially the independents) were firmly behind the insurers, the Fund grew out of the intransigence of both.

53. The annual income through premiums earned by British insurers is somewhere around the £600,000,000 mark. Interviews.

continuing conservatism of many maritime states and particularly of such states as Japan, Greece, and the U.K., all of which suffer serious chronic pollution. When states have had important conflicting industrial interests, these have almost always dictated national policy. The response has been to avoid the *direct* costs of pollution control and suffer the *indirect* costs of a polluted environment.[54]

It is not that these governments or their industries oppose higher environmental standards on principle. It is simply that environmental costs have not traditionally been an internal cost of the economic process—in the terminology of the economist, they have been a "technological externality," a social cost. To internalize these costs is an expensive and unsettling task, especially at the competitive international level where one state's loss is another state's gain. As a consequence, only domestic pressure from within the state has motivated governments—and maritime governments in particular—to advocate costly environmental initiatives: "almost nowhere has the movement on behalf of the human environment originated within governments. Particularly in the more highly developed nations, concern for the environment has been thrust upon often reluctant governments by popular demand."[55]

Public political pressure is the counterbalance to economic interest. Such pressure was at the root of British environmental policy in 1954 and 1962, and it is certainly the source of recent American environmentalism. Unfortunately, the structure of most domestic policy-making does not encourage the acceptance of environmental values. In the maritime states, policy is formulated by the relevant maritime ministry with little input from environmental agencies. This has certainly been the case with the United Kingdom, France, the Netherlands, and Greece.[56] Moreover, individual politicians have been very restricted in their ability to force environmental values on a reluctant government. The strict party line of the parliamentary system has had this affect in the United Kingdom (and Canada), and there are additional impediments in other countries.[57] The ability of the United States to avoid these problems has to a great extent fostered its environmentalism (see further on). Once environmental demands are felt, the concern for competitive advantage can have a positive environmental impact. As

54. In order to achieve the least-cost alternative, even where some measures are taken, the strategy has long been to shift the *locus* of the pollution (through *zonal* prohibitions, regional accords, or "special areas") but not to stop it.

55. Caldwell, *In Defense of Earth*, p. 3.

56. Unfortunately a detailed analysis of all the key states was not possible. Sufficient interviews were conducted on the states listed to support this conclusion.

57. In France, for example, by article 40 of its constitution, no private bill can have adverse national financial consequences. This was cited to the authors as having a substantial dampening impact on the influence of the Assembly on governmental policy after the *Olympic Bravery* grounded off the Brittany coast. Interview.

the government finds itself under strong compulsion at home to set higher domestic standards, it naturally seeks to impose them as well on the world industry. This has been the story of environmental politics at IMCO.

The determinants of the policies of these WEO maritime states are different from the sources of their influence. Influence is the ability of one actor either to alter another's behavior or simply to achieve desired outcomes. Influence is not a rigid thing: there are many levers of power but their efficacy varies with the issue, the situation, and time.

In the early days IMCO was truly a "shipowner's club." As Keohane and Nye would put it, power in the organization could be explained almost totally in terms of the "issue structure."[58] The countries with the relevant commercial interests prevailed significantly because marine pollution was not a highly politicized question and the large majority of the world's non-maritime states either did not belong to IMCO or, if they did, paid little attention to it. At the 1954 Conference and again in 1962, those fifteen states with the greatest share of world tanker tonnage were dominant. Indeed, with the Soviet bloc, they constituted the majority of the total IMCO membership until the mid-1960s.

Throughout this period, the structure of influence was that of a highly dominant coalition of the maritime "regulars" who, for the most part, wanted few changes in the status quo.[59] Certainly there were differences within the maritime group (after all, the United Kingdom did put forward technical proposals that were rejected). However, policies were, by and large, determined by technocrats; negotiations were controlled by a small group of congenial neighbors; and no government was significantly concerned about the outcomes. Until 1965, the basic pattern was of a few like-minded governments operating at a *low bureaucratic level* through *functional ministries* sharing a similar *organizational ideology*.[60] What differences there were never involved higher levels of conflict—they were basically contained within the ministries responsible for shipping. States acted primarily as *unitary actors* with little or no pressure by public groups or the organization itself. Indeed, so low was the contribution of IMCO, that the secretary-general feared for its very survival. Only after the mid-1960s did things begin to change. At that time, industry was becoming directly and independently involved in pollution prevention. The developing countries were joining the organization in increasing numbers. The environment was

58. Keohane and Nye, *Power and Interdependence*, particularly pp. 49–54.

59. Cox and Jacobson, (*Anatomy of Influence*, p. 23), classify five rough patterns of organizational influence: "(1) unanimity, (2) one dominant coalition, possibly led by a dominant actor, (3) polarization between two rival coalitions, (4) a large number of alliances, none of which dominates, or (5) crosscutting cleavages on different issues with no general pattern." Until the late 1960s, IMCO basically mirrored the second pattern.

60. See ibid., pp. 402–405.

evolving into an issue of greater political prominence. And a new coastal coalition was beginning to emerge. Yet even with these strangers wandering into its halls, IMCO was still very much a club for the old boys. Its foundations had been well constructed and the house rules set. Within its walls the structure of influence would not change easily.

The influence of the traditional maritime states over the creation of international environmental regulations grows largely from two sources: their control of the provision of shipping services and their control of the IMCO organization. Each reinforces the other and would be much diminished without it.

The control by the West Europeans and Japan of such an enormous share of world tanker tonnage has been of no small consequence in ensuring them an important voice in marine environmental rule-making. Without their acquiescence, new rules would be but scraps of paper. As is so often heard at IMCO conferences, the nonratification by the maritime states of a new convention would be fatal to it. Indeed, to ensure that this has continued to be the case, the entry-into-force of the environmental conventions often has been made conditional on their acceptance by a substantial portion of the maritime states. The 1973 Convention, for example, must, by article 15 (*i*), be accepted by states representing 50 percent of the world's merchant shipping tonnage before it will become international law (see Chapter 8).

This power is, however, conditional on the cooperation of the maritime group *as a whole*, as a state's possession of considerable tanker tonnage is not itself a reliable source of influence outside the maritime bloc. Shipping states are very competitive with one another and, as we have seen, each has a particular mix of specific interests. With fragmentation of this group, recalcitrant states would find themselves either imposing a competitive disadvantage on themselves (as would have happened had the United Kingdom imposed its demands in 1954 and 1962 for a total prohibition onto its own ships) or cutting themselves off from large segments of the market (which could happen if one refused to comply with higher standards set by other states for vessels entering their ports). The influence provided by the control of tanker tonnage is an influence conferred on the group as a whole when faced with external challenges. Certainly, the possession of a large fleet does provide great prestige and a potential for influence, but as a source of influence for an individual state it is, alone, most uncertain.[61]

The availability of the IMCO forum as a place to execute and legitimize collaboration is, therefore, crucial. Indeed, the possession of high tonnage

61. The position of influence here should be contrasted to the great power Norway and the United Kingdom have over matters involving the insurance industry which is so heavily concentrated in these two states.

without access to the IMCO decision-making process severely constrains a state's influence. Panama and, until 1970, Liberia were ineffective actors in the environmental negotiations despite their great tonnage. Even today, Japan's influence is not proportional to its shipping capabilities because of language and cultural barriers.[62] It is the ability of the maritime states to participate effectively and in great numbers at IMCO that is the most important source of their influence on marine environmental regulations.

As indicated in Chapter 3, the developed Western states have been regular participants in all of IMCO's environmental committees and conferences. Both in frequency of attendance and in size of delegation, this largely maritime group surpasses the Soviet bloc and, of course, the developing countries. In the early conferences, they constituted the numerical majority (or very close to it) and, according to the IMCO constitution, they were by far the controlling majority on the agency's two most important policy-making bodies: the Council and the Maritime Safety Committee. Today, their numerical domination of the conferences has passed, and their formal command of the organization's control organs has disappeared.[63] But the high level of informal control has continued.

In practice, the developed maritime states frequently still constitute a majority in the regular committees of IMCO and, with their large delegations, they are far more effective than their developing, nonmaritime counterparts. Moreover, effective participation requires great expertise, a resource which these states possess in abundance. As Jack Plano has written, "The most basic dimension in any field of functional cooperation relates to the competence of decision-makers in comprehending the nature and scope of the technical problems to be solved."[64] Not only is the nature of the problem itself hyperdependent on scientific evidence and technological solutions, but as we have seen, technical arguments themselves are an instrument of bargaining. Eugene Skolnikoff has noted that "In organizations dealing with technological subjects, technical competence can be a major source of influence. Inevitably, the nation providing the most detailed studies and supporting data is likely to win most of the battles."[65]

62. Over the years this has also applied to the Italians. As they could often communicate only in French, they were frequently led into alliances with France. Despite the existence of translation services (English, French, Spanish, and Russian), the language barrier is still an important one in informal negotiations or working groups.

63. One of the American motivations in demanding that the new Marine Environment Protection Committee (MEPC) be open-ended was to overcome criticism by the developing countries that the organization was a "shipowner's club." The Maritime Safety Committee will soon become an open-ended committee as well. However, as Table 5 shows, without the interest or resources to participate, the developing countries are still vastly outnumbered in the open committees as well.

64. Jack Plano, *International Approaches to the Problems of Marine Pollution* (ISIO Monograph) (Brighton, U.K.: University of Sussex, 1972) p. 31.

65. Eugene Skolnikoff, *The International Imperatives of Technology* (Berkeley: Institute of International Studies, University of California, 1972), p. 120.

The concentration of technological expertise is not totally in the maritime bloc, especially with the defection of Canada and the United States in the late 1960s. But the overwhelming weight of it is there. As an "initiator"[66] in the environmental area must have a large technological research base, the United States, the maritime states, and the oil and shipping industries have controlled the portals of change. The great majority of states are an audience to whom the technologically powerful actors address themselves. The audience does not define the alternatives.

This maritime influence has manifested itself in another, more subtle way in that the entire "organizational ideology" of IMCO reflects the interests of this group. What is expected, indeed, permissible, for an organization is often an unstated, unconscious, unchallenged, and very powerful "ideology." With the Intergovernmental Maritime Consultative Organization, the very goals that were set from the beginning were those of the European shipowners. The agency was simply to provide this group with a *passive forum* for the *voluntary* coordination of policies on the noncommercial aspects of shipping (that is, it was to be an "intergovernmental" and "consultative" body). This was so much the case that, when two outsiders —Liberia and Panama—attempted to take their due places on the Council in 1959, they were rebuffed. It was also the policies of these states that prevented IMCO from ever taking up its economic functions. IMCO was not meant to challenge or control the maritime states. Instead it was only to assist its client states to resolve their problems *en famille*.

The combination of heavily European expert participation with a long-standing ideology of deference has produced an organization that has been greatly influenced by an "elite network" of individuals representative of conservative maritime interests. It is difficult to define exactly either the structure or importance of such a network, but its existence and impact at IMCO are undeniable. At the top, there is what might be called an "organizational elite" composed of those individuals who attend the meetings of the Council and Assembly and who help control the overall direction of the agency. Insofar as the network affects IMCO's environmental affairs, it involves a *legal* component (centered in the Legal Committee) and a *technical* component (centered in the MSC and MEPC). Individuals who are part of the elite group centered in the Council are also often members of the legal and especially the technical elites, but there is no rigid pattern on this, as the issues and interests differ at the various levels. On the legal and technical levels, new individuals may occupy places of great prominence because of their expertise in a particular area.

The elite networks at IMCO are necessarily based on maritime and not environmental expertise. Members of them tend to share certain professional and cultural traits. Personal expertise and diplomacy, an adherence

66. Cox and Jacobson, *Anatomy of Influence*, p. 13.

to the technical/commercial organizational ideology, and a long familiarity with either the private maritime or IMCO structures are prerequisites. Most of the individuals achieving such prominence have had a lifetime's work in the maritime field, and many have been important actors at IMCO for a decade or more. Virtually the entire elite is, as with other international organizations,[67] drawn from the wealthy WEO states or their industries. The developing countries are noticeably absent.

The existence of a powerful elite has both beneficial and detrimental effects. On the positive side, it facilitates the smooth, orderly flow of business. For example, the creation of the International Fund was greatly assisted by the informal efforts of a very few individuals who had established their prominence in the Legal Committee since it was created in 1967.[68] The chairmen of important IMCO or conference committees are usually drawn from the elite, and their ability to steer the debates is crucial to success. An often-mentioned source of some difficulty at the 1973 Conference on Marine Pollution was that the expected chairman of the plenary was rejected in favor of a symbolic but untried newcomer from a developing country. The rejected candidate, Lindencrona of Sweden, had been chairman of the subcommittee that had supervised the preparatory work. Had a similar fate befallen the chairman of the MSC, Dr. Spinelli, in his bid for the chairmanship of Committee II at the 1978 TSPP Conference, agreement would certainly have been less likely. That such an elite network exists is inevitable, and it does contribute to specific achievements.

At the same time, by its very nature the elite network reinforces the status quo. Drawn from those individuals who reflect the traditional biases of the established maritime community, the elite is not receptive to dramatic change. At the organizational level, many of them are, for example, more interested in ensuring that the new developing members do not "rock the boat" than in actively seeking new ways in which the organization might respond to their needs. Likewise, on the issue of the environment, the gradualist, maritime-oriented approach of the elite group reinforces the influence and legitimacy of the traditional maritime states and impedes the transition of nonmaritime power into effective influence.[69] As a whole, the elite continues to constitute a tight group of intimidating competence and authority which enhances the influence of the WEO nations in IMCO.

67. Ibid., p. 408.

68. The four persons identified to the authors were Sweden's representative, Ulf Nordenson, the Netherlands' van Rijn van Alkemande, Switzerland's Walter Muller, and OCIMF's Claude Walder.

69. The list of prominent individuals repeatedly mentioned in connection with IMCO's environmental negotiations (both legal and technical) bears this out. The group includes representatives of British, Norwegian, Swedish, French, Dutch, Italian, Belgian, and Greek nationalities as well as some from private industry.

Other sources of influence are available to states within this developed Western grouping. One of the most interesting derives from these states' political and organizational ties in other areas. The colonial influence of the United Kingdom and France has largely lost its former potency, although France does have, in the words of one delegate, "an inordinate amount of support from its former colonies," particularly from the Francophone African delegates.[70] On another level, the Western European states consult on environmental matters in the European Economic Community (EEC), but this has had minimal impact at IMCO. At the 1978 TSPP conference, the EEC observer was able to convene regular meetings of EEC countries, but to no real effect.[71] With Canada, Japan, and the United States, these states also discuss shipping policies within the OECD. Here, however, their concern is almost totally economic, and the focus is often on their shared interests in contrast to those of the UNCTAD "Group of 77." To date, neither the EEC nor the OECD has been very relevant to IMCO's work. This may change, however, especially as the developing nations show more interest in the organization.

At the same time, the regional interests of some European states have led them to conclude a number of local agreements in the North, Baltic, and Mediterranean seas and to coordinate their ratifications.[72] For one subgroup within this Western group, the Scandinavian states (Norway, Denmark, Sweden, Finland), their regional collaboration has had a profound impact on their influence at IMCO and on IMCO's environmental regulations. Unlike the larger group of European states, these four states have made a genuine effort to resolve differences among themselves. Indeed, before the 1973 Conference, the agreements among them overrode a specific national directive from the Swedish parliament.[73] This coordination has been institutionalized at a very high level with the creation of a Nordic

70. Interviews. This point was widely repeated.

71. On EEC shipping policies, see Couper, "Shipping Policies." The EEC has recently issued a study, *State of the Environment: First Report* (Luxembourg: EEC, 1977). Its concern for marine pollution focuses on land-based pollution, ocean-dumping, and the Law of the Sea Conference. There is scant mention of tankers. The EEC has become much more involved after the *Amoco Cadiz* incident. Regarding the OECD, its participation at the 1978 TSPP was a unique event motivated by the special economic circumstances of the time.

72. See the Agreement Concerning Cooperation in Dealing with Pollution of the North Sea by Oil, 1969 (initiated by F. R. Germany); the numerous special areas in the 1973 Convention; and the recent agreements through UNEP on the Mediterranean (see the Barcelona Protocol for the Prevention of Pollution of the Mediterranean Sea by Ships and Aircraft, 1975).

73. Officials interviewed commented that the Swedish Parliament passed a resolution in 1973 favoring a total ban on ship discharges. However, this environmentalism manifested itself in other forums such as the Seabed Committee and General Assembly. At IMCO, Sweden follows the lead of the dominant interests in the Scandinavian group—those of the independent shipowners.

Council and Committee of Ministers. On a lower level the functional ministries in each state participate in joint committees to draft legislation for all the states, and this has resulted in the conclusion of a Nordic Maritime Code and in a high degree of functional regional cooperation.[74] Internationally, their submissions to IMCO and to foreign governments are often identical and their negotiating positions well coordinated. Only at the 1978 TSPP Conference was a rift visible within this group, and only then as a result of the very special nature of their economic situations at the time. Generally, however, the preparation and execution of their policies is so well coordinated that their influence in the debates has been greatly augmented, especially through their ability to formulate successful compromise proposals.

Developed Nonmaritime. Apart from the members of the maritime coalition and the U.S., there are several WEO states without major shipping interests who have adopted a notably "coastal" orientation. These are the non-European states of Australia, Canada, and New Zealand. Without major involvement in the international shipping and oil industries and without close regional ties to the maritime powers, these states are free to concentrate on protecting their coastal interests. As with the developing states, however, even the environmental "commitment" of the developed coastal states is intimately connected with their larger economic or political objectives, particularly with the whole range of maritime jurisdictions.

Canada, despite its avowed environmentalism, reveals this attitude in its essence. That country's early radicalism at IMCO in 1969 and later at UNCLOS III was only partially motivated by a concern to protect the quality of its coastal environment and resources. By 1969, Canadian domestic environmental policy was beginning to develop, and Canadian representatives at IMCO were concerned at the slow pace of negotiations following the *Torrey Canyon* incident. However, like the more recent American challenges at IMCO, the impetus for Canadian policy came from internal public pressure, pressure motivated largely by jurisdictional and not environmental concerns. The basis for these concerns was the planned Arctic voyage of the American supertanker *Manhattan*, a voyage that aroused far greater fears of loss of sovereignty than of environmental destruction. A firm response was required from the government and a unilateral assertion of jurisdiction under the blanket of environmental protection was the perfect answer. With this unilateralism looming, the 1969

74. The agreements are numerous: the Convention on the Protection of the Marine Environment of the Baltic Sea (1974); Convention on the Protection of the Environment between Denmark, Finland, Norway, and Sweden; Agreement concerning Cooperation in Measures to Deal with Pollution of the Sea by Oil (1971); Nordic Environmental Protection Convention (1974).

IMCO Liability Conference offered an excellent opportunity to demonstrate the government's profound—if new-found—environmental commitment. This change in domestic Canadian policy was the early mainspring for the coastal revolt at IMCO. To achieve their objectives, control of policy was effectively shifted from the Ministry of Transport to the Department of External Affairs. As long as jurisdictional goals were at stake, control remained there, with Canadian environmentalism retaining its high pitch. With the resolution of most of the jurisdictional issues at UNCLOS III, Canadian interest has waned, although the delegation is still advocating an extension of coastal state powers. Not surprisingly, therefore, at the environmentally important but jurisdictionally insignificant TSPP Conference, control of the Canadian position reverted to the Ministry of Transport, with External Affairs only peripherally involved on minor legal issues. As a result, policy in 1978 was much more moderate (Canada opposed retrofitting segregated ballast tanks), although the new lower profile did have the unexpected side benefit of allowing the delegation to play a useful mediatory role. In summary, even for the coastal state, environmental values have their uses—and their limits.[75]

In contrast to the maritime group, the sources of influence available to the developed coastal states are meager. Despite their relatively high status within the global political system, these states are without important levers of influence when confronting their WEO brothers at IMCO. Certainly they have the expertise and resources to participate fully in the agency's meetings—and they have done so with some diplomatic skill. But their small fleets (especially those involved in international trade) combined with their substantial dependence on overseas trade renders them extremely vulnerable to pressure. In both the IMCO and UNCLOS forums their ambitions are dependent on the policies of other groupings—especially the developing countries and the United States.[76]

With the limited influence of the Canadians at IMCO, it is not surprising that their challenge to the maritime states in 1969 and thereafter was only moderately effective. It was, as will be seen, up to the Americans to produce the significant legal changes. With its threat of unilateral action and its "exaggerated offensive" Canada was, in 1969, able to tap successfully its one source of potential influence—an alliance with the developing

75. For a far more thorough exposition of this analysis, see R. Michael M'Gonigle and Mark W. Zacher, "Canadian Foreign Policy and the Control of Marine Pollution," in Barbara Johnson and Mark W. Zacher, eds., *Canadian Foreign Policy and the Law of the Sea* (Vancouver: University of British Columbia Press, 1977).

76. Canada has worked hard with the developing countries and the United States to achieve its objectives. This has been evident at the 1969 and 1973 IMCO conferences and at the UNCLOS III, where it sought to forge a diplomatic alliance with the developing countries.

countries.[77] Combined with the differences within the maritime group itself, this resulted in a successful compromise on the civil liability issue. But its attempts to impose even higher standards at home were not so successful. Its unilateral assertion of jurisdiction in the Arctic has, interestingly enough, largely achieved the government's aim of maintaining sovereign control vis-à-vis the United States, which has been relatively helpless to control the outcome on the *specific* issue. At the same time, unilateralism on shipping regulations does require a high degree of invulnerability which Canada, unlike the United States, does not possess.[78] Its dependence on foreign ships and insurers has forced it to implement a virtual emasculation of the contentious liability provisions of the Arctic Waters Pollution Prevention Act, just as later happened to similar legislation in the State of Florida.[79] Moreover, at UNCLOS III, the country's demands for broader coastal state powers were not very successful. The Arctic has been given a special status, but the coastal state's environmental jurisdiction has otherwise been restricted, not expanded.

The United States. The United States is a unique actor within the IMCO context. Until the late 1960s it could, by the policies it pursued and the size of its shipping tonnage, be classified as a maritime state. Since that time, it has been both the most radical of the environmental actors and among the most conservative. On many issues it has pursued the policies of a coastal state. Indeed, it has led the coastal movement to secure higher technical standards. Yet, all appearances to the contrary, it has continued to share many of the policies of its developed maritimes allies. These policies can be explained only partly in terms *intrinsic* to the IMCO environmental issue. To a great extent, they result from a factor *extrinsic* to purely commercial

77. As an outsider to the elite network, the government ostensibly chose this aggressive policy to put the maritime bloc on the defensive. Interviews.

78. Despite the greater political power of the United States, it was ineffective on the specific jurisdictional issue, to which the Canadians had a greater commitment. As a result, the unilateral passage of the Arctic Waters Pollution Prevention Act with its 100-mile pollution control zone has probably secured the waters of the Canadian Arctic archipelago as "Canadian" in the face of the perceived American challenge. On the environmental thrust of the legislation, however, the far greater power of the London insurers on this specific matter resulted in a two-year postponement of the implementation of the legislation. Even then, the act was proclaimed only with regulations that recognized practices acceptable to the London insurers. See M'Gonigle and Zacher, "Canadian Foreign Policy," pp. 117–121. On the operation of power in such situations as this, see Keohane and Nye, *Power and Interdependence*, especially their section on Canadian-American relations, pp. 165–218.

79. Despite Canada's criticism of the restrictions in the IMCO Liability Convention, the Arctic Water Pollution Prevention Act includes strict, not absolute, shipowner liability, and under the act's regulations the level of liability is equivalent to that in the IMCO Convention but with *additional* restrictions (see reg. SOR/DORS/72-253). Finally, when the act was proclaimed (after a two-year hiatus), the provision requiring evidence of financial responsibility was not put into effect. For a review of the brief life of Florida's Oil Spill and Pollution Act (1970), see "Florida Pours Oil on Troubled Waters," *Seatrade*, May 1974, p. 16.

maritime concerns, the desire of the United States to protect its global economic and political hegemony.

Perhaps the most basic aspect of American policy is the subservience of its shipping policy to the country's larger economic needs. This is most evident in the American role in the creation and maintenance of the much-criticized flags-of-convenience.[80] These flags arose after the Second World War when American shipping interests, with the—at least—tacit approval of the American government, sanctioned an arrangement which would retain American-*owned* vessels under "effective control" by the government in time of crisis while allowing the shipowners to *register* them in foreign countries—primarily Liberia and Panama—in order to avoid costly domestic construction, tax, and crew regulations. These regulations had been seen as potentially fatal to the American fleet, and the flags-of-convenience provided a way out. As the United States already possessed a strong economy, the profits that might have resulted from retaining a domestically registered fleet were insignificant in comparison with the larger benefits to be gained from the use and control of a larger but less expensive foreign-registered fleet. Even today Liberia and Panama are, with almost half their fleets owned by American interests, little more than "godsons of Uncle Sam."[81]

The United States must, in consequence, bear a large measure of responsibility for the environmental policies of these flag-of-convenience states. Clearly, therefore, there is a great similarity in its policies with those of Western European shipowning states. The environmental policies of both are heavily constrained by their economic goals. For the Europeans, their specific interest in maintaining a profitable, domestically registered fleet affects their policies at IMCO. For the United States, a profitable domestically registered fleet is not a specific interest that requires it to take a protective environmental stance at IMCO. At the same time, its general interest in maintaining a strong domestic economy has caused it to sanction an arrangement that has had profound environmental consequences.

80. For more information on the flags-of-convenience, see Boleslaw Adam Boczek, *Flags of Convenience: An International Legal Study* (Cambridge, Mass.: Harvard University Press, 1962); Erling D. Naess, *The Great PanLibHon Controversy: The Fighting over Flags of Shipping* (Epping, Essex: Gower Press, 1972); Robert Rich, "Flags of Convenience—or Necessity?" *Oceans* 10 (March-April 1977), pp. 22–27; and *New York Times*, February 13, 1977, p. 1; February 14, 1977, p. 14; and April 18, 1977, p. 51.

81. Statement cited in Robert Engler, *The Politics of Oil* (New York: Macmillan, 1961), p. 179. With many ships registered through dummy corporations, the figures are possibly even higher than reported. The Panamanian fleet, of which the United States owns 42 percent has been used to facilitate American oil trade in the Caribbean and South America. The United States owns 35 percent of the Liberian fleet (see footnote 93). As these two fleets together account for 97 percent of the total flag-of-convenience tanker fleet, American involvement in it is not small. *World Shipping under Flags of Convenience* (London: H. P. Drewery Shipping Consultants, July 1975).

This concern for protecting its larger global interests has brought the United States into alliance with the European maritime powers on many issues. Both have similar interests in the future of IMCO as a whole, and the United States is integrally involved in the organizational elite at the Council level.[82] On technical environmental standards, there is a significant divergence of interests and less collegial participation, but on jurisdictional matters their policies are very similar. To an even greater extent than for the Western Europeans and Japanese, American opposition to "coastal special measures" at the 1973 IMCO Conference and other coastal powers at UNCLOS III has flowed from the important objective of retaining freedom of navigation for military and commercial vessels. The retention of this freedom has been so strongly advocated by the departments of Defense, Transportation, State, and Commerce that it has become an inviolate principle in the American government's approach to ocean management.[83] Any legitimization of unilateral extensions of jurisdiction—despite its potential for dramatic impact on marine pollution control—is to be avoided. American support for port- rather than coastal-state enforcement also grew directly from this concern. Interestingly, this policy has been recently weakened by the adoption of legislation that implements significant coastal state authority.[84]

With these larger goals protected, the Americans are easily able to demand higher technical standards for all shipping. What shipping it does retain at home (still 3.00 percent of the world tanker fleet) is engaged largely in coastal trading and would not, therefore, suffer a severe competitive disadvantage by being exposed to higher domestic standards.[85]

82. The American representative to the IMCO Council is a retired Coast Guard officer, Rear Admiral R. Edwards. He is also a consultant to the American Institute of Merchant Shipping and was openly critical of U.S. environmental proposals in 1978. At the Council he has been a pillar of strength for the developed, maritime position. In 1977 he was elected to his fourth term as Council president.

83. "The basic goal of the U.S. pollution articles is to protect U.S. navigation interests by preventing coastal states from asserting and enforcing vessel pollution standards in their economic zones" ("The Third United Nations Law of the Sea Conference: Report to the U.S. Senate by Senators Pell, Muskie, Case and Stevens," February 5, 1975; cited in Nancy Ellen Abrams, "The Environmental Problem of the Oceans: An International Stepchild of National Egotism," *Environmental Affairs* 5 (1976), p. 28).

84. American support for port-state enforcement would have meant that its own maritime transport would not have been jeopardized by extended assertion of coastal authority by other states while still allowing the United States to protect its own shores by controlling the standards of ships coming to American ports. However, the passage of the Clean Water Act (33 U.S. Code S.1321) extends American jurisdiction over many aspects of marine pollution out to the edge of the 200-mile fisheries zone. Despite much pressure to have the offending provision (sec. 311) repealed, it now seems as if it will be implemented through a series of new regulations.

85. Robert E. Osgood et al., *Toward a National Oceans Policy: 1976 and Beyond* (Washington, D.C.: Government Printing Office, 1976), p. 76. In fact, some of these ships do engage in

Imposing higher standards would also not affect the flags-of-convenience more adversely than any other shipping, as the standards would be imposed equally on all foreign shipping. Indeed, the United States has a special interest in raising the standards applicable to the convenience fleets in order to avoid the widespread environmental stigma associated with their use. In return, Liberia has proclaimed its willingness to become a "blue-chip" flag-of-convenience.[86]

For the most part, economic and political interests extrinsic to IMCO have been responsible for those aspects of American policy that have had adverse environmental consequences, but two other general political traits have propelled its environmentalist stance at IMCO on many issues. First in importance is the democratic, pluralistic structure of its society. As we have seen, not only are environmental values incorporated by governments only when they are "thrust upon [them] by popular demand,"[87] but pollution controls have been forthcoming at IMCO only when these governments have initiated them. From 1954 to the present day, all environmental initiatives at IMCO have been a direct result of democratic governments seeking international solutions to problems being loudly mooted at home.

With the exception of truly national issues (such as the *Manhattan* voyage was for Canada), we have seen that most governmental structures are not particularly receptive to environmental interests. In the United States, however, popular democratic participation on behalf of environmental protection has been refined to an art. Until the 1960s, American policy as a maritime state was largely unchallenged, but at that time the environmental movement as a whole was making great strides in the United States and this could not help but affect the policy of the government toward the international regulation of marine oil pollution. The Santa Barbara oil spill directed attention to this pollutant, but given the existing bureaucratic and

---

some international transport so are not completely immune from the need to safeguard their competitiveness.

86. The exact nature of the American benediction on the Liberian operations is unclear. Liberian officials repeatedly talked of informal State Department contacts and close consultations. American officials, of course, deny formal government support for the Liberian operations, but the problems involved in shifting away from a reliance on it are clearly enormous. Moreover, except for the environmental problems, the relationship works well. On the environmental front, the situation with the convenience fleets has undoubtedly improved. Until the early 1970s these fleets were certainly inferior to most others. In 1974, an OECD study showed that their records were still substantially below those of the OECD states ("Report of the OECD Ad Hoc Group on Flags of Convenience," Paris, March 14, 1974). Many would argue that this is still so. However, under pressure, Liberia in particular has attempted to improve its standards, and the results of the American "Tanker Boarding Program" would seem to indicate that convenience flags are overall no worse than others. Greece has a particularly bad record. For this Coast Guard analysis, see TSPP/CONF/INF.4.

87. Caldwell, *In Defense of Earth*, p. 3.

commercial interests, the change in policy was not an easy one.[88] It has been the existence of an independent legislative branch that has made the difference.

On the specific issue of maritime environmental protection, the independence of the American legislature can be readily appreciated. First, by not being bound into a rigid party line, independent politicians such as senators Muskie and Magnuson have been able to develop their own national environmental constituencies, from which they draw much of their support. Combined with this, legislative authority within the United States is widely diffused, allowing politicians actually to pass private bills. As a result, with an environmental political base and the capability of making law, congressional power to affect governmental environmental policy has been very real. Indeed, for the 1969, 1973, and 1978 IMCO conferences, threats of legislation had a central role in affecting American policy. In addition, actual legislation passed by the Congress, notably the Water Quality Improvement Act (1970) and the Ports and Waterways Safety Act (1972), revolutionized the role of environmental considerations in U.S. maritime affairs.

Another piece of domestic American environmental legislation has also affected the structure and processes of the United States government in a unique way. By requiring that all governmental agencies include environmental considerations in their planning (and issue "environmental impact statements" to prove it), the National Environmental Policy Act (NEPA, 1969) has guaranteed a prominent role for the environment in governmental planning. To ensure this, the Council of Environmental Quality (CEQ) was created by NEPA in the Executive Office to serve as a continuing prod to government departments. Acting with presidential authority, CEQ has had a substantial impact, particularly before the TSPP Conference (especially in the preparation of the Carter initiatives), but also at the 1973 Conference (where the head of CEQ, Russel Train, was one of the two chief U.S. representatives). This particular bureaucratic arrangement provides the United States with a unique environmental asset. In addition the United States also has an environmental agency, the Environmental Protection Agency (EPA), created in 1970. It too has some effect as a bureaucratic weight on behalf of the environment, but like so many environmental ministries in other countries, its impact is muted by its lack of operational authority in the field of maritime regulation. This remains under the

88. With its broad international concerns, American environmental interests confronted the Department of the Interior (concerned to protect the profitability of the oil industry—see Engler, *Brotherhood of Oil*); the Department of Defense (concerned to protect free navigation); and, of course, the Department of Transportation and Coast Guard (concerned not to dislocate shipping practices).

auspices of the Department of Transportation. As a result, EPA has often been merely a "frustrated observer."[89]

One final aspect of the American legislative process deserves mention: the very strong legislative committee system. With their powerful investigative authority, hearing after hearing has been held on maritime environmental issues where otherwise unknown officials have been called forward to explain and justify both their policies and the results of IMCO conferences. As the lead role at all IMCO conferences still resides in the Coast Guard, this has been most useful in ensuring that other than "maritime" values are considered in the American position. Indeed, with the great opportunity for the Coast Guard to make "transgovernmental" alliances, this high level of accountability has ensured that the Coast Guard has used its contacts not to undermine governmental policy but to achieve instead widely acceptable solutions.[90] Moreover, with policy discussed so openly, special industry interests often have been given short shrift. These committees have provided a receptive channel for public environmental input into the government.

The second factor which has greatly facilitated the American advocacy of environmental values at IMCO is its ability to afford the costs. That is, not being a "maritime" state in the sense of the European shipowners, American policy makers are not hindered in their environmental policies by the specific cost implications of the proposals for their industries. Without this barrier, incorporating the general costs as costs to the consumer has posed little concern at all. It is only in those cases where the specific costs of environmental improvements have not presented a barrier that the general ability of the developed states to pay for environmental values has become important. At this point, the high level of economic development of these states can be a significant aid in achieving higher environmental standards.[91]

Turning to the nature of American influence, it is quite clear that its power in IMCO surpasses that of any other state. Its ability to secure its preferences is grounded in the size of its market, its wealth, and the

89. Interviews.

90. Although the Coast Guard has been widely assailed within the United States for its "anti-environment" policies (such as for allegedly failing to promulgate high enough standards under the Ports and Waterways Safety Act), most groups in the American bureaucracy praise its international activities at IMCO. Knowing the need to "sell" the results of IMCO conferences to the Congress, the Coast Guard naturally must seek a solution that is acceptable both domestically and abroad.

91. One commentator has observed that the "present outcry over environmental quality is as much a function of American economic well-being as it is of the actual state of the environment." Donald R. Kelley et al., *The Economic Superpowers and the Environment: The United States, the Soviet Union and Japan* (San Francisco: W. H. Freeman, 1976), p. 29.

worldwide reach of its industrial machine. On the environmental issue, the United States demonstrates what Keohane and Nye refer to as "asymmetrical interdependence."[92] As a user of the world's maritime services, the United States ranks at the top. It is a major ocean trader and its ports must be available for most foreign shipowners which wish to remain competitive. But unlike Canada or almost any other state, the United States is also relatively *invulnerable* despite its dependence on foreign trade. It has a large domestically owned tanker fleet which provides the country with considerable independence, as it is subject to American "effective control" in time of crisis. Moreover, this fleet is registered worldwide, giving American interests additional leverage over the policies of other governments.[93] At the same time, the country is not dependent on this fleet to produce domestic revenue, and it can impose additional costs on it without great concern. Indeed, American economic strength is so great that many have even urged the country to build all its required tankers at home! With such great *specific* leverage and *general* economic security, American power at IMCO is unrivaled. That power is, with growing domestic commitment, increasingly being transformed into influence which is challenging the European maritime states.

Against the diplomatic influence of the traditional maritime states within IMCO, the threat of unilateral action outside the organization has been the most powerful American weapon in affecting the negotiations. Even before the first major American initiative began in 1970, potential American unilateral action had some impact on the Civil Liability Conference. The existence of draft legislation for oil pollution liability (later the Water Quality Improvement Act of 1970) "was immensely desirable because, when the Brussels Conference convened in 1969, it was fully aware of the bare minimum terms to which the United States might have been prepared to go with an international rather than a unilateral approach."[94] As we have seen, however, the United States has undertaken only two major environmental initiatives, the first culminating in the 1973 Conference on Marine Pollution and the second in the 1978 TSPP Conference. In 1973, it

92. Keohane and Nye, *Power and Interdependence*, pp. 11–19.

93. The extent of American shipownership is truly astounding. Following is a list of countries in which the United States owns a portion of their nationally registered fleet: Panama (42.3 percent), Liberia (34.8 percent), United Kingdom (27 percent), Netherlands (25.6 percent), West Germany (24.6 percent), France (18.6 percent), Italy (9.1 percent), Norway (2.1 percent). This amounts to some 485 tankers or 81.6 percent of all U.S.-owned tanker tonnage. Of this, 41.6 percent is registered in Liberia. *World Shipping under Flags of Convenience* (London: H. P. Drewery Shipping Consultants, July 1975).

94. Allan Mendelsohn, "Ocean Environment and the 1972 United Nations Conference on the Environment," *Journal of Maritime Law and Commerce* 3 (January 1972), p. 393. As a former member of the State Department, Mr. Mendelsohn had been involved in the early IMCO negotiation for this convention.

was the threat of unilateral action under the Ports and Waterways Safety Act that provided the United States with its greatest source of leverage. In 1978, it was the threat of such action by the Congress and the President that provided the United States with its great influence.

With the combination of a high domestic policy commitment, great levers of influence, and relative invulnerability on all levels, the American threat of unilateral action has carried weight.[95] Virtually all the recent advances in international regulations for oil pollution control have resulted from American pressure. At the same time, the United States is not omnipotent and, pitted against the combined weight of the United Kingdom and most maritime states as well as the oil industry, their most costly environmental proposals have been rejected. For one thing, the costs of actually undertaking unilateral action are of a different magnitude than those involved in simply threatening it. The unwillingness, therefore, of the United States actually to proceed alone has meant that the changes it has achieved have, in the end, been those that were acceptable within the larger organizational context. At the same time, it has been the American power and threats to act outside IMCO that has kept the environmental issue moving within it.[96]

Soviet Bloc

Quite visible differences exist at IMCO between the Soviet Union and the United States, but these differences mask an even more basic similarity. The structure of their shipping interests, their policies on a variety of environmental proposals, and their influence in the marine environmental field differ, but for both states these reflect their overriding positions and objectives in the global political system.

Although only in a reactive way, the Soviet Union and its East European allies since the 1950s have been, with one exception, consistent supporters

95. The combination of domestic policy commitment, external levers of influence, and relative invulnerability (asymmetric interdependence) is prerequisite to a successful use of the threat of unilateralism and fundamental regime change. Where these three items are present, the usual source of control within the specific IMCO environment, shipping, becomes a source of vulnerability. The rejoinder to the maritime state's threat of "nonratification" is, therefore, the powerful oil importer's call for "unilateral" action. This is the creative tension within IMCO and the international industry in general that provides the impetus for progress.

96. This reliance on threats to act unilaterally outside the institution reflects Keohane and Nye's analysis that power is not always fungible; "underlying power resources may be immobilized by norms and political processes so long as the regime remains in place" (*Power and Interdependence*, p. 147). For the United States, its power resources are to a great extent immobilized by working through the normal IMCO channels. They are fully realized in this area only by unilateral action. The distinction here is also between power as "control over resources" and power as "control over outcomes."

of more stringent *technical* regulations. In 1954 and 1962, they backed the measures put forward by the British, and in the 1970s they have supported most of the American environmental initiatives. This policy orientation flows from the nature of the Soviet shipping interests, which in turn reflect the internal political and economic characteristics of the Soviet Union and its bloc.

Most important is the "closed" nature of its economic system. In the East-West struggle, the Soviet/East European sphere always has been tightly closed and isolated, and this is evident in its shipping practices. In the first place, the Soviet tanker fleet and the negligible tanker tonnage of the Eastern European countries operate largely between their own ports on the Black and Baltic seas.[97] These areas are semi-enclosed and very vulnerable to oil discharges. Moreover, because their vessels are state-owned and trade largely with other Comecon countries, there is no fear of loss of "competitive advantage" to Western states. As a result, the Soviet Union has been able to react favorably to environmental initiatives which would cost it little in competitive terms while improving the quality of its marine environment. Only with the very costly American proposal to retrofit segregated ballast tanks has the USSR been seen to flinch at costs. Even then, it did not publicly oppose the proposal.

On the other hand, this favorable situation for its fleet has not resulted in the USSR taking an initiating role at IMCO. The supposedly "socialist" economic character of the country has not influenced its general attitude toward the environment, and maritime shipping is no exception.[98] Instead, the closed undemocratic nature of the Soviet state has fostered a reactive approach to its participation at IMCO. At home, the country has not been motivated by popular demands which would encourage the government to advocate higher pollution control standards at IMCO. Given the propensity for environmental concerns to be forced upon governments and IMCO only by public interest, the bureaucratic Soviet system has simply reacted to external pressures at IMCO,[99] while at home waiting until "environmental disruption causes other state interests [within the country] . . . to lose as much as those in favour of greater exploitation of the environment stand to gain."[100]

97. *The Role of Comecon Oil Tankers* (London: H. P. Drewery Shipping Consultants, 1975). Only 17 percent operate outside Comecon or domestic trading, although this is expanding somewhat of late with increased trade to some developing countries.

98. Numerous commentators have pointed this out. See William Ophuls, *Ecology and the Politics of Scarcity* (San Francisco: W. H. Freeman, 1977), pp. 203–206; Falk, *A Study of Future Worlds*, p. 402; Kelley et al., *Economic Superpowers*; Marshall Goldman, *The Spoils of Progress: Environmental Pollution in the Soviet Union* (Cambridge, Mass.: MIT Press, 1972).

99. For an interesting comparison to the actions of the USSR in other international organizations, see Cox and Jacobson, *Anatomy of Influence*, pp. 409–414.

100. Goldman, *Spoils of Progress*, p. 273.

Soviet environmental policy is, therefore, largely a product of attributes associated with its larger political concerns. This is unlike the European powers, the individual policies of which vary with their specific maritime interests, largely unrelated to their more general global ambitions. However, this is similar to the sources of American policy. At one level, this manifests itself in the desire of Soviet delegates to further détente by establishing good relations on specific issues of importance to the United States. That this was a factor at both the 1973 and 1978 conferences was attested to by numerous delegates.[101] On specific issues, Soviet policy also reflects its larger East-West interests. Indeed, on jurisdictional issues, the East-West rivalry is most germane to Soviet policy. Like the United States, it too has opposed any jurisdictional initiatives at IMCO and the law of the sea conference that could affect its freedom of military navigation. The USSR has, however, gone substantially further in also opposing virtually all attempts to facilitate the detection and prosecution of violations by vessels in foreign ports. This intransigence is based on its fear of any regulations that could pry open the lid on its secrecy and control or result in prejudicial treatment by the authorities or courts of non-Communist states.[102] On enforcement issues, their policies have been very close to those of the major Western powers and Japan, but unlike them the Soviet concerns have been political, not economic.

Within the IMCO bargaining context, the influence of the Communist states has been quite modest. Few IMCO members use their tankers (although Soviet and Eastern European countries import little oil from outside their group). At the same time, the USSR's technical competence and status as one of the world's "superpowers" has given it some weight in IMCO, especially with the United States on those law of the sea issues of mutual interest. Most important, however, has been the fact that, within the IMCO forum, the USSR controls up to nine votes: its own, those of the six East European IMCO members,[103] and those of the ByeloSSR and UkSSR which have gained admission to IMCO conferences as members of the United Nations. On issues of minor concern to the USSR little control has been exerted, but the votes have been dutifully assembled on matters of interest. As a result, they have been instrumental in bringing about the defeat of many jurisdictional and enforcement proposals.

101. Interviews. The validity of this is hard to verify, although it is consistent with events.

102. This defensive attitude has been clear at all conferences. But it was perhaps most explicit in 1969 in the discussions on the Intervention Convention. There it opposed any strong coastal state power of high seas intervention and even introduced some very restrictive proposals (such as *requiring* flag state notification of all circumstances and *requiring* the coastal state to return the crew after a maritime casualty). Recently they have reluctantly bowed to an expansion of in-port inspection and port-state enforcement, but in the latter provision have insisted on the right of flag-state preemption.

103. Bulgaria, Poland, Czechoslavakia, Romania, Hungary, and East Germany.

Developing Countries

A division of growing significance to IMCO is that between the developed and developing countries: the "north-south" split. Until recently, IMCO was totally dominated by the WEO and Soviet bloc groups. It was truly their agency. Of late, however, there has been an increasing shift toward the developing countries: their proportion of the membership and their role in the governing organs has increased; their participation in the specialized committees and particularly in the conferences has grown; and the agency has created and has agreed to "institutionalize" the Technical Cooperation Committee. Especially under the most recent secretary-general, C. P. Srivastava of India, there has been a greater orientation toward the problems of development within the organization as a whole. This reflects trends that are evident in all United Nations agencies.

The developing state grouping encompasses states with varied characteristics. In fact, as a collectivity, the "Group of 77" has been little in evidence at IMCO. On financial matters or issues of technical assistance, there is certainly a shared perspective, but lack of interest has meant that these issues have been addressed only in the particular committee designated for them (see Chapter 8). At conferences, the Group of 77 seldom collaborates to secure particular accords. It was certainly in evidence at the 1973 Conference, but only because jurisdictional issues were being discussed prior to the opening of UNCLOS III. At the 1978 TSPP Conference, the Group of 77 did not meet prior to the conference and, at the conference itself, met only once at the request of India. Only on explicitly "political" issues—most notably, election to Council—does the Group of 77 work closely together.

In general, the interests of these states are so varied and their participation so unpredictable that patterns are difficult to discern. To an extent there is a growing regionalist identity among many of these states, but this is still an incipient development. Within the IMCO context, it is most convenient to classify developing nations into one of three categories: flag-of-convenience, developing maritime, and developing nonmaritime. Even these groupings are rather arbitrary. The small group of flag-of-convenience states are readily identifiable and do play a crucial role in the global shipping industry, but the other two groups are often difficult to distinguish. After all, even those with the largest tanker tonnage fall below 1 percent of world grt, while many of the others aspire to build their own shipping fleets. These two groups will be discussed together.

Flag-of-Convenience. There are numerous flags-of-convenience,[104] although together Liberia and Panama account for about 97 percent of the total

104. These include Liberia, Panama, Cyprus, Somalia, and Singapore. See footnote 80.

tanker tonnage of the group. The distinguishing characteristic of these states is, of course, that their registered tonnage is owned by nonnational interests attracted to the state by lower crew costs, taxes, and shipping standards and less rigid enforcement. As a result, the vessels registered in these states have been great contributors to maritime environmental pollution of both the accidental and operational variety.[105] With one exception, they are largely nonparticipants in IMCO's environmental negotiations.

The one exception is Liberia, and with approximately 90 percent of the convenience tanker fleet (30 percent of the world fleet), it is an important exception too. Liberian policy at IMCO is unlike that of the other convenience flags because of its unique situation. First, as previously mentioned, the Liberian flag is an American creation. The operation is run financially by the International Trust of Liberia, which is 80 percent owned by the International Bank of Washington, D.C. It is run administratively by the Liberian Deputy Commissioner of Maritime Affairs (an American citizen) and a staff of seventy. Its offices are in Virginia. The deputy commissioner's senior deputy is a vice-president of the International Bank of Washington, D.C. Liberian shipping policy is strongly influenced by the Liberian Shipowners Council, which is composed of sixty-two major international shipowners, and by the Federation of American-Controlled Shipping, whose members control 90 percent of American-owned tonnage registered in Liberia and Panama. Although American interests can be identified as accounting *directly* for only 35 percent of Liberian tonnage, Liberia is truly, in the words of American Assistant Secretary of Commerce for Maritime Affairs Robert Blackwell, "a phantom maritime power that was created by American businessmen."[106]

Until the late 1960s, Liberia was not concerned with the marine environment. Its policies are, however, highly sensitive to American governmental and corporate interests, and as environmental concerns came to the fore in the United States, these policies began to change. Moreover, the need to tighten standards became unavoidable after 1970 when two inadequately manned Liberian ships, the *Pacific Glory* and the *Allegro*, collided in the English Channel. With this collision, Liberian shipping came under attack as never before. Soon after, it was decided that Liberia would try to become a refuge only for well-run vessels. In the words of Liberia's IMCO legal representative, Frank Wiswall (an American), Liberia would become a "blue-chip flag-of-convenience."

As a result, in recent years, Liberia has entered fully into all levels of IMCO negotiations on the environment. Like the other maritime states, it

105. See footnote 86. Because of their great quantities of registered tonnage, they seem at times to be the only polluters!

106. Quoted in two excellent articles on the subject by John Kifner in *New York Times*, February 13, 1977, p. 1, and February 14, 1977, p. 14.

frequently opposes environmental initiatives in the conferences, although its acceptance of segregated ballast and the "special measures" compromise at the 1973 Conference and its silence on the retrofitting issue at the 1978 TSPP Conference reflect its greater sensitivity to the need to accommodate environmental interests. Most importantly, Liberia is concerned to be *seen* to accept the highest international standards and has, as a result, ratified almost all the IMCO conventions. Moreover, to enforce them, it has created its own worldwide inspection service. In some areas, Liberia's record has improved as a result.

Until the last few years, Liberia's influence on environmental regulation was minimal. The general disdain by other maritime governments for flags-of-convenience, its own position as the environmental villain, and its limited interest in IMCO regulations militated against its achieving a prominent role. Even today, as a "phantom power" that is still very much of a *bête noire* to many governments, its leverage outside IMCO is slight. Within IMCO, however, its standing and influence have greatly improved with the adoption of its more respectable policy and with its concerted efforts to put forward delegations with a high level of diplomatic and technical competence. Moreover, in the face of the American environmental challenge, the Liberian interest in reasonable international standards is akin to and not competitive with that of the European powers. In this debate the Liberian voice can be most effective when in harmony with those of the maritime state coalition. Yet, as we have seen, American interests can usually call the Liberian tune.

Developing Maritime and Nonmaritime. Apart from the flag-of-convenience nations, the developing countries occupy minor roles in the world shipping industry, and their share of tanker tonnage is considerably below that of dry-cargo tonnage. They have voiced their aspirations to develop their maritime interests—especially within UNCTAD's Committee on Shipping—but they have made only minor inroads to date. Even countries like Brazil, Argentina, Venezuela, and India, which have had modest shipping industries since the 1950s, have not been able to achieve significant increases in their positions. Their delegates may talk more like shipowners than most of the representatives of other developing nations, but in fact they too are only aspirants to maritime status.

The overriding concern of the developing nations in international environmental negotiations has not been the quality of the marine environment. This they view as largely an issue for the rich. Rather, they have been primarily worried about the effect which pollution regulations could have on the future of their shipping industries and on their economic development in general. With limited capital resources and largely secondhand vessels, they have frequently proclaimed their opposition to regulations which would force them into uneconomic retrofitting of old vessels and would raise the cost of new. Not only might such costs curtail their maritime ambitions, but they would require a diversion of resources from higher priority objectives. As one observer has noted,

> Capital investments for a cleanup for developing nations involve a direct tradeoff with increased employment opportunities or increased exports or other economic growth objectives which are necessary simply to sustain life. The quality of life issues for non-affluent countries will tend to take lower priority than the more immediate issues of economic survival.[107]

One way for the developing countries to avoid such costs has been through their advocacy of a "double standard" whereby requirements for environmental protection would not be as stringent for themselves as for the wealthier industrialized nations. The desirability of taking the level of economic development into account in setting environmental standards was accepted in the Stockholm Declaration on the Human Environment, and it has been basic to the developing state approach ever since.[108] The quite reasonable justification for the principle is that:

> The developed states realized their greatest period of economic growth under conditions of completely free, laissez-faire access to international water and international air. This provided them with a decided cost advantage. Now, if environmental constraints are to be imposed on all states, the developing states during their period of major economic growth will not enjoy that cost advantage. Indeed, it can be and is argued that their further economic growth will be penalized by the earlier growth of the developed states, and that therefore, the situation of an abuse of rights may be extant. . . . Political reality demands that if developing states are to be asked to order their economic growth within established environmental limits, they should be compensated for the economic handicap thus imposed.[109]

107. Harold B. Malgren, "Environmental Management and the International Economy," in A. V. Kneese, S. E. Rolfe, and J. W. Harned, eds., *Managing the Environment* (New York: Praeger, 1971), p. 54.

108. Principle 23 of the Declaration states, "Without prejudice to such criteria as may be agreed upon by the international community, or to standards which will have to be determined nationally, it will be essential in all cases to consider the system of values prevailing in each country, and the extent of the applicability of standards which are valid for the most advanced countries but which may be inappropriate and of unwarranted social cost for the developing countries."

109. E. W. Seabrook Hull and Albert W. Coers, *Introduction to a Convention on the International Environmental Protection Agency* (Kingston, Rhode Island: Law of the Sea Institute, Occasional Paper no. 12, 1971), p. *v*.

Although a recognition of varying "capabilities" was included in UNCLOS III text, the double standard provision there is surprisingly not a strong one.[110] Nevertheless, were it to be invoked in IMCO, its implications would be enormous.[111]

Despite the great concern expressed by the developing countries at the Stockholm Conference that the environmental issue not be used to impede their development, the Group of 77 has failed at IMCO to push for provisions that might actually benefit them. Although the "double standard" has occasionally surfaced, it has been quickly quelled and discussions of technical assistance have been surprisingly absent from conference negotiations.[112] Here they have lacked the interest, the expertise, the bureaucratic continuity, and the resources to participate effectively. Cultural and language barriers have exacerbated their situation so that they have tended to ignore such matters or be swayed by the "initiators" from the developed world. On some issues there has been a fair measure of agreement with the developed maritime states. For example, they have supported these states in their standard inclusion of "grandfather clauses" to exempt existing vessels from new regulations.[113] However, at the 1973 Conference and to a lesser extent in 1978, they were notably silent on the issues of more costly ship design, construction and equipment standards for new vessels, and new requirements for port facilities. In fact, several participants in the technical committee negotiations at these meetings commented on how few developing state delegates had a grasp of the technical issues. Despite their clear articulation outside IMCO of general economic objectives, their low level of interest in environmental problems, lack of expertise and poor participation have produced a lobby that is sporadic and ineffective.

There have been some issues on which the developing states have supported proposals with important environmental implications. Sometimes this support is a product of their willingness simply to "go along" with developed state delegations they trust. Often, however, it is because these states do recognize the necessity for compromise, for achieving a solution

110. Informal Composite Negotiating Text, U.N. doc. A/CONF.62/WP.10, July 15, 1977, art. 195(1).

111. For one thing, the double standard would exaggerate any competitive disadvantages inherent in future regulations, thus providing greater disincentives to their acceptance.

112. Discussions of a "double standard" at the TSPP Conference went nowhere, and when the Economic and Social Commission for Asia and the Pacific raised the prospect of a regional double standard, the IMCO secretary-general dissuaded them from this tack. Interviews.

113. Considering the enormous implications of applying new regulations to existing ships, such clauses are a *sine qua non* to improved standards, although they do extend the time before a new convention will have a noticeable environmental impact. With the enormous tonnage now on the seas and the decrease in future construction, these clauses have come under increasing challenge as retrofitting has become more attractive to environmentalists.

acceptable to the broad base of the IMCO membership. This was particularly evident at the TSPP Conference, where India played a constructive role in achieving the support of most developing states for the final package of new regulations.[114] At the same time, the developing states have occasionally supported new environmental initiatives as a matter of conscious government policy. After all, many African, Middle Eastern, and Asian states are located near heavily used shipping routes, and some specific measures would obviously benefit them. When they do participate in this way, they provide a solid numerical basis for the coastal contingent.

At the conferences on liability and compensation in 1969 and 1971, large numbers of the developing states backed higher standards because as coastal states with minor shipping interests they were most often the victims of other people's spills and they would be the major beneficiaries of higher compensation for them. Even states like Brazil and India, with nascent shipping interests, supported the revisions in order to assure compensation lest there ever be a major spill adjacent to their coasts. Of far greater interest to these countries has been the question of *jurisdiction* over environmental protection. This is properly a law of the sea issue, but the 1973 IMCO Conference was strategically timed so that it could affect some of the jurisdictional issues. It was this that drew so many developing countries to the 1973 Conference. Led by a phalanx of law of the sea negotiators, these states played an active and important role in the debates. Many of them backed extended coastal jurisdiction, but their policy had little to do with a strong environmental concern. Rather, they saw this extension in terms of its impact on their general prestige and offshore economic interests and on their other law of the sea goals. Not really so different from many of their developed state counterparts, the environment is rarely a factor in their calculations.

Because of their lack of interest and resources, the developing states have had very modest influence on environmental regulation at IMCO. On liability and jurisdictional questions where they participated with interest and in numbers, their votes did have an impact. On technical issues, they have had almost none. Yet within the IMCO context, these states would seem to have great potential influence if only because of their numbers. Substantial changes would have to take place before this could occur, however. Even apart from their general lack of interest, their undeveloped bureaucracies would inhibit them from building up the sort of continuous, organized representation that is so crucial to effectiveness. Moreover, despite some

114. Especially as the alternative to IMCO agreement (that is, American unilateralism) might be even more costly, the developing states are interested in achieving a convention. As their actions at the conference are rarely subject to serious scrutiny at home, developing state representatives have the flexibility to support a variety of proposals.

sense of common purpose, there is great diversity among the developing states. As they develop further, this diversity will likely increase both along special interest and regional lines.[115]

In conclusion, it is evident that the policies and influence of states in environmental negotiations stem from a variety of national characteristics which are both intrinsic and extrinsic to the oil pollution problems. Maritime or coastal, developed or developing, the pattern is a complex one, but one where only tangible political and economic pressures have meaning. In such a maze, environmental values have been slow to make themselves apparent.

NONGOVERNMENTAL TRANSNATIONAL ACTORS

In the international politics of the environment, states are neither cohesive or alone. Their power is divided internally, and increasingly it is being shared externally as well. As many commentators have pointed out, the number of international "nongovernmental organizations" (NGOs) has increased markedly since the Second World War, and particularly since the early 1960s.[116] Especially since 1967, this trend has been mirrored at IMCO, where a number of the organizations—notably the International Chamber of Shipping (ICS), the Comité Maritime International (CMI), and the Oil Companies International Marine Forum (OCIMF)—have had an important effect on IMCO's environmental responsibilities.[117]

The ICS, like the CMI, is an "umbrella organization" representing a large number of constituent national associations. Founded in 1921, it encompassed thirty national shipowner associations in 1978, eight of which were from developing countries (including Yugoslavia and Liberia). The ICS is dominated by the Western maritime states, as might be expected considering its designation as "a body representing shipping interests which operate on the basis of free enterprise."[118] Similarly, the CMI includes forty national "maritime law associations"—largely from the developed Western countries.[119]

115. Regional identification has definitely been in evidence among the developing states at IMCO. Regional groupings would include the Arab states, French-speaking Africans, Latin Americans, and ASEAN members. However, the impact of this has been minimal to date, given the very uneven and *ad hoc* participation.

116. Werner J. Feld, *Non-Governmental Forces and World Politics* (New York: Praeger, 1972), and Kjell Skjelsbaek, "The Growth of International Nongovernmental Organization in the Twentieth Century," in Keohane and Nye, eds., "Transnational Relations and World Politics," p. 420. Feld notes an increase from 795 NGOs in 1950 to 1,321 in 1960 and 2,296 in 1970 (p. 117).

117. In 1960 and 1966 there were eleven NGOs with consultative status at IMCO. This had increased to twenty-one by 1971 and to forty-five by 1978.

118. ICS Constitution, art. III(1)(*a*).

119. In 1974, only four developing states—Argentina, Brazil, India, and Mexico (all with shipping interests)—were represented at its important Hamburg Conference.

The Oil Companies' International Marine Forum (OCIMF) is structurally different from the ICS or CMI in that its members are individual companies and not national associations. Prior to 1969, the views of the oil industry were channeled through the ICS, but as the work of IMCO gained political salience and the interests of the ship- and cargo-owners diverged, a separate oil lobby was thought to be necessary. Originally, OCIMF was composed of eighteen oil companies of largely American, Japanese, and European nationalities (including all of the "seven majors"), but by 1978 over forty-five companies accounting for 80 percent of world oil shipments were represented.[120] The American companies are, by far, the most numerous and account for a large percentage of the oil moved by sea. Hence, they exert tremendous influence over OCIMF policies.[121]

Nongovernmental organizations are important vehicles for the expression of industry interests, and they operate in three principal ways: by affecting policy at the national level, by affecting policy through direct lobbying at the intergovernmental level, and by providing private international alternatives.[122]

Action at the national government level is a longstanding method of operation for both the ICS and the CMI, with their many national shipping and maritime law associations. The national associations include some of their countries' most eminent maritime experts, who carry great authority and legitimacy when making representations to their governments. At the same time, these experts can call on their international counterparts for advice and assistance—a strategy which has been used with great efficacy whenever the industry has been threatened by unilateral state action.[123] Moreover, by passing information about the impending legislative actions of a state to various foreign associations, foreign governments more sympathetic to the welfare of the industry are encouraged to exert pressure on the state in question. This is an oft-repeated pattern.

Unlike the ICS and CMI, the insurance lobby organization at IMCO, the International Union of Marine Insurers (IUMI), serves basically a

120. Some of the companies—such as Shell—represent numerous individual companies worldwide (Shell has over a hundred). They are drawn from both the private and public sectors. Most of the 20 percent of shipments not represented is carried by state-owned corporations from either the developing or Soviet grouping.

121. The chairmanship of the General Committee has alternated between Americans and British. One of the three vice-chairmen and six of the twelve elected members have been American. At the 1978 TSPP Conference, four of the seven OCIMF representatives were American.

122. For some of the permutations and combinations of strategies in the international airline business, see Robert L. Thornton, "Governments and Airlines," in Keohane and Nye, eds., "Transnational Relations and World Politics," pp. 550–552.

123. Shipping and insurance experts were flown in to address Senate committees in both Canada and the United States when, in 1969, each contemplated unilateral legislation. Their impact has been quite significant in dissuading states from passing unilateral or commercially damaging legislation.

peripheral information-gathering function. Because the marine insurance industry is largely concentrated in a single country, the United Kingdom, the insurers have a minimal need for international coordination and operate largely on their own initiative (see Appendix 2).[124] OCIMF actively lobbies many national governments, while also coordinating the activities of its constituent corporate membership.[125] While all states respond in varying degrees to the arguments of these organizations, one country, the United Kingdom, has sometimes acted as the national spokesman for these interests at conferences over the last ten years.

The environmental lobby has, with one exception, not been sufficiently well organized transnationally to affect the policies of national governments. The exception was the early activities of the British-based ACOPS which, to a limited extent, functioned as an international environmental body. Its pressure on the United Kingdom and its own multinational conference in 1953 had a substantial impact in motivating governments to convene the 1954 London Conference, and its 1959 Copenhagen and 1968 Rome conferences may have had some impact on foreign governments. Since that time, however, it has functioned largely as an elite pressure group operating almost solely within the United Kingdom.[126] Even there, its influence is now quite modest. In the 1970s two other environmental organizations have sought to influence IMCO deliberations, although the only effective one—the Center for Law and Social Policy in the U.S.—has only a national base. Representing many large environmental organizations within the United States, the Center has participated actively in congressional hearings and committees, has instituted lawsuits under domestic American environmental legislation, and has had representation on the American delegations to IMCO conferences and UNCLOS III.[127] The

124. The impact of this complex industry, especially in the area of liability and compensation, has been profound in influencing both national and international policies. As a result, a special section (Appexdix 2) has been included to explain its operation. In addition to its own independent activities, the insurers do cooperate within the interindustry CMI in its "private" legislative activities. One commentator has alleged that only two individuals in London control the entire process: "their conclusions as to capacity and cost determine . . . the limits and terms of liability that can be negotiated in international conventions." Allan J. Mendelsohn, "Maritime Liability for Oil Pollution in Domestic and International Law," *Environmental Law Review*, 1970, p. 418.

125. One example of this coordinating power was explained to us by an OCIMF participant. Lebanon was contemplating new legislation which a resident oil company—ARAMCO—opposed. ARAMCO notified OCIMF, which then brought in the other oil companies in a successful transnational campaign to criticize and change the legislation.

126. For more information on the work of ACOPS, see copies of its *Annual Report* (10 Percy Street, London, U.K.).

127. The director of the organization, Richard Frank, reported to one Senate committee that the Center spoke for 250,000 voters from the Sierra Club, Friends of the Earth, Audubon Society, Environmental Defense Fund, and the National Resources Defense Council. These

other organization, Friends of the Earth (FOE), became the first international environmentalist body to obtain consultative status at IMCO. Despite the existence of a London office and national chapters in many Western nations, it has been unable to devote large resources to IMCO negotiations. Its attendance at meetings is sporadic, and it has not coordinated the lobbying of national governments by its constituent members.

Besides dealing directly with national governments, a second strategy for transnational organizations is *lobbying at the intergovernmental* (that is, IMCO) *level*. When issues are being discussed or decided in the agency, the commercial organizations have performed useful functions in bringing matters to the attention of their national constituents and, more importantly, in participating in the IMCO decision-making process itself. This they have done to such an extent that there are few, if any, other agencies where nongovernmental actors are as active. With this end in mind, these organizations have often created structures parallel to those at IMCO. The International Chamber of Shipping, for example, has committees on marine safety, marine pollution, maritime law, and insurance. Policy positions are prepared through joint consultation, approved by the body as a whole, and submitted to IMCO (and recently to UNCLOS III as well). ICS Secretariat officials attend IMCO meetings as observers, interjecting their views into the discussion and lobbying individual delegates. Other ICS members often attend as representatives on their national delegations.[128]

OCIMF is likewise an extremely important lobbyist on environmental matters at IMCO. Its committee structure also reveals a clear parallelism to IMCO's, producing coordinated oil industry policies on such problems as tanker construction, watchkeeping and training, reception facilities, load-on-top, and the legal implications of pollution. As a result OCIMF is a very active participant in IMCO bodies, submitting detailed technical papers and participating at length in the debates. Its influence in the Legal Committee, the Marine Environment Protection Committee, and in all the conferences since 1970 is commensurate with this expertise and activism. Prior to the 1971 Fund Conference it prepared an excellent paper, *Fund in Action*,[129] to which all conference delegates paid tribute and which had an important impact on the final Fund Convention. It has produced a plethora of studies on other, more technical matters. On many of the issues, there is

---

other groups lobby politicians directly as well. In 1977, Mr. Frank was appointed director of the National Oceanic and Atmospheric Administration, the key U.S. oceanic agency.

128. These industry representatives literally pepper the delegations of the maritime states. Other countries (such as Canada and the United States) have also included such officials on their IMCO or UNCLOS delegations in order to provide them with expert advice—only to find them openly lobbying against their positions. On the composition of IMCO delegations, see Harvey B. Silverstein, *Superships and Nation-States*, pp. 51–59.

129. LEG/CONF.2/INF.2 (October 25, 1971).

a high degree of coordination with the ICS, as the oil industry wants the cooperation of the independent shipowners in order to implement its policies as widely as possible.[130]

Transnational coordination and lobbying at the intergovernmental level are functions the oil companies take very seriously. Robert Engler, the renowned student of the international oil industry, refers to the "private government of oil,"[131] and Anthony Sampson comments that oil companies "were institutions that had appeared to be a part of world government [with] their supranational expertise [that] was beyond the ability of national governments."[132] In the environmental field, these assertions are certainly borne out. In its organization, OCIMF is a parallel to IMCO, while the International Petroleum Industry Environmental Conservation Association (IPIECA) works with the United Nations Environment Program (UNEP) with consultative status there and at UNEP's parent body, the United Nations Economic and Social Council. Moreover, anticipating the creation by UNCLOS III of an International Seabed Authority, the oil industry has already created its own Exploration and Production Forum (E & P Forum).[133]

In contrast, the environmentalists have exercised minimal impact within the IMCO forum. Almost by their nature, international agencies tend to rely most on actors that can both offer the most expert commentary and affect the implementation of their regulations.[134] Environmental lobbies at IMCO have been able to fill neither of these prerequisites. With *very* limited technical and financial resources and the absence of strong national constituent organizations, their international lobbying is at a distinct disadvantage. Indeed, although FOE was accredited to IMCO in 1972, it submitted its proposals on the 1973 Convention only the day before the meeting and, even then, only with the assistance of an environmentalist accredited to the U.S. delegation. However, with the increasing political importance of the

130. The joint ICS-OCIMF publications, *Clean Seas Guide for Oil Tankers* and *Monitoring Load-on-Top*, were put out at least partially to demonstrate to IMCO the potential effectiveness of the industry-inspired regulations. But where their interests diverge (for example, independent versus oil-company-owned tankers or ship- versus cargo-owner), each will seek alliances with national governments against the other. For example, although the ICS and OCIMF had agreed on the appropriate apportionment of liability between the ship- and cargo-owners prior to the 1971 Conference, the ICS withdrew its support. OCIMF then accused the ICS of a "doublecross" and gained the support of its allies, Britain and the Netherlands. As a result, much of the conference was a pitched battle between these two states and the independents led by Norway.

131. Engler, *Brotherhood of Oil*, p. 6.

132. Anthony Sampson, *The Seven Sisters* (New York: Bantam, 1975), pp. 6–7. See especially pp. 6–18.

133. Interviews. Regional organizations such as CONCAWE (Committee on Clear Air and Water in Europe) have also been created.

134. This is the major thesis of Anne T. Feraru, "Transnational Political Interests in the Global Environment," *International Organization* 28 (winter 1974), pp. 31–60.

oil pollution issue in the last few years, FOE has been able to devote more resources to the issue and has been a more regular attendant at the meetings of the MEPC. But, even now, its participation is largely that of an observer rather than an active lobbyist.

The American environmental lobby, especially the Center for Law and Social Policy, has had an indirect impact. It has been intimately involved with the development of American policy at IMCO as well as at UNCLOS III. Perhaps the best "idealist" prescription for the environmental component of UNCLOS III was produced by an official at the Center,[135] and its members have actively worked on behalf of environmental issues with developing countries at UNCLOS III.[136] But its potential bases of support (like those of other environmentalist bodies) rest in the wealthy, pluralistic, democratic countries of the West (and particularly the United States), and most of its activities have been directed there.

A final and quite influential role of the transnational bodies has been the *provision of their own alternatives* for adoption by governments or as substitutes for intergovernmental action. In this area the oil industry (and OCIMF) has been particularly active and adept. The load-on-top system was devised and implemented by the oil companies despite its obvious incompatibility with the 1962 Amendments and was only later accepted by governments. TOVALOP and CRISTAL were implemented prior to the IMCO conferences of 1969 and 1971, and for many years these industry agreements were the only ones in force. Likewise, many oil companies had implemented crude-oil-washing before the TSPP Conference (although as a commercial, not an environmental, measure), and their willingness to accept it had a decisive impact on the conference.

Another body which has been very active in generating proposals for international regimes has been the Comité Maritime International—the longstanding private international organization for the "unification of maritime and commercial law, maritime customs, usages and practices."[137] Only after an intense debate did it reluctantly cede its role to IMCO in drafting the Civil Liability Convention in 1969. With the growing political significance of many aspects of maritime trade—especially through IMCO and UNCTAD (the latter with its Committee on Shipping)—the CMI is rapidly being outpaced in its role as an independent drafter of conventions. Today, it operates largely as an auxiliary organ to IMCO's Legal Committee. Many have warned that its continued viability is in doubt.[138]

135. Robert Hallman, *Towards an Environmentally Sound Law of the Sea* (London and Washington: International Institute for Environment and Development, 1974).

136. Interviews.

137. CMI Constitution, art. I(1).

138. For a brief history, see Albert Lilar and Carlo Van den Bosch, *Le Comité Maritime International, 1897–1972* (Belgium, 1972). At its 1974 Hamburg Conference, it heard a

The NGOs—particularly the ICS and OCIMF—have had a crucial effect on IMCO's environmental law-making. They have, first of all, increased the influence of national shipping and oil interests on the positions of their governments. Many national associations or firms do not have the resources to acquaint themselves with matters being discussed at IMCO and to evaluate the impact of different proposals on their economic goals. Their participation in the transnational bodies provides them with such an understanding and facilitates their ability to put forward detailed and systematic positions to their governments. On occasion they can also call on experts from other countries to discuss complex issues with their governmental officials.

In addition, the impact of many national industries on IMCO deliberations is increased by their adoption of common policies, as this promotes a unified stand by the nations in which they reside. IMCO's chief legal officer, Thomas Mensah, has expressed their contribution to IMCO's law-making process in this way:

> Accepting advice from and relying on the expertise of specialized bodies is a major and now established part of the institutional arrangements within IMCO. It has worked well mainly because, even though these bodies, especially the industrial and commercial concerns among them, are profit-making bodies and hence generally interested financially in the schemes and measures being discussed, they are all international organizations in which the competing (sometimes even conflicting) interests of component national bodies have first been harmonized and accommodated. The result is that it is very rare, if ever, that one finds their advice or proposals wholly unacceptable to the majority of governments in the respective IMCO organs and bodies.[139]

However, to the environmentalist, this reliance is seen from a different perspective. Instead, it is alleged that IMCO depends "on industry to define the scope and content of its inquiries and only promote[s] and adopt[s] marine pollution initiatives if they are acceptable to and indeed often accepted by industry."[140] As competitive industry is, by its nature, committed to a minimum of regulation and interference, too heavy a reliance on it must certainly reduce the speed of change.

Apart from enhancing the influence of particular industrial interests, NGOs have also augmented the influence of the developed WEO states,

---

warning from an IMCO Secretariat official that without an expansion of its political base—that is, the inclusion of more developing countries—it was a doomed organization. Others also pointed to the need for it to recognize the growing convergence of public and private law as a result of the importance of many "private" issues for public interests.

139. Thomas Mensah, "The IMCO Experience," in John Lawrence Hargrove, ed., *Law Institutions and the Global Environment*, pp. 242–243.

140. Eldon Greenberg, "IMCO: An Environmentalist's Perspective," *Case Western Reserve Journal of International Law* 8 (winter 1976), pp. 147–148.

since these organizations "all have their origins in advanced Western countries." As such they have increased the "asymmetric relationship" between developed and developing nations.[141] Moreover, within these bodies, those particular national groups with the resources, interest, and voting power can come to dominate the institution itself. The ICS is certainly heavily influenced by the British and to a lesser extent by several other European members, and OCIMF by the five American majors. These institutions are often, therefore, convenient vehicles for the policies of a few national interests both at IMCO and with other governments. Like the subsidiaries of the multinationals which carry their parents' message abroad, these organizations can be yet another means of reinforcing the dominant perspective. Transnational relations may make a complex tapestry of international politics, but if they do, a few colors still stand in bold relief.

THE IMCO FORUM

The Intergovernmental Maritime Consultative Organization was not created as an environmental regulatory agency. When the IMCO Convention was drafted in 1948, neither the rise of the ecology movement nor the notorious developments in tanker pollution was anticipated. No mention of pollution was made in the agency's charter.[142] But events took a different course, and today IMCO's environmental role is one of its most important. Indeed, it has been the continuing conference diplomacy at IMCO in this one area that has set the pace for the international legal protection of the marine environment. It is important, therefore, to understand how the evolution of an environmental legal regime under IMCO has been affected by the agency's formal and informal political processes.

Looking at its *formal* structure, it is clear that the agency was created simply to facilitate the cooperation of likeminded European maritime states and not to channel opposition against them. This is not surprising, for:

> If an organization is to have functions that might affect significant values, those in control of these values will generally demand structural and procedural devices to ensure for themselves the means of exerting special influence. How far they will press their demands and how successful they will be will depend on the configuration of forces they face at the time.[143]

141. Keohane and Nye, eds., "Transnational Relations and World Politics," pp. 736–739. This Western concentration reflects the level of economic development of these countries as well as their pluralistic ideology and highly developed transportation and communications system (p. 737).

142. The pollution function was officially added to the IMCO Constitution in amendments accepted by the Assembly in 1975.

143. Cox and Jacobson, *Anatomy of Influence*, p. 3.

Given the agency's orientation, its structure was intentionally hierarchical so as to place effective policy control in two bodies with tightly restricted memberships, the Council and the Maritime Safety Committee. For many years, this ensured that the organization was run by a limited number of specially qualified states. With the expanding membership and changing expectations of the organization, this particular *formal* control has disappeared. All IMCO's organs have broadened their membership base. In fact, all IMCO bodies (except the Council) are now open to all agency members, including the new Marine Environment Protection Committee. Most importantly, with the November 1977 elections to IMCO Council, control of that central organ has now passed to a majority of developing countries.

Even with this loosening of the formal authority of the maritime states, they have retained considerable informal control. Even in those committees where membership is unrestricted, the extent and nature of actual participation by the developed maritime states far surpasses that of all other members. IMCO is a rule-creating agency and, as such, is dependent on regular meetings of its committees to prepare resolutions or conventions in draft form. This routine procedure is the real lifeblood of IMCO. It is at these meetings with their limited participation that the basic transgovernmental and transnational elites are formed. Personal contacts are renewed on a regular basis; issues are evaluated and alternatives proposed; strategies are agreed on. As these meetings are of a more technical, preparatory nature, representation is largely by individuals from the shipping ministries and industries. The drafts produced tend to reflect their more gradualist, less environmentalist perspective.

The impact of this system on the conferences that follow is profound. The relatively small number of maritime specialists who participated in the preparatory meetings continues to exercise great authority at the conference. In fact, they have often expressed a feeling of frustration and resentment at the many new participants—outsiders—who have been unfamiliar with and often hostile to the draft convention. This is frequently manifested in relation to states that are discussing the issue for the first time, but it is also evident in the hostility that often exists between the regular transport representatives and the delegates from the many foreign affairs ministries. While the former group sees their work as "technical" and "nonpolitical," the latter have been accused (with some justification) of politicizing debate by introducing extraneous issues.[144] As a result, the

144. The antipathy between "technical" and "political" actors at IMCO is one of the omnipresent features of IMCO's work. This attitude was typified by one representative of a maritime transport ministry who, on reading the reference in an earlier draft of this chapter to a "phalanx" of law-of-the-sea negotiators, commented wryly, "A phalanx is an organized company of soldiers. I would prefer a more accurate description of the LOS negotiators, such as horde, mob, swarm (in the sense of locusts), or gaggle."

conferences do not provide a receptive forum for the participation of individuals who are not a part of the already established transgovernmental network.

In this situation, it is not surprising to find that a proposal which has not been introduced in the preparatory committee has very little chance of being accepted at the conference. This is, of course, especially so for proposals which would introduce more stringent environmental controls. Because of the significant amount of work it represents, there is almost an aura of sanctity surrounding the entire draft convention itself which militates against substantial changes in its provisions. A government which seeks bold changes must, therefore, enter the process early, in the preparatory committee. Only those with a strong commitment will be willing to persevere through the long negotiations that must ensue. For those other delegations which seek to strengthen the convention only at the conference itself, they are under a heavy burden, especially as the intense atmosphere of the conference setting itself mitigates against discussion of new proposals for which delegates are unprepared.[145] The potential of the conference is limited. This is the unavoidable, practical reality of IMCO's legislative process.

IMCO's Secretariat has done little to alter the existing political processes. Formally, the organization and its officials have no power. The agency was conceived to be merely a "consultative" forum to facilitate intergovernmental cooperation and not an agency with an independent control function. As a result, the Secretariat was simply to perform a "housekeeping" function: "to maintain all such records . . . prepare, collect, and circulate the papers, documents, agenda, minutes, and information that may be required."[146]

In initiating action within the agency (that is, in identifying and investigating problem areas and setting the agenda) the Secretariat has had little influence. Virtually all the environmental conventions and nonbinding Assembly resolutions have come from either "previous commitments" contained in earlier conventions or from delegation proposals.[147] This governmental control is a common pattern within international organizations,

145. The former British Cabinet minister Richard Crossman found a similar situation when he tried to assert ministerial control over a large bureaucratic government department. So carefully are policies prepared within the bureaucracy before they surface at the ministerial level that "the whole job is precooked . . . to a point from which it is extremely difficult to reach any other conclusion than that already advanced by the officials in advance." Cited in "The Crossman Diaries," *Sunday Times* (London), February 10, 1975.

146. IMCO Convention, art. 2.

147. This has certainly been the case with environmental issues, but one student of IMCO has found this to be the case with *all* of IMCO's work. See R. McLaren, "An Evaluation of a Secretariat's Political Role," Ph.D. dissertation, University of Pittsburgh, 1973, pp. 153–156 and 187–194. Instances of a convention resulting from a "previous commitment" are the holding of the 1962 Conference as a result of resolutions passed at the 1954 Conference, and the 1971 Fund Conference following from the 1969 resolution.

where initiatives may result in new rules of law which have great practical consequences.[148] In IMCO this means that state policy and power is all-important in getting issues before the organization.

So heavy is IMCO's reliance on its members and the relevant industries that occasionally the agency has been almost totally circumvented, at least in the early stages of developing proposals. The genesis for the important LOT amendments occurred totally outside the organization, which was involved only after the proposal had been thoroughly developed *and implemented* by industry in cooperation with a number of important maritime states. Much of the work for the 1971 Fund Convention was done outside IMCO by an informal network of industry and government representatives. And, most recently, the United States developed domestically an entire package of new regulations which in effect it simply asked the agency to ratify. Much of the package was accepted, and what was not was replaced with requirements that had again already been developed or implemented by the oil industry.

The focus of this study has been the entire "issue area" of marine oil pollution. IMCO has been only one actor in it. IMCO has, however, remained the central focus of decision making, not only because it alone can create international legal regulations but because

> it has most fully reflected the real power relations bearing upon the issue area. . . . Moreover, in remaining the central point of decision, IMCO has changed its nature, from being a relatively exclusive "maritime power club" to becoming an organization open to a wider range of influence amongst which the maritime powers continue to be predominant. In effect, the maritime interests seem to have been able to make concessions to other interests without giving away anything really vital to their own.[149]

In all of the environmental negotiations, IMCO was ultimately involved to a substantial degree. Yet if IMCO's attractiveness stems from its being a reflection of the "real power relations bearing upon the issue area," there are strict limits on the contribution the organization can make. IMCO can, to use an analogy, service the machine but not redesign it. The organization has made its substantive contribution in its role as the locus of negotiations and source of legitimacy. Its independent contribution has been

148. As Cox and Jacobson have noted, "The greater the immediate practical consequences are in material terms, the more predominant influence is likely to be exercised by the governments—and also by those private associations with a large stake in the outcome" (*Anatomy of Influence*, p. 389). Very few of IMCO's products are simply "hortatory" statements. Most, if successfully implemented, would impose substantial costs.

149. We are indebted to an anonymous reader for this valuable perspective. We would, however, include the United States within the phrases *maritime powers* and *maritime interests* in order for this statement to be strictly accurate.

extremely limited. However, with the creation of the Group of Experts on the Scientific Aspects of Marine Pollution (GESAMP) in 1969, IMCO has had access to a research body which has produced many useful studies on marine pollution. GESAMP itself has no independent initiating role but has operated well on behalf of IMCO at the latter's request.[150]

With respect to the Secretariat, McLaren—in a study based on an examination of IMCO documents issued during the 1960s—found no noticeable contribution by the Secretariat in initiating topics, delimiting the scope of the problem, fact finding, or suggesting alternatives. Instead, officials functioned in administrative and facilitative capacities. This they do exceedingly well and have been justifiably praised as "efficient facilitators." That they did not undertake a broader role is explained by McLaren to be the result of four factors: their small numbers; the existence of a plethora of experts in many national delegations; the lack of an official executive function for the organization; and the personalities of the Secretariat officials themselves.[151] The first three factors have strongly circumscribed the Secretariat's role to date. Especially where no official benediction is given to initiative and in fact is disapproved of, members of a secretariat are naturally cautious and nonassertive.

This image of IMCO Secretariat members as completely non-political has never been as extreme in reality as that just portrayed. And in recent years they have unquestionably begun to exert greater—although still quite limited—influence. IMCO's first two secretaries-general came from European maritime states and felt little inclination to challenge their expected roles.[152] IMCO's third secretary-general, Colin Goad, was also from a maritime state (the United Kingdom), but he came to the office soon after the *Torrey Canyon* incident when, according to his own observation, the organization began to "take off." At this time, his role did expand, but, as he noted, it was confined to attempts to ensure the survival of the organization as a whole within the United Nations system and to persuade states of IMCO's growing utility.[153] Particular environmental objectives did not figure in his activities.

Under IMCO's fourth secretary-general, C. P. Srivastava, the Secretariat has begun to move in new directions. As one not schooled in the European maritime tradition, Mr. Srivastava could not be expected to be

150. See Chapter 3 for a discussion of GESAMP.

151. R. McLaren, "An Evaluation of a Secretariat's Role Within the Policy Process of an International Organization—The Case of IMCO" (Ph.D. dissertation, University of Pittsburgh, 1972), pp. 225-235.

152. These were Ove Nielson (Denmark), 1959-1962, and Jean Roullier (France), 1963-1967. Colin Goad (United Kingdom) was the third secretary-general, 1968-1973.

153. Interviews. The *Torrey Canyon*'s timing was certainly fortuitous for IMCO, as it was under great challenge at the time from the newly created UNCTAD and its visibility in ECOSOC was languishing.

concerned with the European maritime interests alone. Yet as an Indian national who achieved great success as head of the Shipping Corporation of India, he has been held in great respect by the maritime community. In addition, having successfully chaired UNCTAD's Committee on Liner Conferences for two years prior to his arrival at IMCO, he was well acquainted with the political aspects of maritime affairs, particularly those concerned with maritime development. The augmentation in the Secretariat role in recent years has unquestionably been furthered by Mr. Srivastava's occupation of the secretary-generalship. But, to a much greater extent, it has been due to other developments—the growing importance of marine safety and marine pollution control, the creation of new IMCO bodies and new memberships for old ones, the growing assertiveness of the developing states, divisions among the developed states, and increases in Secretariat personnel. Many of these developments had been going on for some time, but they have accelerated in recent years as never before.

The first goal of the Secretariat is to preserve and expand the political responsibilities of the organization, and this has not changed in the recent period. At the same time, there has been a diversity of new opportunities to pursue it. At the Law of the Sea Conference, the IMCO Secretariat members have been very visible in their quest for enlarged responsibilities for the agency, and they have been vigilant in protecting IMCO's area of jurisdiction from any encroachment by UNEP or UNCTAD. Within IMCO, the paramount responsibility has been, as ever, to demonstrate the agency's usefulness by concluding successful conventions in all fields and by enhancing the organization's attractiveness to its members. This has led the secretary-general into some political activism, especially in the period prior to the TSPP Conference. In the face of possible unilateral action by the United States, Mr. Srivastava was able to plead the cause of IMCO in an attempt to dissuade the United States from undertaking any action that could harm it. In this he had the support of the entire maritime community and so was not, in any way, risking attack. However, insofar as he did restrain the United States, his actions had an impact on the shape of the final outcome. With his particular background, Mr. Srivastava has also been eager to enhance IMCO's appeal by improving its work in the field of technical assistance, an area in which the agency had long been weak. The activities of the Technical Cooperation Committee in maritime training and development have been financed by UNDP, UNEP, or voluntarily, so their expansion has not impinged financially on the IMCO membership. Finally, the secretary-general has been eager to improve IMCO's attractiveness by showing that it works—that its conventions are ratified. This he has actively promoted through numerous diplomatic initiatives. His tactful demarches in this area have apparently not elicited hostile reactions and have, in any event, been supported by the Council. The first goal of preserving and enhancing the organization has, in summary, occasionally led the secretary-general into ''political'' areas, but none that have had serious consequences for the ''real power relations'' among states in the maritime area or for environmental protection.

The second goal of the Secretariat has been to promote the *successful* formulation of conventions. Here its greatest contribution has unquestionably been to *facilitate* the discussions—as it did so well recently by organizing a large, intergovernmental conference on only nine months' notice. In this function, the Secretariat has occasionally strayed into the realm of substantive politics—into the negotiations themselves. For example, not only did the secretary-general work hard before the 1978 TSPP Conference to forestall American unilateralism and achieve an international solution acceptable to the maritime states as well, but he served as an effective bridge to many of the developing nations. His task was to convince these countries that their interests lay not in rejecting new environmental standards but in ensuring that an acceptable compromise was reached. Before the conference, he was also active in ensuring that useful chairmen would be appointed to the delicate jobs of overseeing the committee debates. Within the debates themselves, Secretariat officials take a very quiet but nonetheless important role. Although they never openly advance their own proposals, they will often seek to prevent problems from arising from the proposals of others.[154] Often they will explain the ''technical'' implications of unsound proposals to their proponents or, while not arrogating to themselves the role of defining the ''global interest,'' will elucidate constructive proposals that are being misunderstood or unfairly attacked. As well, they can function as peacemakers trying to restrain extreme behavior, explaining one state's apprehensions to another, or advising a state on how best to pursue its goals so as to minimize conflict. These are small but useful additions to the negotiating process. And in seeking to ensure that the conventions produced are ratified, the secretary-general, in particular, has helped to give substantive meaning to what might otherwise be purely paper achievements.

In conclusion, the role of the Secretariat is primarily to preserve the organization while facilitating interstate agreements. Its activities do occasionally stray into substantive political fields but only tangentially and never in a way that challenges the powers of the member states. Despite its central access to the many networks of transgovernmental and transnational interaction, it seldom has used its position to build its own special

154. Occasionally Secretariat officials will put forward specific proposals, but only surreptitiously through either their own national delegation or a sympathetic one.

interest alliances or to advance its own policy preferences. It certainly has not done so on environmental issues.

In addition to those constraints on IMCO's role as an environmental agency is the external expectation that IMCO remain a limited purpose organization. Most importantly, the organization is not expected to involve itself in altering the basic jurisdictional regime of the oceans. On one level, proposed legal regulations have been opposed whenever they might set legal precedents for other areas of maritime law. At a much more basic level, however, IMCO's work is confined by the interdiction that it not affect the prevailing laws of the sea. Each attempt to deal with such issues at IMCO has failed. Also, with the probable termination of UNCLOS III and the absence of a continuing body to revise the environmental law of the sea, the future work of IMCO will likely be even more confined in dealing with changing global needs than it has in recent years. To a large degree, the revision of relevant jurisdictional and enforcement provisions will depend not on IMCO but on the ability to convene yet another law of the sea conference. Developments in maritime environmental law will have to await demands for change in other unrelated issues in the law of the sea.

Not only is IMCO restricted by the external oceans regime but, like all international organizations, its effectiveness in promoting environmental protection is also limited by the resistance of member states to limitations on their independence. This resistance is very real, whether derived from "habit, parochial nationalism [or] the 'autonomy illusion' that by retaining national control of policy, one retains the substance of power."[155] As Skolnikoff has noted, "The general attitude of governments in this era of nationalism is, and will almost surely continue to be, characterized by a jealous guardianship of national prerogatives."[156] This has certainly manifested itself many times at IMCO. One instance of this has been the continuing opposition of any regulations that could affect a state's internal legal system. This had a decided impact on the British proposals in 1962 and 1967 and the American proposal in 1973 for reversing the burden of proof for discharge violations. It has similarly affected proposals to set an international scale of penalties and to allow IMCO or coastal states to monitor flag-state prosecution proceedings. Not only does this often hamper the possibility of achieving improvements in specific regulatory provisions, but

155. Robert Keohane, "International Organization and the Crisis of Interdependence," *International Organization* 29 (spring 1975), p. 359.

156. Skolnikoff, *International Imperatives*, p. 117. Also see John G. Ruggie, "Collective Goods and Future International Collaboration," *American Political Science Review* 66 (September 1972), pp. 874–893. Ruggie has written (p. 878), "the propensity for international organization will be determined by the interplay between the need to become interdependent upon others for the performance of specific tasks, and the general desire to keep such dependence to the minimum level necessary."

it confronts at a basic level the long-term requirements of an effective global environmental protection system.

In conclusion, as a forum IMCO has served as a valuable focus for a large transgovernmental and transnational network and has generated many important accords. At the same time, IMCO has been restricted formally and informally, internally and externally, and has been hindered from developing as an *independent* environmental actor. As such, it is like most international organizations—a mirror of the pattern of interests and influence of its governmental and nongovernmental participants.

CONCLUSIONS

Environmental pollution is a new danger on the international political scene, but it has not evoked a new response. Instead, attempts to protect the global environment have been as buffeted by political bargaining as have attempts to regulate, say, the terms of international trade. If, as the functionalists claim, there is a "common good" to be perceived in technical areas, the experts on maritime pollution have done a good job at hiding it. Partisan interests prevail.

The history of the negotiations at IMCO points to a multiplicity of factors which have affected the development of international environmental regulations. First, of course, is the nature of the problem itself. Anything short of visible, immediate catastrophe is equivocal and invites an equivocal response. The burden of proof is overwhelmingly on the environmentalist and not on the polluter. Pollution is but a side effect of a larger, expanding global economic and technological process, and this process resists restriction. Each state and each industry must protect its own specific competitive interests—despite the fact that pollution ignores boundaries and is a global problem demanding a cooperative response. At a very basic level, our economic and ecological systems are in conflict.

Within international organizations, states are the ultimate decision-makers and their decisions reflect a variety of interests and influences. In the case of oil pollution regulations, some of these interests and influences spring from factors intrinsic to the issue of maritime oil transport while others seemingly bear no relation to it. Of course, those states possessing industries affected by environmental regulations (in this case, states with shipping, oil, or insurance interests) will seek to protect them. The impact of these intrinsic interests can be seen most clearly in the *technical* regulations which bear directly on them. Here the policies of the shipping nations diverge from those of the nonmaritime states in general. However, some particular issues have significance beyond their impact on pollution-related industries. At this point, extrinsic concerns may intrude into the negotiations. This has been so particularly with *jurisdictional* issues that states see as affecting their larger economic and military ambitions. Here too environmental concerns may be jettisoned in the protection of other interests.

Most interesting, however, is the relationship between environmental policies and the larger (extrinsic) pattern of world power and allegiance. In general, the environmental problem, in its source and in its solution, is a problem for the developed, democratic states. It is Europe, Japan, and the United States that are the major polluters. Yet it is also these "WEO" states, with their vast wealth and democratic governments, that have provided the environmental initiatives for the world. And it is from this group that the many transnational organizations have come which are of significance to the environment both within and without IMCO.

Other states are not disturbed by the international oil pollution problem or are, at best, reactive to proposals about it. The Soviet Union only reacts to IMCO's environmental activities, and even then its reactions are ultimately a reflection of its closed and isolated position in the larger structure of world political alignments. The developing countries pay little attention to IMCO, and when they do it is their identification with the necessities of economic development, not oil pollution, that conditions their policies. IMCO provides a focus for global involvement, but the issue is still one for the Western rich.

In this situation, IMCO's independent role is a limited one. For years, the agency has been little but a forum for the Western shipowners. They set the tone for its work, and they defined the scope of its undertakings. Recently, with both a growing split between the United States and Europe and generally increased opportunities for participation by all states through new bodies such as the MEPC, the potential for an augmented contribution by the organization and its Secretariat has become apparent.

Chapter VIII

The Implementation of Conventions: From Treaty to Law, From Law to Compliance

During a recent IMCO meeting a national shipping official stood in the lobby of IMCO's London headquarters gazing at the glass-encased model of a vessel presented by one of the member states. Like many more distant observers, he mused, "I'm not sure what relationship there is between that convention we're discussing upstairs and what those ships are going to be doing on the oceans in the future." Almost without exception, the sentiments of this experienced official regarding the difficulty of translating written conventions into maritime practice have been echoed by other experts in the field. Unquestionably the gap between the conference hall and the ship at sea is immense.

The process of translating convention norms into maritime practice has two basic stages. First, the convention must attain the formal status of international law. This requires that the treaty be ratified by a specific number and/or group of states. Second, vessels and states must comply with the legal strictures they have accepted. That is, the law must be enforced. Each of these two processes is strewn with hurdles and impediments—the drafting of an international convention is by no means the end of the political process.

FROM TREATY TO LAW

The process of translating international conventions into law is slow and uncertain. Only six of IMCO's nine accords on oil pollution have entered into force in the past two decades and, even then, only after a hiatus of about six years each.[1] Three agreements—the two 1971 Amendments, and the 1973 MARPOL Convention (including the 1978 Protocol)—have not, despite many years since their drafting, attained the status of international

1. The 1954 Convention became law in 1958, the 1962 Amendments in 1967, the 1969 Amendments in 1978, the 1969 Intervention and Civil Liability Conventions in 1975, and the 1971 Fund Convention in 1978.

law. The delays and uncertainties involved in turning conventions into mandatory laws undermine the very basis of IMCO's "legislative" process. Yet in contrast with the energy and publicity surrounding the creation of a new convention, this massive obstacle has been inadequately discussed or confronted.

Ratification is the acceptance of an international agreement as binding domestic law by the government of a state.[2] The mechanics of *entry-into-force* vary with each convention or amendment as determined by the participants at the particular drafting conference. Generally, however, a convention will enter into force when it has been ratified by an agreed number of states, including a certain proportion of those states with a special interest in the subject matter. For example, the 1971 Fund Convention (article 2) provides that the convention shall enter into force after ratification by "at least eight states" which have transported by sea "during the preceding calendar year a total quantity of at least 750 million tons" of oil.

In the five pollution prevention conventions considered in this study, the number of ratifications required is quite small: 1954 Convention, ten; 1969 Intervention, fifteen; 1969 Liability, eight; 1971 Fund, eight; 1973 MARPOL, fifteen.[3] However, the provision requiring acceptance by a number of states with a special interest in the subject matter has ensured that the actual entry-into-force of most conventions has effectively been controlled by a small group of maritime states.[4] Indeed, the requirement of the Fund Convention that the convention would enter into force only when the ratifying states account for 750 million tons of oil moved by sea was found to be extremely restrictive. The stipulation was intended to ensure that the convention would come into force only if the largest oil importers (United States and Japan) ratified it with the acceptance as well of the large European importers (Italy, France, the Netherlands, the United Kingdom,

2. The terms *acceptance* and *ratification* will be used interchangeably to indicate a state's consent to become party to a convention. The term *acceptance* is often used "to connote ratification, accession, succession or any other form by which a state expressed its consent to become a party to a treaty." Oscar Schachter, Mohamed Nawaz, and John Fried, *Toward Wider Acceptance of U.N. Treaties* (New York: Arno Press, 1971), p. 91. For a discussion of different procedures employed by states, see pp. 93–131.

3. The entry-into-force provisions for the respective conventions can be found as follows: 1954—article 11, 1969 Intervention—article 11, 1969 Liability—article 15, 1971 Fund—article 40, 1973—article 15.

4. The ten states referred to in the 1954 Convention also had to include "five governments of countries each with not less than 500,000 gross tons of tanker tonnage." This gave control to about ten maritime states, while a similar provision in the Liability Convention focused control on about the top fifteen shipowning states. As there were no adverse implications for maritime interests in the Intervention Convention, no special provision was inserted beyond the need for fifteen ratifications. It should be pointed out that it is the nonmaritime states that usually demand a high absolute number of ratifications in order to prevent a few maritime states from bringing a convention into force.

and West Germany).[5] This stipulation long stalled the convention's entry-into-force.[6]

It is quite possible to criticize these provisions to the extent that they unnecessarily impede the early implementation of agreed conventions. Yet with all the conventions before 1970, the special interest provision had no tangible effect, as the earliest ratifiers were the maritime states (see Chart 17). However, the figure used in the 1971 Fund Convention was undoubtedly an impediment,[7] and it may be questioned whether the formula used in the 1973 Convention is not too high. Four years after the completion of the MARPOL Convention, it had received only three of the fifteen ratifications needed, so was not near to entry-into-force in any event. But there is an additional requirement in both the 1973 Convention and the 1978 Protocol (which has now superseded the 1973 MARPOL Convention by incorporating it) that the ratifiers must also possess 50 percent of total world grt. This gives almost a veto power to the top four or five shipping states that account for well over half the world's merchant tonnage. Clearly this provision may be excessive, and such provisions should be carefully controlled by other IMCO members in the future.

At the same time, it cannot be denied that such special controls do have their place. Acceptance of a convention by a number of states is an obvious prerequisite to its functioning effectively. That this number should also include the major actors in the field is not only realistic but desirable. Without these actors, the new rules would, in practice, be quite meaningless. Like the development of customary international law, the law must be accepted by those with the greatest interest. More importantly, no major maritime state would ratify a convention imposing significant competitive disadvantages on itself except where its main competitors also accepted the measures. By including the safeguard requiring a number of the major competitors to accept the new measure before it becomes effective, a substantial *dis*incentive to early ratification is removed. In a voluntaristic legal system such as exists internationally, these safeguards are unavoidable.

Much room for improvement does exist, however, for not only do the

5. See Appendix 1, Table C. These countries accounted for about two-thirds of world oil imports by sea. The actual stipulation in the convention (sec. 10) refers to oil "received" in its ports, not imported. The intention was to set a figure (750 million tons) that would be about 50 percent of the expected annual oil movement by sea in a few years. The effect was to concentrate control of the convention's entry-into-force in a very few states.

6. The Convention required the ratification of France to push the tonnage total from 633 million to over 750 million in 1978. The need for this single last ratification caused a delay of two years.

7. The reason it was set so high was to spread the cost of a spill and because of the certainty that an Assembly would be created to administer the Fund immediately upon the convention's entry-into-force. Large oil importers wanted to ensure that the Assembly could not be set up before they were a party to the convention.

provisions for entry-into-force need to minimize impediments but they need to maximize incentives for ratification as well. In this latter regard, some important progress has recently been achieved. In the 1971 Amendments, the 1973 Convention, and the 1978 SOLAS and MARPOL protocols, actual dates were stipulated when the provisions on ship construction would apply even *before* the legal entry-into-force of the agreements. This has meant that the conventions had a wide practical impact soon after they were drafted, thus reducing the costs involved in ratification.[8] Similar to this is the fact that many states—including the United Kingdom[9]—have in the past implemented domestic legislation incorporating the provisions of new conventions long before they were accepted internationally. This was done on a widespread basis in the 1970s in order to allow the 1969 Amendments and the load-on-top system to be implemented. The 1978 TSPP Conference went so far as to accept two resolutions encouraging states to implement the provisions on existing ships before they legally come into force.[10] This is a major development which could dramatically expedite the implementation of new international environmental regulations if powerful countries like the United States take advantage of it.

Another incentive to wide ratification has been created by stipulating that contracting parties are to apply the provisions of a new convention to *all* vessels whether or not the flag government is itself a party. This has the practical effect of ensuring that all states comply with the convention—once it has come into force—putting any *non*complying states at a competitive disadvantage vis-à-vis the contracting parties. This was done in the 1969 Liability Convention, where a contracting party is obligated to require full insurance for all vessels entering its ports and not just for those accepting the convention. Noncomplying vessels stand to lose access to a substantial portion of the available market. Not surprisingly, the result has been that virtually all major international traders have accepted the convention since it came into force in 1975. Those that have not (only Italy and Finland) comply with it in any event. The 1973 Convention also requires con-

8. The stipulation limiting tank sizes in the 1971 Amendments and in the 1973 Convention (annex I, reg. 24) and the requirement for segregated ballast [1973 Convention, annex I, regs. 13 and 1(6)] set such dates. As a result, despite the fact that these instruments have not yet entered into force, vessels have been built to the new specifications.

9. The Intervention Convention was, for example, incorporated into the United Kingdom's Prevention of Oil Pollution Act (1971). This was four years before the convention came into effect. Japan and Sweden did it at the same time.

10. Resolutions 1 and 2 of the 1978 Conference provide that the requirements be implemented "to the maximum extent without waiting for the entry-into-force of the instrument" (in the case of the MARPOL Protocol) and "by" a certain date (in the case of the SOLAS Protocol). TSPP CONF/12, Res. 1 and 2.

tracting parties to apply the convention to the vessels of nonparties "as may be necessary" to prevent their being treated more favorably.[11]

Such improvements as these are important for expediting the practical implementation and legal acceptance of new international controls. Others must be adopted as well, and one of the most important must be in the area of *amending* existing conventions. Progressive amendment is a far more expeditious and efficient method of legislative development than is the creation of new conventions. This is, of course, the method of domestic legislation, and it allows governments to focus only on those areas most in need of improvement. It is especially necessary at the international level, where national suspicions result in endless examination of all aspects of a new convention in order to ferret out all their "national implications." Given the national protectionism and inertia of modern bureaucracies, this is not a minor consideration.

The absence of a reasonable amendment procedure has been one of the greatest impediments to successful work by IMCO in the pollution prevention field. The 1969 and 1971 conventions have completely omitted any provision for amendment, requiring instead a full-scale diplomatic conference to alter them. This is quite unnecessary and inefficient. By contrast, the provision for amendment in the 1954 Convention is so onerous as possibly to be even worse than no provision at all. That convention stipulates that an amendment can be legally incorporated into the convention only after two-thirds of the contracting parties have ratified it. Despite the fact that the load-on-top system was the basic operating procedure at sea even before the 1969 Amendments were drafted, the cumbersome amendment requirements prevented the 1969 changes (and the LOT system) from becoming international law until 1978. As more and more states ratified the 1954 Convention, more states were needed to attain the two-thirds requirement to amend it.

A serious obstacle to the rapid development of the law has been, therefore, the absence of a fast, efficient amendment procedure. This has hindered IMCO's ability to make an independent contribution to the law's

11. At the 1973 Conference the U.S. urged that such application to the vessels of noncontracting states be made obligatory, but many states either opposed it or preferred some flexibility. However, an article finally was accepted which did require parties to apply it "as may be necessary to ensure that no more favorable treatment is given to such ships" [art. 5(4)]. It is interesting that one source of opposition to these proposals has been from the Soviet bloc. Due to their minority position in the international community, they strongly desired to restrict international law creation to explicit contracts between states. The developing states sympathized with this position especially as the measure obviously would be directed against them. This caused the defeat of the *mandatory* provision. The developed maritime states will undoubtedly apply article 5(4) (once they have accepted the convention and it is in force) despite its voluntary nature in order to protect their competitive positions.

evolution, especially when IMCO is compared to other international organizations.[12] The 1973 Convention (as incorporated in the 1978 Protocol) has somewhat alleviated this problem with the introduction of the "tacit acceptance" procedure. This provides that two-thirds of the states at a meeting of the "appropriate" IMCO body—the MEPC—can (where they also account for 50 percent of world grt) amend the annexes of the convention, with the amendments to come into force automatically for all parties.[13] All parties will have at least ten months to refuse to accept the amendments, although if the amendments have already been accepted by two-thirds at the IMCO meeting, it is unlikely that they would then be rejected. Although the "explicit acceptance" procedure is retained for the convention articles, the introduction of tacit acceptance for the technical annexes is an important structural change for, as Contini and Sand have pointed out, "What will be crucial for future environmental management is the capacity of the system of norms to respond adequately to a constant change of situations, including crisis situations, and to accelerating technological progress."[14]

Ultimately, however, the languid progress of IMCO's pollution conventions is not a structural problem of the entry-into-force provisions but a political problem of national opposition or indifference. Opposition to new conventions is not surprising, as the conventions are never to everyone's satisfaction. For example, for the reasons they espoused during the conferences, France continued for many years to oppose the 1971 Fund Convention and Canada, the 1969 Liability and 1971 Fund conventions. Such opposition on particular policy grounds is to be expected from numerous states for each new convention. In general, however, it cannot be said that the cost implications of new measures should have produced major policy impediments to rapid ratification. The 1954 Convention and its 1962 Amendments required only minor expenditures by the shipping industry and minimal enforcement effort by contracting states yet took four to five years to enter into force. Likewise, it was six years before the Intervention and Liability conventions were implemented even though, by 1971, it was apparent that the only real cost of either convention, pollution liability

12. For a comparison to other "quasi-legislative" international agencies, see C. H. Alexandrowicz, *The Law-Making Functions of the Specialised Agencies of the United Nations* (Sydney: Angus and Robertson, 1973), especially pp. 60–78 and 152–161. ICAO has been particularly successful at achieving acceptance of its conventions and their regulations. As one delegate commented to the authors in referring to ICAO's greater success, "Politicians fly in aircraft—hence the safety of civil aviation. Nobody important travels in tankers."

13. Art. 16(2)(*f*)(*i-iii*) of 1973 Convention and art. 6 of 1978 MARPOL Protocol.

14. Paul Contini and Peter H. Sand, "Methods to Expedite Environmental Protection: International Ecostandards," *American Journal of International Law* 6 (1972), p. 39. On IMCO's new amendment procedures, see A. O. Adele, "Amendment Procedures for Conventions with Technical Annexes: The IMCO Experience," *Virginia Journal of International Law* 17 (1977), p. 201.

insurance, was readily affordable. Only the 1971 Fund and 1973 MARPOL conventions had significant ramifications.[15]

The real problem of national ratification is not opposition but indifference. A report to the International Law Commission noted quite accurately,

> *political reasons*, or more often fears concerning the possible repercussions of certain rules on particular situations, may explain delay in ratification or accession, or even failure to ratify. But in most cases the reasons why a State delays transmission of the instrument formally establishing its consent have nothing to do with any real opposition, either on principle or on a particular point.[16]

Unlike the adoption of a convention, its ratification is not a "collective action" but "a series of individual actions," each of which must first overcome the massive "inertia of the political and administrative machinery of the modern state."[17] A very different set of actors will often be involved than those who actually negotiated the convention. Indeed, codification will likely require substantial coordinated effort by a number of ministries and the legislature. These bureaucracies and legislators are already overburdened by pressing domestic responsibilities, and they are under little pressure and no deadline for initiating long and complex codification procedures. Many officials (especially those from the developing countries) do not even understand the technical requirements or how to implement them.

In this situation, only those states with a motivation to ratify will do so. Not surprisingly, therefore, the developing countries, despite their predominance in the total IMCO membership, have ratified IMCO accords in no greater numbers than the much smaller group of WEO countries.[18] It has actually been the ratifications of the WEO group, and particularly the major shipping states, that have been responsible for bringing the conventions or amendments into force.[19] In the past, the developing countries

15. The 1971 Fund Convention requires extensive international collaboration through the creation of a Fund and a Fund Assembly. The 1973 Convention required many technological improvements to existing and future tankers and demanded the installation of shore reception facilities by coastal states. These latter cost disincentives were exacerbated by the fact that states had to accept the requirements for oil reception facilities (annex I) and those for chemicals (annex II) simultaneously. This problem has largely been removed with the 1978 MARPOL Protocol.

16. Robert Ago, "The Final Stage of the Codification of International Law," International Law Commission, 20th session, U.N. doc. A/CN.4/205/Rev.1, July 29, 1968, part. 10 and *passim*. Also see Schachter et al., *Toward Wider Acceptance*.

17. Ago, "Final Stage," pars. 6 and 11.

18. Overall, the WEO states have provided 47 percent of the ratifications, the developing states 48 percent, and the Soviet grouping 5 percent. This pattern is comparable to that discovered in an analysis of fifty-five U.N. multilateral treaties. See Schachter et al., *Toward Wider Acceptance*, pp. 27-28.

19. Of those ratifications that brought measures into force, fifty-six came from the WEO group, thirty-four from the developing states, and three from the Soviet bloc.

have followed the lead of the developed states, although this pattern has been changing in recent years.[20]

Within the developed "Western European and Others" grouping, it is the large maritime powers that are the best and earliest ratifiers. Although they value the environmental benefits these conventions do confer on them,[21] their interest is overwhelmingly in achieving uniform international shipping standards and in preventing the unilateral alternative. The Scandinavian shipowners and the United Kingdom have the best ratification records. The other maritime states, although not quite so enthusiastic, have also been instrumental in achieving the acceptance of the international standards. If the shipping states are to have access to a wide variety of markets, they must all "join the club."[22]

On the issue of ratification, the problem lies not with the European maritime nations but with the developed nonmaritime states. In general, their ratification records are abysmal. Except for New Zealand, even the leaders of the coastal state group at IMCO have not accepted many of the conventions they fought so hard to achieve. The United States and Australia have ratified little, and Canada has completely ignored the four conventions that have been drafted since the *Torrey Canyon*. Canada's failure here reflects its steadfast opposition to the limited jurisdictional and geographical scope of these conventions.[23] But, like the United States, its attitude also springs from its continuing mistrust of IMCO and its international control, and from a predilection for the unilateral alternative. Unquestionably, the threat of unilateral action by these states has been a most powerful motivating force for environmental protection at IMCO, but these states have worked hard and successfully to gain acceptance for a wide variety of new measures. Their failure to accept the 1969 Liability

20. In the case of the 1954 Convention, the WEO nations constituted ten of the eleven countries that brought the treaty into force; the 1962 Amendments, fourteen out of twenty-one; the 1969 Intervention Convention, nine out of fifteen; the 1969 Liability Convention, only five out of fourteen; and the 1969 Amendments, only nineteen out of thirty-eight.

21. Japan's ratification of the 1954 Convention was, for example, at least partially attributable to its government's inability to prosecute discharges which had occurred adjacent to its coasts. A Japanese official commented that the government was not only unable to send reports of discharges outside the territorial sea to flag states for prosecution and hence deter such polluting activities, it was also faced by claims of ship officers whose vessels discharged oil within the territorial sea that they did not think such discharges were illegal because Japan was not a party to the 1954 Convention—and this argument was sometimes accepted by Japanese courts.

22. The parallel between extent of ratification and conservatism in the conferences is not perfect. While most maritime states are eager to accept new conventions in the interests of uniformity, some states (such as Greece and Panama) seem to seek to avoid stricter regulations right up to the point of their entry-into-force.

23. Canada's fear is that any acceptance of these conventions would prejudice its claim to a larger coastal jurisdiction. This is not necessarily so, especially as other countries with broad coastal claims (such as Brazil and Ecuador) have ratified the Liability Convention.

Convention and their part in stalling the implementation of the 1971 Fund Convention are unnecessarily obstructionist.[24] Such failures to ratify treaties undercut their reputation and bargaining power at IMCO and undermine the organization's legislative process. If the environmentalist coastal states are to be taken seriously in their future demands at IMCO, they must not only be willing to pay their own costs for implementing new measures that they demanded (such as the 1973 MARPOL Convention and its 1978 Protocol), but they must positively encourage their entry-into-force for all.

This obligation is especially great when one considers what a minor role most other states play in the ratification process. Most Communist states, for example, have virtually ignored IMCO's environmental work. As the only state in Eastern Europe with a large tanker fleet, the Soviet Union has ratified those conventions that will allow it to travel freely to the ports of any IMCO member, but it has certainly not encouraged their entry-into-force. Except for Poland, no other East European country has followed the Soviet lead.

In general, the developing countries have played an equally inactive role, although some have taken more interest in recent years in IMCO's environmental conventions. The flag-of-convenience countries which seek access to the ports of the developed world—Liberia,[25] Panama, and the Bahamas—have accepted the major conventions, although many of the developing maritime states such as Brazil, India, Venezuela, and Argentina have not. For these latter states the additional costs implied by a widespread acceptance of IMCO conventions are to be avoided. (This is not always possible when agreements such as the 1954 Convention have attained a wide measure of acceptance.) With less concern for the environment and rather ad hoc shipping policies, their ratification patterns are quite haphazard.

Ratifications by nonmaritime developing states are equally infrequent and sporadic. Thirty-eight of IMCO's developing country members have ratified no conventions at all. When others have accepted new regulations, a number of diverse factors—such as the presence of an interested official in a maritime affairs post or a general desire to "join the club" and appear as

24. No national scheme could possibly be as efficient as the international Fund. Indeed, the proposal to establish a "Super Fund" in the United States is grossly inefficient, as it would result in collection of revenues to a limit of $200 million *in advance* of any spill. The Canadian "Maritime Pollution Claims Fund" already had amassed over $40 million by the end of 1977 and yet had been used only once for a very small claim. National schemes such as these are both extravagant and unnecessary.

25. Liberia's sweeping ratification of five agreements in 1972 corresponds with its adoption of its "blue-chip" flag-of-convenience policy. Likely, the sudden ratification in 1976 of many conventions by Panama and the Bahamas also reflects the increased environmental awareness of other states.

Chart 17
Ratifications by IMCO Members (to November 1977)[1]

WEO	1954 Convention	1962 Amendments	1969 Amendments	1971 (Tank Size) Amendments	1969 Intervention	1969 Liability	1971 Fund	1973 Convention
Norway	57	63	71	74	72	75	75	
Sweden	56	63	72	72	73	75	75	
U.K.	55	63	71	74	71	75	76	
Denmark	56	64	71		70	75	75	
Finland	58	66	74	74	76			
France	57	63	72		72	75		
Germany (F.R.)	56	64			75	75	76	
Japan		67	71		71	76	76	
Netherlands	58	63	75		75	75		
Belgium	57	66	73		71	77		
Canada	56	63	72	74				
Greece		67	76	75		76		
New Zealand		71	76		75	76		
Switzerland	66	66	77	77				
Spain	64		76		73	76		
Iceland	62	66	70					
Italy	64		75	76				
Malta		75	75	75				
U.S.	61		73		74			

USSR and bloc	1954 Convention	1962 Amendments	1969 Amendments	1971 (Tank Size) Amendments	1969 Intervention	1969 Liability	1971 Fund	1973 Convention
USSR		69	71	76	74	75		
Poland	61	63			76	76		

Developing states	1954 Convention	1962 Amendments	1969 Amendments	1971 (Tank Size) Amendments	1969 Intervention	1969 Liability	1971 Fund	1973 Convention	Number of Ratifications
Tunisia		73	73	73	76	76	76	76	7
Bahamas		76	76	77	76	76	76		6
Liberia	62		72	72	72	72	72		
Syria		68	75	75	75	75	75		
Algeria	64		76	76		74	75		5
Lebanon		67	72	72	75	74			
Yugoslavia		74	76	76	76	76			
Dominican Republic	63		77		75	75			4
Fiji		72	72		72	72			
Jordan	63	64		72				75	
Panama	63		76		76	76			
Philippines	63	65	73	73					
Egypt	63	63	72						3
Ghana	62	64	76						
Ivory Coast		67		72		73			

																		Malagasy	65	*65*	*71*						
																		Mexico	*56*		77		76				
																		Morocco		68			74	74			
																		Saudi Arabia		71	*71*	*75*					
																		Senegal		72			*72*	*72*			
																		Surinam		76	77		75				
Australia	62		*73*						Bulgaria		76	77						Argentina		76	*76*						2
Ireland	*57*	*65*																Chile		77			77				
Israel	65	*66*																Ecuador					76	76			
Portugal		67				76												Kenya		75						*75*	
																		Kuwait	61	*63*							
																		Libya		72	*76*						
																		Nigeria	63		*77*						
Austria		75																Brazil						76			1
																		Cuba					76				
																		India		74							
																		Uruguay		75							
Turkey									Czechoslovakia									Venezuela	63								
									Germany (D.R.)									Thirty-eight others[2]									0
									Hungary																		
									Romania																		

NOTE: Italicized dates indicate that the ratifications have helped or will help bring that instrument into force. Ratifications of the 1954 Convention occurring after the 1962 Amendments came into effect (for all parties to the 1954 Convention—art. XVI) are entered in the 1962 Amendments column.

1. All conventions and amendments except the minor 1971 Amendment pertaining to the Australian Great Barrier Reef have been included. Several non-IMCO members have ratified IMCO Conventions: Monaco—1954 Convention in 1970, 1969 Amendments in 1975, 1969 Intervention in 1975, and 1969 Civil Liability in 1975; South Africa—1959 Liability in 1976; and the People's Democratic Republic of Yemen—1954 Convention in 1969. In the case of the Netherlands Antilles, the 1954 Convention was extended to it in 1962, the 1969 Amendments in 1975, and the 1969 Intervention in 1975. In the case of Hong Kong, the 1954 Convention was extended to it in 1969, the 1969 Intervention in 1974, and the 1969 Liability and the 1971 Fund in 1976.

2. Bahrain, Bangladesh, Barbados, Burma, Cambodia, Cameroon, Cape Verde, China, Colombia, Congo, Cyprus, Equatorial Guinea, Ethiopia, Gabon, Guinea, Haiti, Honduras, Indonesia, Iran, Iraq, Jamaica, Korea, Malaysia, Maldives, Mauritania, Oman, Pakistan, Papua-New Guinea, Peru, Sierra Leone, Singapore, Sri Lanka, Sudan, Thailand, Trinidad-Tobago, Tanzania, Zaire.

"good environmentalists" when this can be achieved at no cost—have been important. At the same time, however, there does appear to be an active environmental interest, particularly in recent years, by the coastal states of Africa and especially the Arab states on the Mediterranean Sea and Persian Gulf. These states have important coastal interests and no real shipping concerns, and their ratifications have been important in bringing into force those agreements (such as the 1969 Amendments and Liability Convention) with definite coastal benefits. As with other developing states, however, it is most unlikely that they will devote any resources to enforcing the agreements they have accepted.

The codification process is one of the weakest links in the chain of international law creation. However, as one recent study has pointed out, there is much that an international organization and its secretariat can do to expedite a convention's acceptance.[26] The International Labor Organization, for example, requires states to bring its agreements before their legislatures within a year and to report and explain any actions taken or not taken.[27] IMCO has certainly not gone this far, although in recent years it has begun to devote greater attention to the problem.[28] The secretary-general has set improved ratification as one of his prime goals and was diplomatically active in the campaign to bring into force the 1969 Amendments. IMCO's greatly expanded technical assistance program (see further on) has also been useful in this regard, and the MEPC now gives some consideration to the status of IMCO's environmental conventions at one meeting every year. At those sessions, governments are expected to explain their policies.[29] In addition, the 1978 Conference on Tanker Safety and Pollution Prevention has set target dates by which all members should ratify both the 1973 MARPOL Convention and its Protocol and the 1974 SOLAS Convention and its Protocol. All states not meeting these targets must submit explanations on their failure to do so to the secretary-general.[30] These measures represent a valuable beginning of the effort to *internationalize* the codification process, although much remains to be done if

26. Schachter et al., *Toward Wider Acceptance,* pp. 62–79. The authors refer to disseminating information, giving advice about ratification, revising the treaties, and providing technical assistance as four main techniques. The provision of technical assistance will be discussed further on.

27. Ibid., pp. 57–60. The League of Nations recommended that a new conference should be convened automatically after a set period if the required ratifications had not then been received (p. 71).

28. The formal functions conferred on the IMCO Secretariat by the entry-into-force provisions give little more than administrative powers such as receiving the instruments of ratification, collecting information of relevance to ratification, and circulating relevant documents. Until recently, little more was done.

29. MEPC VII/WP.7, par. 7.

30. TSPP/CONF/12, res. 1 and 2.

states are really to be accountable for their actions on the accords they conclude.

In conclusion, the whole ratification process provides a shorthand account of the low deference states give to the requirements of an international legal order of environmental protection. The provisions that have been designed to bring new agreements into legal force have incorporated only minor incentives to ratification while sometimes positively discouraging the convention's implementation. Clearly more conducive provisions are necessary both to bring conventions into force and to facilitate their amendment. Even with such provisions, however, states sometimes reject negotiated agreements such as the 1973 Convention because its costs are undeniably high. More often, it is simply that national indifference prevails. Without a major concern for the quality of the marine environment, states are just not interested in committing themselves to an international agreement that imposes costs or political restrictions. Only an active organization committed to a real international solution can counter this.

FROM LAW TO COMPLIANCE

"It is very easy for states to sign a piece of paper and ratify conventions, but there are great differences in what they do about it." For former IMCO Secretary-General Sir Colin Goad, "the fundamental problem of controlling oil pollution is enforcement."[31] The evidence supports this assessment. All too often, accession to conventions has been a public relations ploy, "window dressing" to pacify domestic groups or foreign states when governments and their industries have really only wanted to continue business as usual. Efforts at implementation have clearly been inadequate.

Most commentators have treated this problem far too cursorily in their concern for the formulation of the rules themselves.[32] But enforcement is "the crucible of law, the test of its reality."[33] And it is not an easy or simple problem. Indeed, it requires consideration of a number of perspectives: (1)

31. Interview. A different opinion has been set forth by the head of IMCO's Legal Division, Thomas Mensah: "The case of a state deliberately failing to take action at the state level in clear acknowledged violation of its international undertaking is the exception rather than the rule." "The IMCO Experience," in John L. Hargrove, ed., *Law, Institutions and the Global Environment* (Dobbs Ferry, N.Y.: Oceana Publications, 1972), p. 241.

32. Some have addressed this issue: William T. Burke, Richard Legatski, and William W. Woodhead, *National and International Law Enforcement in the Ocean* (Seattle: University of Washington Press, 1975); A. V. Lowe, "The Enforcement of Marine Pollution Regulations," *San Diego Law Review* 12 (1975), pp. 624–643; Donald E. Milstein, "Enforcing International Law: U.S. Agencies and the Regulation of Oil Pollution in American Waters," *Journal of Maritime Law and Commerce* 6 (1975), pp. 273–285.

33. W. Michael Reisman, "Sanctions and Enforcement," in Cyril E. Black and Richard A. Falk, eds., *The Future of the International Order, Vol. 3: Conflict Management* (Princeton, N.J.: Princeton University Press, 1971), p. 275.

the character of the regulations themselves; (2) the rights and practices of surveillance, inspection, and prosecution; and (3) the monitoring of this entire process. Here again, as in the formulation of the conventions themselves, there are reasons explaining the successes and failures of enforcement strategies, reasons which must be understood in order to design a more effective system.

The *characteristics of the regulations* have had a major effect on compliance. The nature of the discharge standards and the existence or absence of requirements for installing the requisite technologies are crucial to success. Particularly influential is the nature of the discharge regulations, since it affects the ability of national officials to detect and prove violations. Until January 1978 when the 1969 Amendments came into force, an illegal discharge was defined as one in which the oil content exceeded 100 parts per million (ppm). This rule was accepted in 1954 and reaffirmed in 1962, despite the fact that it is virtually impossible to prove violations of the rule outside of harbors. Even photographs from low-flying planes could not establish whether the concentration of oil in an effluent exceeded the limit.[34] Short of admissions of guilt by the masters of ships—a very infrequent occurrence—it was impossible to prove violations by vessels at sea. The acceptance by shipping interests of the extension of the prohibition zones from 50 to 100 miles at the 1962 Conference was unquestionably made easier by this consideration.

The 1969 Amendments and the 1973 Convention have altered these regulations and should facilitate the detection of violations. Within 50-mile coastal zones the 1969 Amendments state that tankers should discharge only an "effluent . . . [which] if it were discharged from a stationary target into calm water on a clear day, would produce no visible traces on the surface of the water." This is the "clean ballast" criterion (defined as 15 ppm in the 1973 Convention), and it means that photographs from surveillance aircraft are now acceptable evidence of violations at sea. In addition, the 1969 and 1973 tanker regulations require that the *total* discharge of ballast contain no more than 1/15,000 of the cargo-carrying capacity. This too is an important development for it means that, for the first time, in-port inspections can control the quantity of allowable discharge. Were a reliable "black-box" monitoring device to be developed, the port-state control of both the concentration *and* quantity of discharges would be further enhanced.

34. In 1967 the Warren Springs Laboratory of the U.K. government investigated this problem. It concluded that photographs could not determine whether discharges were above or below 100 ppm—something which was generally recognized by government prosecutors. Its report stated, however, that it might be possible to prove whether discharges were above 50 ppm. Interviews.

Clearly, then, the potential for a more rigorous enforcement system now exists.[35] But it must be decided just what evidence is acceptable to courts, a problem on which the MEPC has recently been working. With clear criteria detailing the characteristics of acceptable evidence, investigating states will be more certain as to what the outcome of prosecutions (whether by the port or flag state) *should* be. The determination of the commercial forces opposed to more rigorous enforcement must not be underestimated, however. Their ability to challenge the evidence must be reduced.

Even with enforceable discharge standards, compliance is not assured. Shipowners and governments must be required to install the technologies (both on the vessels and in the ports) that will allow these standards to be complied with by vessels and enforced by governments. The regulatory system for nontankers under the 1954 Convention and its amendments has always been a sham simply because of its weak requirements for the installation of oily-water separators on vessels and for reception facilities in ports. Shipping interests in many countries have long opposed such requirements, which demand capital expenditures and delays in ports, and governments have opposed any obligation to install the reception facilities. As a result, in 1954 states accepted an obligation to build facilities in "main ports" only and then purposely left "main port" undefined, thus emasculating the requirement. In 1962, the requirement was completely abolished. Moreover, as vessels entering ports without adequate reception facilities were explicitly permitted to ignore the discharge rule altogether, massive noncompliance was guaranteed. Officials were concerned with apparent, not real, progress.

Similar deficiencies existed in the regulatory scheme for tankers. The participants at the 1962 Conference did not require the reception facilities and oily-water separators which were necessary if new vessels over 20,000 tons were to implement a complete prohibition. The effect of this was to place all responsibility for financing facilities on the oil industry. It, of course, also wished to avoid such expenditures and so developed its own pollution-control system instead—load-on-top (LOT). When the major oil companies with the backing of the shipping industry sold the new system to governments in the late 1960s, they claimed that on-board technologies—

35. The new standards for nontankers are not as rigorous as those for tankers. According to the 1969 Amendments and the 1973 Convention, nontankers are not to discharge an effluent in which the oil content is greater than 100 ppm and are not to discharge more than 60 liters per mile *anywhere* in the oceans. Violations of these standards at sea are impossible to prove. The 1969 Amendments also state that they should discharge any oil "as far as practicable from land"—a meaningless requirement. The 1973 Convention prohibits nontankers from discharging any oil above 15 ppm within twelve miles from land—a standard whose violation can more easily be verified.

particularly oily-water separators and monitoring systems—were unnecessary, excessively expensive, and, of course, not developed. The IMCO member states accepted these claims, and no such requirements were included in the 1969 Amendments. Not surprisingly, the industry was wrong, and the absence of these technologies undermined both the operation and enforcement of the new system. Once again, the concern with costs prevailed over the desire to provide an effective system.

The 1973 Conference significantly improved this situation. By requiring segregated ballast tanks on *new* tankers (over 70,000 dwt), a major source of discharges was removed. On existing ships (both nontankers and tankers), oily-water separators, slop tanks, and monitoring systems were now required. In 1977, the MEPC set down technical specifications for these devices, but even now reliable monitoring devices have not been developed. A monitoring system does exist for black, crude oils, but it is not effective (it fails to meet MEPC specifications) and is not "tamper-proof." The situation with white, refined oils is even worse, so a special exemption for them was included in the 1973 Convention. There is doubt as to whether the industry is really concerned to develop these devices, a situation which, if true, would necessitate governmental involvement.

The 1973 Convention also includes extensive requirements for reception facilities for nontankers and for tankers on short voyages and in special areas. The MEPC has also set out criteria by which port authorities might assess the nature of the facilities they require, and it has recommended procedures for vessels to report inadequate facilities to their flag states and for flag states to forward the information to IMCO.[36] The entire system appears impressive. But some governments—especially developing ones—argue that the requirements are not obligatory, and others still exhibit a great deal of reluctance to build the facilities. A very serious problem continues to exist in the Mediterranean, as many states are unwilling to finance the facilities required for special areas. The formal regulations may be adequate to control oil pollution, but the commitment of governments still is not.

Next along the path to enforcement is the *detection of infractions*. The first step in any inspection system is to survey a vessel before it goes to sea and at set intervals thereafter. This is to ensure the ship's initial and continuing compliance with *design*, *construction*, and *equipment* standards. These inspections are generally carried out by classification societies such as Lloyd's Register of Shipping, the American Bureau of Shipping, Bureau Veritas, and Nippon Kaiji Kyokai. Under the Safety of Life at Sea Convention (SOLAS), these bodies issue a Cargo Ship Safety Construction Certificate

36. MEPC VII/WP.2, Add. 1, annexes II and III.

and a Cargo Ship Safety Equipment Certificate. Under the 1973 Convention for Prevention of Pollution from Ships, they will issue the International Oil Pollution Control Certificate. A few governments such as Canada and the U.S. also inspect and issue government certificates to their own flag vessels. This is, however, not a common practice and is, in any event, usually limited to controlling equipment, not design or construction.

Most surveys by the classification societies are thorough and the judgments of their surveyors accurate. This is not always the case, however, and a significant minority of vessels—especially older ones—obtain certificates which they should not.[37] This is, in part, due to the fact that classification societies compete for business and hence do not always carry out rigorous surveys since they might lose business to other societies.[38] Through such bodies as the International Association of Classification Societies (IACS), the major societies do try to control such occurrences, but the competitive nature of the system continues to encourage them. Other problems exacerbate this situation. For example, some "nonexclusive" surveyors in out-of-the-way ports are not regular employees of the societies and are neither well-trained nor very rigorous in their inspections.[39] Others at repair yards are under pressure to make sure that vessels are not detained so as to delay their schedules.[40] Moreover, the classification societies are themselves run by committees composed largely of shipowners who have a vested interest in ensuring that their vessels are not kept out of operation by unfavorable reports. It is at least partially for this reason that a few states do demand that their flag vessels also be inspected and granted government certificates. All experts agree: more rigorous inspections by the societies are fundamental to future attempts to improve ship safety and prevent accidental as well as operational pollution. Governments and IMCO must assert greater control over the quality of the work. However, with

37. For example, the American Bureau of Shipping gave the *Arrow* an "A1" rating six days before it grounded off the coast of Nova Scotia in February 1970. According to Canadian officials it had a number of deficiencies—including no diesel auxiliary and no reliable radar.

38. Government officials from several countries have commented on this fact. One official said that he heard a shipowner tell a surveyor of a classification society that he would transfer responsibility for issuing certificates for his vessels to another classification society if his vessel were not "cleared" immediately. For an analysis of weaknesses in the classification-society inspection system (including competition among societies), see Trevor Lones, "IMCO Takes Action over Substandard Ships," *Seatrade*, August 1975, pp. 63, 65.

39. For an excellent criticism of "nonexclusive" surveyors by an individual who has worked as a classification society surveyor, see Commander J. W. McCurdy, U.S. Coast Guard (Ret.), "Flagging the Unsafe Liberian Flag Ships," *United States Naval Institute Proceedings* 10 (June 1977), pp. 102–103.

40. A government official who had observed classification societies' surveyors at ship repair yards said that the surveyors tend to exhibit a stronger loyalty to the repair yards than to the societies. In particular, they hesitate to delay a vessel's departure in order to have certain deficiencies corrected if it will delay the schedule of the repair yard.

representatives from these societies frequently members of national delegations and in attendance as nongovernmental observers, resistance to such endeavors should not be underestimated.

Once vessels have received the appropriate certificates, it has largely been up to governments to detect infractions of ship standard regulations. Yet the SOLAS Convention and the 1973 Convention state that government surveyors should board a vessel to inspect it only if they have "clear grounds" for believing that its construction, design, or equipment does not meet the standards of the conventions and their certificates.[41] This provision is a substantial deterrent to more rigorous inspection, especially when combined with the costs of the inspections and the pressures from the shipowners against them. Even so, the regulations have not always deterred officials from undertaking "spot checks." These are, however, sufficiently infrequent that they have been tolerated despite the fact that they can lead to costly delays when deficiencies are discovered and their rectification required before departure from port.[42]

In general, therefore, there is little policing of the maintenance of ship standards between regular inspections. Since significant deterioration can occur—especially on older vessels—frequent ad hoc inspections by port authorities are important for preventing intentional and especially accidental pollution. At the 1978 TSPP Conference some progress was made on this front by requiring contracting parties to have their flag ships surveyed either annually or on an unscheduled basis. While this regulation is an improvement, it does return control to the flag states and the classification societies. The societies, in particular, will hesitate to challenge the judgments of the previous surveyor or to cross shipowners by instigating an ad hoc inspection. Port states must ensure that the new system works before they abjure their responsibility in this field. Indeed, to encourage its effectiveness, they should retain their option to inspect at will.

The detection and deterrence of *discharge* violations can be accomplished both by surveillance of coastal waters and ports and by inspection in ports. In general, surveillance by states of their coastal waters outside ports has been infrequent or nonexistent.[43] The cost of regular air and sea patrols is great so that states have relied almost completely on reports of aircraft and

41. Art. 19 of the SOLAS Convention and art. 5 of the 1973 Convention.

42. Many shipowners (especially European ones) indicated at the 1978 TSPP Conference a willingness to have unscheduled inspections by the classification societies. They preferred such a system to inspections by government officials or more frequent (and costly) surveys by the classification societies. Some also realized that the existence of some substandard vessels encouraged harmful unilateral action by governments against all ships.

43. Interviews. Even the U.S. Coast Guard engages in little surveillance outside the harbors. *Marine Environmental Protection Program* (Washington, D.C.: U.S. Coast Guard, Dept. of Transportation, August 1975), pp. 53 and 101–112.

vessels engaged in other duties. The incentive for mounting such surveillance programs also has been undermined by the great difficulty of obtaining evidence which will be accepted in court as proof of a violation. The 1969 Amendments should reduce this disincentive, but problems still remain. Criteria setting out clear evidentiary standards have not yet been set down by the MEPC. Even with such criteria, the high costs of surveillance and the impossibility of its being conducted at night will severely mitigate its effectiveness. If detection and deterrence of discharge violations are to be improved, it will have to come through inspections in port.

No such right of port-state inspection for discharge violations was conferred under the 1954 Convention, or even under the 1969 Amendments. It is, however, included in the 1973 MARPOL Convention and will take effect when that convention is implemented. That will greatly enhance port-state powers at *oil-loading* terminals, especially in conjunction with the rule restricting total discharge quantities to less than 1/15,000 of cargo-carrying capacity. The acceptance in 1978 of crude-oil-washing will enhance port-state inspection even more, as that system depends on enforcement at *oil-discharge* ports in the oil-importing countries. These latter states are much more likely to conduct such inspections than are the Middle East oil exporters. Even so, for the new inspection powers and the crude-oil-washing system to work, resources will have to be expended to expand national inspectorates, and this is an uncertain prospect.

With the lack of authority under existing rules for any state to carry out in-port inspections for discharge violations, that task has fallen by default to the oil industry. According to OCIMF, one-third of all tankers using oil industry loading terminals have been inspected.[44] These inspections have improved the use of the LOT system, although nonutilization is still a widely documented phenomenon. Since 1972, the industry has offered to provide the information discovered to the oil-exporting states for referral to the flag states for prosecution under the 1969 Amendments. No interest has been shown by these states despite the entry-into-force of the amendments.[45] Such a system of relying on the industry to police itself is not the most reliable but, to this day, is the only system available. Only with the introduction of crude-oil-washing will the possibility for real policing exist. Then it will be up to government inspectorates in oil-importing states to make the system work.

Beyond surveillance and inspection, the next element in the enforcement

44. MEPC/Circ. 17.

45. OCIMF initially put forward the proposal in 1971, and the responses to the IMCO inquiry were meager (MSC XXIV/19; MP XIII/3.1). The same offer and poor response occurred again in 1974 (MEPC II/3; MEPC Circ. 17; MEPC III/10). In 1974 OCIMF claimed that one-third of all oil tankers were being inspected (MEPC Circ. 17).

system concerns *prosecution of violations*. Traditionally, coastal states have had the right to prosecute vessels for infractions in their internal waters and territorial seas, and flag states have had jurisdiction over their vessels on the high seas. Flag states have, of course, not been eager to devote resources to prosecutions, and many have not been very diligent in responding to the accusations of coastal states.[46] For example, between 1967 and 1977 Canada did not receive replies to 49 percent (39 out of 80) of the communications it sent, and in only 21 percent of the cases (17 out of 80) were there convictions.[47] The ICNT has altered the traditional flag state/coastal state regime with the introduction of port-state enforcement. Under the new provision, port states will be able to initiate proceedings against a vessel in its ports regardless of where the detected infraction occurred. This will apply to violations in coastal waters outside the territorial sea, to the high seas, and to any violations referred to it by other countries. However, a flag state's right to "preempt" the port-state prosecution has also been included in the ICNT, and this could substantially reduce the impact of the new powers. Flag states will, however, be forced to initiate proceedings if they wish to prevent the prosecution of their vessel in the foreign port state. By increasing the frequency of prosecutions, violations will more likely be deterred.

Regardless of what rights of prosecution states are granted, it is only if they are exercised and meaningful sanctions imposed that respect for the law will be enhanced. In the past officials of all countries have hesitated to prosecute ships, whether out of indifference, lethargy, or a concern not to offend the shipowners. The crews of ships are well aware of the variation in enforcement policies, and they have acted accordingly. Moreover, courts have imposed quite modest fines (an average of $1,200 in the U.S., U.K., and Canada). Even their deterrent value is limited by the fact that they are covered by P & I insurance.[48] IMCO conferences have discussed the issue of the level of fines in the past. The opposition of shipping interests to higher penalties and the extreme antipathy of governments to any external restrictions on their sovereign jurisdiction to legislate and apply sanctions has prevented the establishment of any international scale of penalties. The

46. Virtually all government officials have voiced this opinion. See MEPC/Circ. 17 (1974) and subsequent reports by states to MEPC on states' enforcement activities and the results of reports to flag states. An impediment to some prosecutions has been the inadequacy of evidence.

47. MSC/MEPC/WP.12.

48. A shipping industry official commented that the elimination of such coverage would definitely improve compliance. The possibility of getting all states to make such coverage illegal is, however, not very good. Another official suggested that an effective strategy would be for the P & I Clubs to increase the deductible so that shipowners would have to pay most fines or a portion of them. Again, the political problems involved in securing such a change are considerable.

issue is to be discussed again within the MEPC, yet no substantive agreement can be expected from such discussions. They should continue, however, if only to prod governments to set reasonably effective penalty levels themselves.

Fines are a very important sanction which can be imposed on vessels, but they are not the only type. Costly delays in ports while investigations are conducted and deficiencies corrected are even more effective. In fact, some officials regard keeping a vessel in port as their most effective sanction, since the costs to shipowners can be considerable and are not covered by insurance. There have, however, been significant differences among states regarding their willingness to detain vessels and their interpretation of the stricture that they not cause "undue delays."[49] Those with significant fleets and strong shipping lobbies have been particularly reluctant.

International conventions explicitly recognize only the role of *states* in applying sanctions, but private industry also plays a crucial role. Industry also has the possibility of employing positive sanctions (inducements) in addition to negative sanctions (penalties).[50] In the past, oil companies have been reluctant to impose penalties on tankerowners whose vessels did not comply with the LOT system. They considered charging them for any oil lost or canceling a charter arrangement when their utilization of LOT was inadequate, but in the end they rejected such a punitive strategy. Until 1977 the oil companies did not even make the hiring of tankers contingent on their previous pollution prevention and safety records. Consequently, tankerowners and masters had no reason to worry that their future ability to secure contracts would be influenced by their concern for environment protection. This has begun to change with the *Argo Merchant* and other tanker accidents around the U.S. in late 1976 and early 1977. Some charterers are now beginning to inquire about the safety, pollution-prevention standards, and history of a vessel and to make chartering arrangements contingent on such information.[51] It is still too early to judge the extent and efficacy of this program. Its voluntary nature does limit its likely impact.

The oil industry has been more willing to consider *positive* sanctions or inducements for promoting compliance with discharge standards. This has

49. On the reluctance of the U.K. government, see J. A. Sullivan, "Two Cases of the Practical Enforcement of Pollution Standards in Harbours: (*b*) Milford Haven," in Jolanta Nowak, ed., *Environmental Law: International and Comparative Aspects* (Dobbs Ferry, N.Y.: Oceana Publications, 1976), p. 140. The 1973 MARPOL Convention (as well as the 1978 SOLAS Protocol) does provide that shipowners must be compensated if a vessel is "unduly detained or delayed" [art. 7(2)].

50. For this distinction, see David Baldwin, "The Power of Positive Sanctions," *World Politics* 14 (October 1971), pp. 19–38.

51. Regarding the new policies of some major oil companies, see Trevor Lones, "Getting Oil Off Troubled Waters," *Seatrade*, July 1977, p. 135; and *World Shipping Journal*, March 15, 1977, p. 38.

largely taken the form of compensation to independent tankerowners for expenses incurred as a result of using load-on-top. Many oil companies opposed compensation, and it was not until 1976 that it was accepted.[52] Led by Shell, a "voluntary code" for compensation was finally adopted in that year as a way to overcome the evident lack of success with the load-on-top procedure. Not formally accepted by many companies (especially American), it is being used by most large oil firms in the world.[53]

A final and much-neglected enforcement strategy is the *monitoring* by IMCO of national actions to promote compliance. Since governments do not want to have a reputation for illegal or irresponsible behavior, the recording, publication, and discussion of enforcement records could put substantial pressure on them. In the 1954 Convention, the only provision along these lines was one which "obligated" contracting parties to send IMCO information on all investigations and prosecutions of violations which they detected or of which they were notified. This obligation was completely ignored by all concerned until after 1973. Prior to that date, another monitoring provision was discussed by states but was soundly defeated. This was the proposal (by the U.K. in 1962 and by France in 1967) that all reports of violations by foreign flag vessels be sent to IMCO as well as to the flag states. Even apart from the desire not to be publicly pressured into expending resources to investigate and prosecute their own ships, most participants in the negotiations saw the provision as yet another attempt to infringe on the autonomy of their legal processes.

With the 1973 Conference IMCO's members began to expand the monitoring system, and it has been strengthened since then. At the same time many of the new provisions are still only on paper, and a great deal must still be done to take advantage of their potential. The 1973 Convention added the requirement that contracting parties provide information on their reception facilities in ports and the identity of the classification societies which they authorize to issue the International Oil Pollution Control Certificates. Then in 1974 the MEPC began to urge states to fulfill their reporting requirements under the 1954 Convention. Only about a dozen countries—largely from the developed world—have responded.[54] This was followed by a large number of new MSC, MEPC, and Assembly *recommendations* and the launching of several new monitoring exercises by

52. Those opposed were willing to compensate the owners for the carriage of the oil/water residues and any extra canal dues. But they did not compensate them for extra time at sea (especially when going to repair ports) and extra time in ports.

53. *Pollution Prevention Code (Oil Tankers)* (London: International Chamber of Shipping, 1976). The U.S. companies unfortunately cannot sign for fear of being prosecuted under antitrust legislation. There are differences within the U.S. Department of Justice on the question. Collaboration in pollution prevention could be allowed in the future. Interviews.

54. MEPC Circ. 17, VI/16/2, VII/18/1, VII/18/2, VIII/12, Circ. 47.

IMCO organs. These have included a request to forward reports of safety and pollution violations to IMCO as well as to flag states, the requirement that had been rejected in the 1960s. In addition, states are now requested to forward reports on vessel casualties. This information is to be circulated and reviewed regularly.

In recent years, some reorientation in the policies of IMCO members is occurring on this issue of monitoring compliance. Although only an incipient trend, states have at least begun to encourage the fulfillment of reporting obligations, to recommend some new ones, and to commit the MSC and MEPC to improved monitoring programs. These changes could be supplemented by undertaking studies of specific compliance problems such as those sponsored by the Committee of Experts of the ILO,[55] by the dispatch of "implementation teams" such as those used by ICAO which both examine the compliance by governments and provide advice,[56] and by formal "confrontations" with officials of particular states such as those carried out by the OECD.[57] Both the administration of studies and the provision of advice could be sponsored by an IMCO Marine Safety Corps, whose creation was supported by the 1978 TSPP Conference.[58] It is, however, still very much an open question as to how far members will go in criticizing states whose governments, flag vessels, and inspection agencies are not fulfilling convention obligations or are inadequately performing tasks delegated to them. But a framework for IMCO to monitor compliance with the conventions is now in place. It must be used.

Deficiencies in the compliance system have in part been due to a failure to understand the implications of some regulations and the limitations of existing technologies. However, they have also been due to the *policy preferences* of governments and the shipping and oil industries. Most governmental and nongovernmental actors have been unwilling to support regulations and practices which would assure a high level of compliance since they have not wanted to assume the attendant costs. The weaknesses which they built into past conventions assured that they would not have to respect the regulations too assiduously and hence would not have to accept major new expenses. Experience with past regulations and the rise of environmental

55. Robert Cox and Harold Jacobson, *The Anatomy of Influence: Decision-Making in International Organizations* (New Haven, Conn.: Yale University Press, 1973), pp. 112 and 390; Evan Luard, *International Agencies: The Emerging Framework of Interdependence* (London: Macmillan, 1977), pp. 140–142. On the effect of ILO activities, see Ernst B. Haas, *Beyond the Nation-State: Functionalism and International Organization* (Stanford, Calif: Stanford University Press, 1964), p.268.

56. Luard, *International Agencies*, pp. 69–72.

57. Eugene Skolnikoff, *The International Imperatives of Technology* (Berkeley, Calif.: Institute of International Studies, University of California), p. 111.

58. MSC 21/4/Add.3; TSPP/CONF/12, Res. 11.

concern have led to some constructive changes, but progress has been slow. The concerns of industry about passing on costs and losing competitive advantage, together with the competing demands of nonenvironmental interests for attention and capital, have been difficult obstacles to overcome.

As with the development of the convention regulations, the advances made have been due to a number of political developments. Of great importance has been the pressure of the United States and its environmental interests, which have been frustrated by the obvious lack of control over many ships and flag states. In addition, some maritime states and the oil industry have exhibited greater concern with enforcement in response both to the pressures of the environmentalists and to the fear that their continued inaction could lead to unilateral action by the others. As many of the worst vessels were thought to be those of the flags-of-convenience, it was believed that stronger enforcement might actually help the more responsible shipowners by decreasing the competitive advantage of the convenience ships. Finally, it was realized that the problem *was* enforcement and that better enforcement would reduce the continually escalating demands for tighter standards.

The developing countries for the most part have been as indifferent toward the compliance problem as they have generally been toward the character of the technical regulations. Their governments have had more pressing domestic demands on their limited financial and bureaucratic resources. The most serious effect of their attitudes for the compliance issue has been the lack of interest on the part of the oil-exporting states to inspect tankers in their ports. In addition, their failure to build reception facilities in conformity with the obligations in the 1973 Convention will make it virtually impossible for many vessels operating in some special areas or on short voyages to respect the discharge regulations. Without a change in their policies on these enforcement matters, the 1969 Amendments will remain virtually unenforceable, as will important aspects of the 1973 Convention.

In review, it can be seen that improved enforcement of IMCO conventions is a necessary and complicated task. Its successful resolution depends on improvements on a number of fronts. The nature of the regulations themselves, the enforcement process of surveillance, inspection, and prosecution, and the improvement of IMCO's monitoring capabilities—all these require detailed consideration and action. Without such action, the conventions could become merely "paper law" and improvements in the standards merely hollow achievements. With the failure of the Middle East oil exporters even to take an interest in the problem, the responsibility for securing these improvements falls squarely on the shoulders of those states which can make them work—the major Western oil-importing and ship-owning nations.

TECHNICAL ASSISTANCE

In the formulation, ratification, and implementation of IMCO's oil pollution conventions, developing states have lagged far behind the rest of the IMCO membership. These states have no less of a stake in a clean and healthy environment, but their expertise and resources are so limited that often they are technically and financially unable to indulge in the "luxury" of environmental protection. Before turning to the future of oil-pollution-control under IMCO, it is useful to review briefly how IMCO is responding to the special needs of these countries.

The provision of technical assistance has been of increasing importance in IMCO's work. With the expansion of the United Nations Development Program (UNDP) in the early 1970s, IMCO's program expanded as well. Even so, in comparative terms, IMCO's share of the UNDP budget is miniscule—only $1.7 million in 1973. Of all the international agencies, only the Universal Postal Union received less.[59] This amount is, however, supplemented by lesser amounts from the United Nations Environment Program (UNEP) and by some individual states through a "funds-in-trust" arrangement.

Despite its small size, the technical assistance program is an important plank in IMCO's program to build up the maritime industries of the developing states.[60] (In this program, IMCO concentrates on the technical side of shipping services with UNCTAD responsible for the commercial aspect.) As an indication of its importance to the agency, not only has the Secretariat been expanded to include a Technical Cooperation Division, but the agency's Committee on Technical Cooperation has been "institutionalized" in recent amendments to the IMCO Charter.[61] According to the secretary-general, this will make IMCO the first United Nations agency with a constitutional provision of this kind.[62]

Within this field, environmental protection is a concern but the main focus is elsewhere. As the secretary-general commented in a recent interview,

59. *Yearbook of the United Nations, 1973* (New York: United Nations, 1976), p. 327. The Universal Postal Union had $1.6 million.

60. For a complete review of IMCO's activities in this area, see *IMCO's Technical Assistance Program*, a pamphlet published in 1977 by IMCO.

61. Res. A.400(X).

62. A.X/12(*b*), annex II, p. 3. The extent of the present secretary-general's commitment to technical assitance was made clear in a recent speech where he said that it was an "equally fundamental" task to that of setting technical standards. A.X/12(*b*), annex I, p. 2 (August 19, 1977).

> The key is personnel. Ships can be bought abroad, but these countries must build up their own pool of master mariners, engineers, navigators, able seamen. Now developing countries are beginning to set up naval academies. We at IMCO have made clear that they cannot aim at lower standards of expertise than those prevailing in the major maritime states. It would be a recipe for disaster.[63]

Training is the real focus of activity, and with money from both UNDP and the recipient countries themselves, IMCO has assisted in the development of maritime training centers in Algeria, Egypt, Francophone Africa, Nigeria, Indonesia, and Brazil. At Alexandria, Egypt, fifteen Middle East countries have, with IMCO's assistance, recently completed the construction of the Maritime Transport Academy. It will have a student body of approximately one thousand.[64] Other such large-scale projects are found in Brazil (where two centers providing almost all Brazil's maritime personnel needs have been built) and Nigeria. UNDP funds have provided only a small fraction of the total costs of these projects.[65]

The second main thrust of IMCO's technical assistance program has been in the provision of "advisory" services. These are largely also funded by the UNDP. Three "interregional advisers" are based at IMCO headquarters in London to offer assistance in the three areas of maritime legislation, maritime safety administration, and marine pollution. In addition, six regional advisers have been established, two each in Africa, Asia and the Pacific, and Latin America. Combined with the training function, these assistance functions will greatly improve safety, administrative capabilities, and, indirectly, pollution control. In the longer term, in particular, the improvement of various marine administrations will increase the ability of these states to participate directly in IMCO's work of drafting and implementing environmental conventions.

Technical assistance of a specifically environmental nature is minimal. The interregional adviser for marine pollution (financed, interestingly, not by UNDP but out of IMCO's regular budget) has an important role, but the impact he can make is limited.[66] Before the arrival of the marine pollution adviser, UNEP did provide funds for expert advisers to be sent to the

63. "IMCO's Srivastava is 'Mr. Shipping,'" *International Herald Tribune* (Paris), January 30, 1978, p. 9.

64. See A/IX 12(*b*), p. 2; MEPC VII/8, pp. 3–4; TC XI/10, p. 4.

65. The Alexandria center's total cost is about $50 million, with UNDP covering $2.8 million. The Brazil academies will cost about $10 million, with UNDP providing $3.2 million. Of the estimated $42 million for the Nigerian facility, only $157,000 will be spent through a "fund-in-trust" arrangement with IMCO. This national school will be supplemented by regional centers in Ghana and the Ivory Coast to provide for most of West Africa's training needs.

66. On his activities, see MEPC VII/8, pp. 9–10.

scene of several pollution emergencies,[67] and it has assisted in a variety of small projects.[68] UNEP has also provided small amounts of money toward assisting states to understand the implications of ratifying IMCO's oil pollution conventions. A special study was made for Nigeria, and a symposium on problems of implementing the 1973 MARPOL Convention was sponsored in Acapulco in 1976.[69] As UNEP's policy is that its funds are *not* to be used as a substitute for regular program funds in the agencies, its monies have largely gone to promote new initiatives. IMCO has benefited from this in one project, the Regional Oil Combating Center in Malta. As part of the regional pollution control agreement for the Mediterranean (under UNEP auspices), this center was established in December 1976 under IMCO guidance with a $700,000 grant from the UNEP fund.[70]

In conclusion, it can be seen that IMCO's technical assistance program is of a very limited scope. It has largely to depend on the priorities of those states drawing on their allocations from the UNDP budget. With UNEP's reluctance to spend money on regular assistance programs, this circumscribes the extent to which monies are allocated to specifically environmental matters.[71] If assistance for marine environmental protection is to increase, it will have to come through direct national contributions of "funds-in-trust." Norway and Sweden have begun to do this, providing sizable grants for particular *designated* projects. This is the only reasonable route for expanded environmental assistance. It is, after all, neither realistic nor justifiable to expect developing states to allocate a major segment of their development funds for environmental projects when they have more pressing priorities and when, to a very great degree, it is the developed states that are responsible for the problem in the first place. If significant international resources are to be devoted directly to maritime environmental protection through the developing states, this money will have to come through new channels. The continuing opposition of some developed states—such as the United States—to such special funds is, therefore, quite inconsistent with their avowed goal of "cleaning up the world's oceans."

In conclusion, there is a long way to go from the drafting of a new convention to its actual working implementation. The convention must first be

67. These were the sinking of the *Metula* off Chile and the *Trans Huron* off India. The marine pollution adviser visited Colombia and Equador following the foundering of the tanker *St. Peter* in 1976.

68. Such activities include pollution-control training courses, advice and assessments on the state of the marine environment in some areas, regional contingency planning advice, and advice on national legislation. To date these activities have been extremely small.

69. See inter alia MEPC III/15, MEPC IV/5/1, and MEPC VIII/5.

70. See A X/22/1 for a review of this project.

71. On proposed UNEP/IMCO environmental assistance projects, see MEPC VIII/5.

accepted—ratified—by a sufficient number of states to be transformed from an object of speculation into real law. Even then the numerous obstacles to effective enforcement can make the convention mere "paper law." This has been the story of much of IMCO's work. Finally, the convention must be accepted and enforced worldwide, not by just a handful of states.

Conclusion

Chapter IX

The Political Process and the Future of Environmental Protection

Oil is but one pollutant and pollution but one aspect of the environmental crisis. Indeed, the list of present and future environmental threats is a long and daunting one—even without the forbidding prophecies of much contemporary environmental literature. These threats have a common origin in today's complex technological, economic, and political system, and their resolution, if it is to come at all, must contend with that system. Interestingly, as a readily identifiable source largely beyond the prickly barriers of national sovereignty, ship-generated oil pollution is one of the easier of these many threats to control. It was for this very reason that vessel-source pollution was singled out for detailed treatment by the Law of the Sea Conference, while other, more serious land-based sources were made subject to only the vaguest of obligations. Even so, the history of maritime oil pollution control has been fraught with frustration, and it provides lessons of wide applicability.

The primary concern of this study has been the politics of ocean oil pollution control. These politics present an important study in themselves, but they are also a useful example of the political processes underlying many other contemporary international issues. In this concluding chapter, we shall review the central elements of the political process operating to date and prescribe strategies for improved oil pollution control in the future. This latter task is an especially complex one, as it demands a detailed knowledge of the political setting which underlies specific policies and laws. In prescribing strategies for future environmental regulations, it is necessary, therefore, to assess how developments in this political setting might affect progress.

THE POLITICAL PROCESS AND INTERNATIONAL POLITICS

In the past, international political studies have focused largely on interstate relations, and particularly on the competitive struggle for military power and security. As we discussed in the first chapter of this study, recent developments in international politics render this focus no longer sufficient. Today, as a result of contemporary processes of technological and economic expansion, many more international issues are addressed by a wider range of actors than just states. With this diversity, power has been fragmented on an issue-by-issue basis so that outcomes reflect a variety of specific powers, not just that of military force. In such a situation, international organizations also play a much larger role and actually have an impact on events. As recent studies have attempted to show, far greater complexity exists in world politics than has hitherto been addressed.[1]

As with many of the new issues, the oil pollution problem stems from the rapid technological and economic growth since World War II, a growth that has heightened the contacts and sensitivities among societies. The specific development of concern to this study stems from the growing appetite of the developed world for foreign sources of energy, an appetite with obvious environmental consequences. Greatly relied on by both producers and consumers, transported oil is the quintessence of international interdependence. At the same time, oil has unique environmental consequences and its control has special technological and economic ramifications. The negotiations for its control are, as a result, highly specific to the particular issue.

The types of actors involved in the negotiations to control this environmental problem are common to many contemporary international issues. First, of course, state actors have no monopoly. Indeed, nonstate—"transnational"—actors have had significant roles. It is their activity beyond any single national control that provides much of the basic character of the problem. Moreover, as a consequence of their control over information and technologies, their active lobbying within national capitals and at IMCO, and their ability to provide their own solutions, these actors have an important influence over outcomes as well. Second, it would be erroneous to perceive the state actors at IMCO as coherent units. On this issue, it has been one ministry (transportation) that has largely controlled the course of negotiations. These ministries are often in conflict with others within their own governments. Consequently, the whole process is best perceived as a continuing "transgovernmental" one rather than one of unitary states formally consulting at a diplomatic level. The transgovernmental network is centered on the specific concerns of the issue—oil, shipping, and the marine environment.

1. See Keohane and Nye, *Power and Interdependence* (Boston: Little, Brown, 1977). These commentators refer to a "diversification of the agenda," "multiple channels of contact," the absence of a "hierarchy of issues," and the "eroding of hegemony." Their analytical framework highlights important areas of concern for these new issues. See also our Chapter 7, where these various perspectives are addressed.

Because of these characteristics of the issue, the patterns of policies and influences mainly reflect the specific attributes of the issue, not the dictates of the overriding security balance. The policies and alliances of the major actors are thus largely along the lines of shared industrial and commercial interests in the areas affected. Likewise, the influences of these states vary according to their possession of the relevant levers of influence: tanker tonnage, control of the oil market, and maritime expertise. The states possessing these capabilities control the agenda, shape the outcomes, and provide for the pace of implementation and enforcement. Conflict among the major participants usually has been contained within the limits of the issue, and minor actors have shown no interest in making linkages to outside issues where they might possess greater power.

Finally, as in many other new international issues, the role and influence of the international organization itself are crucial. As the focus of the network of transnational and transgovernmental interaction, IMCO exercises a decisive constraining influence on negotiated outcomes. A uniform regulatory regime is unanimously accepted as desirable and action outside the international forum shunned. Moreover, by holding regular meetings, acting as a center of communication offering administrative facilities, and providing legitimacy for the final result, IMCO offers positive incentives to undertaking action within the framework. As a result, although the threat to act unilaterally outside the agency carries great weight, that weight is ultimately brought to bear within IMCO's structure, not outside it. Once inside, the peculiar history, ideology, and process of IMCO shapes the results.

With this high degree of issue specificity, it is clearly inaccurate to speak of "*the* international system" dominated by one set of issues, priorities, and levers of power. At the same time, the specific issue of oil pollution does operate within a larger economic/political/military context, and its political process is not isolated from this extrinsic context. After all, a crucial division within IMCO is that among the WEO/developing/Soviet bloc groupings, a situation remarkably similar to many other international organizations. It is important, therefore, to understand how the politics of this issue are linked to larger military, and particularly economic, concerns.

Although military security still remains the ultimate concern of states, it is undoubtedly correct that military issues have recently been diminished as the central focus of international politics. This situation could quickly change, but a continuing balance of power and the obvious lack of utility of military force for the achievement of many goals does allow other issues greater prominence. Even so, military concerns do still impinge on specific environmental issues, often directly. Indeed, two very important events in shaping the agenda of the oil pollution issue were military conflicts: the

Middle East wars of 1967 and 1973. The 1967 war, with its closure of the Suez Canal, changed tanker routes and the locus of pollution and ushered in the supertanker boom. The 1973 war, with its OPEC oil embargo and subsequent price rise, brought on a decline in the growing demand for oil and the supertanker bust. Moreover, while military force may not be a useful instrument to achieve all goals, where issues have been of consequence to military security, the requirements of continued security for the powerful have dominated the requirements of, for example, environmental protection. In the oil pollution issue, this has been evident in the defeat of all proposals that could possibly affect freedom of maritime military movements and trade in strategic goods. Finally, in a more indirect way, the entire military calculus is based on the needs and defense of a provocatively high level of resource consumption.

In a world where many new issues and actors crowd the political scene, the relevance of military power is, however, limited. What must be stressed is the extent to which economic power has replaced it. Today, economic issues dominate the international agenda,[2] and, in numerous ways, economic power dominates international politics *at all levels.*

This has been clearly so in the politics of oil pollution control. The very character of the problem is a result of the technological and economic growth of the capitalist economies of the WEO states which depend on the oil of the developing nations. Likewise, state policies are molded predominantly by their economic interests, and their ability to shape outcomes reflects the general structure of global economic power. In this light, the nearly monopolistic control of the WEO states at IMCO reflects their *general economic dominance.* Meanwhile, the minor influence of the Soviet bloc derives from its commercial and *economic isolation.* The weakness of the developing states likewise follows from their *economic dependence.* As in other organizations, therefore, it is the wealthy, pluralist states of Western Europe, the United States, and Japan that define the alternatives and outcomes.[3] While the distribution of power may stem directly from resources pertinent specifically to the transportation of oil, these are themselves products of the larger global economic system.[4]

2. One example of the growing importance of economic as opposed to military issues is among the newly independent states. Whereas ''alignment'' was clearly the most crucial perspective of the 1960s, the issue of these nations' economic future—for example, the ''new international economic order''—is stressed in the 1970s.

3. Robert Cox and Harold Jacobson, *The Anatomy of Influence: Decision-Making in International Organizations* (New Haven, Conn.: Yale University Press, 1973), pp. 393-397 and 417-422, show how power in most organizations reflects the large global power structure. Minor variations do occur in specific areas and organizations, but the general pattern is clearly retained across a wide spectrum of concerns.

4. With such a central role for economic power in the political process, a revision of our concept of *political* power is called for. Susan Strange has argued strongly for this revision, seeing ''the structure of the international political economy [as being] determined as much by

With economics as the basis of power, it is not surprising that it is economics that is at issue too. The desire to preserve "competitive advantage" (to maintain one's economic position) is a primary consideration, and solutions are acceptable only insofar as they do not impinge on this larger interest (or do so only on the weaker states). In other ways economic interests intrude as well, such as in the preeminent desire of the maritime powers to prevent coastal states from asserting jurisdiction that could interfere with free maritime transport. Indeed, the basic dichotomy between maritime and coastal states is an economically, not environmentally, defined distinction—most maritime states possess extensive coastlines.

Not only, therefore, does economics profoundly influence the nature of the issue and the policies of states, but it controls the powers of states and possible outcomes as well. Clearly it is necessary in proposing future strategies for better environmental protection to consider this larger political setting. Changes must be worked not only in those political factors which are intrinsic to the oil pollution issue but also in those factors that transcend the particular environmental issue.

STRATEGIES FOR ENVIRONMENTAL PROTECTION

Previous chapters of this study have provided a detailed evaluation of a large number of specific regulations concerning oil pollution control. It is not the purpose of this section to reiterate our views on these matters. Rather, we shall focus on general strategies which can be pursued within and toward the general political setting which has molded and placed constraints on the policies of IMCO participants.

The political setting may be considered as being composed of factors both *intrinsic* to the specific oil pollution issue and *extrinsic* to it. The intrinsic setting includes those political factors specifically related to the issue at hand. For example, the distribution of tanker registries and the nature of the IMCO forum are intrinsic to the oil pollution issue. These are matters often susceptible to short-term changes, and our discussion of future specific policies at IMCO will be considered in this context. In addition, there are extrinsic factors which transcend the specific issue. For example, the high consumption demands that require oil or other energy sources and which give rise to the environmental problem in general is such an extrinsic factor. So too is the general pattern of economic wealth that underlies the distribution of shipping and oil interests. These factors exercise a substantial constraining influence on the specific issue of oil pollution as well as

the distribution of wealth and 'economic' power as it is by the distribution of military and political weight and 'political' power." "Transnational Relations," *International Affairs* 52 (July 1976), p. 337.

on a variety of other issues. A comprehensive strategy of environmental management must focus on efforts to mold these matters as well as on the factors and policies pertinent to specific pollutants.[5]

Intrinsic Developments and Oil Pollution Control

When examining those intrinsic factors affecting the future of rule-making at IMCO, one must consider first the nature of the oil pollution problem itself. Trends in oil consumption point to increased pressure for stronger measures. Although the consumption of oil decreased worldwide for a few years following the OPEC price rise in 1973, consumption is again on the increase. Projections of oil supply and demand indicate that the problem will be with us well into the twenty-first century.[6] In addition, as today's supertankers begin to age and traffic rises with increasing demand, the potential for pollution will rise. Tanker catastrophes, such as that of the *Amoco Cadiz*, are a statistical certainty, and they will continue to be the single most important spur to legislative action. Inadequacies in the regulatory system for operational discharges will also become evident, as they have in the past, and this will provide the impetus for ongoing reform. Finally, as more becomes known about the effects of oil in the oceans, it is probable that the problem will be viewed with increasing seriousness.

It is likely, therefore, that the nature of the oil pollution problem will continue to provide pressure for more and better legislative action. The major determinants of improved regulation are, however, the policies of states. These policies are, in large part, a reflection of states' possession of the relevant commercial interests. Trends in this area seem to point to an increasing environmental conservatism, not environmental concern. It is most important, therefore, to consider and recommend specific future policies in light of these evident trends.

Within the maritime states grouping, beneficial changes in most areas of commercial interest cannot be expected. Some European states (notably Italy, West Germany, and the Netherlands) may continue to drop in their percentage share of world tanker tonnage, but their fleets will still remain sizable, as will their interest in protecting them. In addition, for most of these states their domestic oil industry is central to their general economic welfare. As a result, these states will probably exhibit a continuing reluctance to force new costs onto these industries, even in the face of severe

5. In this discussion, we shall utilize the four analytical headings devised in Chapter 7 for the discussion of relevant political factors: the nature of the problem, the determinants of state policies, the role of transnational actors, and the operation of the international forum.

6. Data on reserves are located in the *B.P. Statistical Review of the World Oil Industry, 1976*, p. 4.

chronic pollution and some serious tanker casualties such as the *Torrey Canyon*, *Pacific Glory*, and *Urquiola*.

Even so, some specific policy changes could be made with little economic disruption. These states could demand greater accountability from oil companies for the behavior of the independently owned tankers which they hire. This could be assured by requiring that specified environmental controls be included in all chartering contracts in combination with a variety of positive and negative sanctions to ensure compliance. Similarly, pressure could be applied to domestic shipowners registering vessels under flags-of-convenience in order to secure better safety and pollution-prevention programs by those states. Finally, the governments of the maritime states would be wise both to utilize their port-state powers to control foreign substandard ships and to demonstrate their own reliability by dutifully fulfilling their responsibilities as flag states. Flag state enforcement has not worked well to date, and it is in the self-interest of responsible maritime nations to show that it can work in the future.

Within the Soviet bloc, maritime interests will be increasing, not decreasing, as the USSR moves out into the world transportation market.[7] One can expect, therefore, a reduced environmentalism on technical issues and a continued conservatism on jurisdictional and enforcement issues. Moreover, given the inaccessibility of the Soviet decision-making process, it will be difficult to affect directly their policy development. Even so, Soviet maritime officials, like their Western counterparts, should recognize that their failure to secure and enforce high environmental standards will lead only to continuing demands for expanded coastal state powers. One can clearly see this occurring in the United States, where recent pollution incidents have provoked the Congress into broadening American powers in the economic zone through the Clean Waters Act.

In the case of the developing states, most will probably feel strong incentives to adopt environmentally conservative policies as they develop their maritime industries and seek to direct maximum resources toward economic growth. Our treatment of these states has been relatively undifferentiated, in keeping with their undifferentiated performance. Policies have been uneven and ad hoc, often dependent as much on the individual representative as on the interests of the country. As they develop economically, individual orientations will undoubtedly become more apparent. This is slowly becoming evident now, and the patterns are not encouraging.

7. This Soviet expansion is discussed in "World Shipping: A Special Report," *International Herald Tribune* (Paris), January 30, 1978, p. 9. The report features a cartoon of a large Russian ship portrayed as a hungry shark. This increased competition will also have the effect of increasing the concern of other maritime states for their competitive position. Such moves are already underway. See, for example, "British welcome for EEC Move on Russian Ships," *Times* (London), June 14, 1978, p. 4.

Virtually all the lead actors—Brazil, India, Mexico, Argentina, Indonesia, and Nigeria—have been maritime-oriented with a concern to protect their expanding shipping interests. Only some Arab states (especially those on the Mediterranean) have manifested, in the past, what might be considered at all consistent environmentalist perspectives. With their imminent maritime development, even this limited environmentalism will be in jeopardy.[8]

The maritime development of these states need not, however, lead to widespread environmental conservatism. Most of the quite considerable pollution from which many of them suffer comes neither from their own flag vessels nor from those ships bringing oil into their ports. While much of the benefit of higher environmental standards will, therefore, accrue to them, most of the costs will not. Additionally, if these states participated seriously at IMCO, they could achieve regulations favorable to their interests. Unnecessary politicization should be avoided if tangible achievements are to be obtained, but their past conduct has shown that, unlike at UNCTAD, this is their intention at IMCO.

What the developing states must do to realize this objective is to utilize their large voting resources. This will, of course, not allow them to dictate results, but it does provide substantial bargaining leverage. With minimal organization and delegation of responsibilities to a few key individuals, the developing countries could participate effectively in IMCO's regular meetings so that their interests would be considered throughout the process. This would also require that they formulate specific proposals on issues (such as technical assistance) that are particularly relevant to their shipping needs. Only through increased participation will they be able to forge alliances with those sympathetic individuals within developed states that could upgrade the developed state commitment to assist their special needs, environmental or other. Moreover, with a serious effort at participation, much experience and expertise could be gained that would better help them to perceive their own needs. Ultimately, it is only the fostering of long, continuous contacts at IMCO that will give them the access to the "elite network" that is so important if their interests are to be translated into achievements.

This increased participation is in the interests of all IMCO members in order that the groundwork be laid for avoiding later confrontation and politicization. As it is presently the maritime voice that is heard from the developing states, it is especially important for environmental interests to solicit the participation of a wide range of developing countries. Only

8. See V. H. Oppenheim, "Arab Tankers Move Downstream," *Foreign Policy* 24 (summer 1976), p. 117. The *Times* article cited in footnote 7 highlights some of the problems the Arab states have had in moving into the world tanker market.

through continuing contact and communication will environmental values be transmitted to developing state representatives. In return, an ongoing discussion will likely make developed state representatives more sympathetic and more responsive to the particular problems of the developing nations in this area.[9]

Despite evident commercial trends, some beneficial environmental policies may still be forthcoming from the developed maritime and developing countries. Realistically, however, it will probably continue to be up to the developed "coastal" countries to lead the way. In the past, Canada has been one of the central environmental actors, but this was often for jurisdictional rather than environmental reasons. With the passing of the Law of the Sea Conference, Canada's environmentalism has diminished. Nevertheless, its coastal interests and large market for shipping could lead it to play an important role in the future. Not having the power to force through its own policies in any event, Canada's most useful role could be as a mediator—between the maritime states and the United States, and between the developed and developing states. In 1978, Canada did act successfully as a mediator between European maritime interests and the United States. Canada has also successfully demonstrated its ability to mobilize the interests of the developing nations at IMCO (in the years 1969–1973), at UNCLOS III, and in the United Nations generally. With its reputation as a conciliator, Canada could play a pivotal role in one of the most basic aspects of future IMCO negotiations, the relations between developed and developing nations.

Environmental leadership at IMCO will, however, probably rest with the United States for the foreseeable future. Trends in commercial interests indicate that the American perspective will continue to have great weight. The rapidly increasing share of world tanker tonnage being registered in Liberia (over which the U.S. does have considerable influence) and the increasing size of the American oil import market will provide even greater weight to the positions of the United States government. In addition, although coastal state power has been reduced by the recent Law of the Sea Conference, port-state power has increased. This will give Canada and particularly the United States even greater influence.

In order for its leadership to be credible and acceptable, American policies will need to be carefully considered and consistent. This will mean demonstrating a willingness to do more than demand new standards from

9. To improve communications *before crises arise* would be in the interests of all members. Not only would it provide a valuable learning experience for developing state representatives, but it would enable the "power holders . . . [to be] responsive to signals from the less powerful without abandoning control of the action to them." Cox and Jacobson, *Anatomy of Influence,* p. 428. The authors of the Trilateral Commission study, *The Reform of International Institutions* (p. 25), make the same point.

IMCO. It must actually implement these when they are produced. It will also mean that the United States must show a concern for the legitimate problems its demands pose for others, particularly the developing states. In this regard, American opposition to special funds-in-trust through IMCO is harmful to the environmental cause. To most developing states, environmental costs are often a luxury and understandably so, given the reluctance of even the most developed states to accept them. Special funds *in addition* to regular assistance are necessary if the United States expects developing states to address this issue seriously. Without such a concern for the legitimate problems of the developing countries, American environmental leadership will rest on power, not legitimacy.

The American leadership on environmental matters also raises the question of its most appropriate strategy. Clearly the threat of unilateral action has been and will continue to be its most effective instrument. By its nature unattractive, unilateral action is nevertheless a fact of life and can be a potent and constructive instrument for overcoming the inertia of the status quo. But it is an exceptional remedy, as it puts one value (in this case, environmental protection) above all others (maintenance of smooth shipping practices, collaborative interstate relations, and so on). As a result, it has been little used in fact and its effect has been, somewhat contradictorily, actually to improve the quality of international rule-making at IMCO, not to replace it. It is certainly naive to suggest, as Noel Mostert has done, that the only "effective means of controlling the pollution and destruction of the seas by tankers . . . is through strong local laws imposed by states to which they sail."[10]

When viewed from this multilateral rather than isolationist perspective, it follows that the threat of unilateral action should not be the first line of attack. Instead, its effectiveness is as a possible threat should international negotiations fail to produce a reasonable solution because of the intransigence of narrow protectionist interests. After all the interests have been fairly and openly assessed, the unilateral option would remain. To adopt such a gradualist strategy would require greater domestic coordination than has occurred previously.[11] As international negotiations have nearly always taken place before such environmental action anyway, this could only increase the effectiveness of the strategy without changing its substance.

10. "Supertankers and the Law of the Sea," *Sierra Club Bulletin* 61 (June 1976), p. 16.

11. There was certainly little coordination prior to the Carter initiatives, which were really a spillover into the international arena of domestic politics. The result was a rancorous situation where the transgovernmental network was used to restrain domestic environmental actors. Congress, in particular, must give greater attention to the international ramifications of its domestic politics. It has, in the past, unnecessarily caused the loss of much international goodwill. For an excellent discussion of this issue, see Joseph Nye, "Independence and Interdependence," *Foreign Policy* 22 (spring 1976), pp. 130, 151–153.

Such an approach has other important consequences, not the least of which is improving the moral and legal justification for any such action and, therefore, increasing its credibility. In this light, unilateralism could be evaluated not as a defiant and impetuous breach of faith but as a legal act resulting from the failure of other states, after sufficient opportunity, to fulfill their obligations. It must be remembered that there is a duty to protect the international environment[12] and that inaction can be as wrong as action. From this perspective, unilateralism can be a contribution to world order, helping to overcome the obstacles inherent in a voluntaristic legal and political system. In such a system unilateralism cannot be condemned without first being understood. Its true legal nature and political role demand much closer attention than it has been given hitherto.[13]

In conclusion, trends in those commercial interests that affect states' roles at IMCO point both to a continued or increasing conservatism for the European shipowning, Soviet, and developing states and to greater power for the environmentalist United States. At the same time, political strategies are evident for all these states that could lead to greater environmental protection with reduced chance of confrontation. It is strongly urged that these policies be pursued.

In addition to the nature of the oil pollution problem and to the attributes and policies of states, the character and interests of transnational actors will continue to exercise an important influence on future international regulation. In particular, transnational commercial interests have had a decisive impact on pollution control—from the creation of the problem itself to the enforcement of private and intergovernmental regulations concluded to control it. Their participation in IMCO deliberations has, therefore, not been without benefits. Not only has it provided IMCO with valuable expertise, but it has made the final product more acceptable in that those most affected by it have had a voice in its formulation. Nevertheless, commercial interests have generally been willing to accept more costly controls only as a result of pressure by governments and environmental groups—and this is likely to continue to be the case. For this reason, the strengthening of national and international environmental organizations is important for future progress.

There are certainly environmental organizations concerned with the oil pollution problem in a number of developed countries. In France, for

12. See the Declaration of the Stockholm Conference, Principles 1, 2, 6, and 7 and the ICNT, article 193.

13. For one attempt at this, see R. Michael M'Gonigle, "Unilateralism and International Law: The Arctic Waters Pollution Prevention Act," *University of Toronto Faculty of Law Review* 34 (1976), p. 180. What is often at stake when unilateralism is threatened is not a breach of international law at all but simply the failure to negotiate an important issue of interest to particular states.

example, massive demonstrations and a national boycott of the cargo-owner were staged following the *Amoco Cadiz* incident. The wider development of these lobbies at various national levels is profoundly important. Not only do they put pressure on the industries and government in the country to raise or enforce domestic standards, but they may prompt governments to attempt to bring other states along with them in order to preserve uniformity and their competitive position. Unfortunately, in recent years it has been only in the United States that environmentalist pressure has had this transnational effect. Even in the United Kingdom, where the one environmental NGO accredited to IMCO (Friends of the Earth) is located, its activities have had little noticeable effect on Britain's entrenched policy. Strengthening these national lobbies should be a high priority even for the governments themselves. Improved transnational environmental coordination on the oil pollution issue is also desirable, although not likely in the near future. The resources of these groups are so limited and uneven and their priorities so varied that little cooperation has occurred. Their influence would, however, be greatly enhanced if they could support continuous representation at IMCO which could acquire the scientific, technological, and economic expertise needed to command respect in the agency's meetings and to communicate the complexities of the issues to national groups.

Finally, many improvements can be suggested for the IMCO forum itself. In the past, IMCO has responded very well to the new challenges thrust upon it. Especially when one considers that pollution control was not one of the purposes set out for it by its founding members, IMCO has, in the decade since the *Torrey Canyon* disaster, developed fairly rapidly. Its output compares favorably with any United Nations agency (especially considering its small size). At the same time, its operation could be greatly improved for the challenges of the future.

In dealing with the oil pollution problem, IMCO has developed little independent authority. Nevertheless, national laws follow international agreements—what is accepted at IMCO sets the pace. It is, therefore, of utmost importance to the long-term control of the problem that IMCO's law-creation process be as efficient as possible. The ad hoc conference was and, to a certain extent, still is its standard method of law-making, and it is not the most effective method available. As a way to deal with new situations it is useful, but as an institutionalized procedure for managing the environment it is deficient. The development of a more continuous, simpler institutionalized process of rule-making and review is a matter of immediate concern. A conference requires much superfluous energy in its preparation; the event itself is a pressurized marketplace that often elevates the semblance of achievement above actual achievement; and the result is an instrument that as frequently as not gathers dust instead of ratifications. Nye and Keohane have referred to the "tired old institution" of conference diplomacy, and Contini and Sand express doubts

> whether traditional treaty techniques will prove to be suitable for meeting the technical requirements of effective "ecomanagement" on the global or regional scale, once international action passes from the declaratory to the operational stage. Environmental problems characteristically require expeditious and flexible solutions, subject to current updating and amendments to meet rapidly changing situations and scientific/technological progress. In contrast, the classical procedures of multilateral treaty making, treaty acceptance and treaty amendment are notoriously slow and cumbersome.[14]

IMCO has moved in this direction with the establishment of the MEPC as the agency's continuous environmental review body and with the introduction of the new process of "tacit amendment" in the MARPOL Convention. The challenge now is for the MEPC to apply itself to the full range of rule-creation and rule-application problems and to utilize the new amendment procedure. Also, there are other quasi-legislative procedures utilized by other international organizations which the members might consider in the future.[15]

As a more efficient legislative body, it would also be desirable to see IMCO's role in ensuring the enforcement of its regulations advanced. Setting standards for evidence of pollution violations, setting standard levels for penalties, and monitoring flag-state enforcements are a few of the many tasks with which IMCO could be charged. This task expansion will not be easy, as it will meet the resistance of states concerned to protect their sovereign authority. However, responsible flag states should recognize that it is in their interest to increase such international accountability in order that the less reliable flag states can be exposed to public censure. Again, the trend in this direction has already begun (see Chapter 8) and needs to be reinforced.

With these basic developments in legislating and enforcing international regulations, IMCO should also expand its own independent programs and activities. Considering the passive administrative role of IMCO's Secretariat in the past, there will be some resistance to such developments, but

14. Paul Contini and Peter H. Sand, "Methods to Expedite Environmental Protection: International Ecostandards," *American Journal of International Law* 66 (1972), p. 35.

15. On the operation of quasi-legislative bodies in the United Nations system, see Evan Luard, *International Agencies: The Emerging Framework of Interdependence* (London: Macmillan, 1977); Charles Henry Alexandrowicz, *The Law-Making Functions of the Specialized Agencies of the United Nations* (Sydney: Angus and Robertson, 1973), especially pp. 61–69, 152–161.

there is no inherent reason why such activities could not occur. Many other United Nations agencies possess large research staffs which frequently submit proposals for future action to their member governments.

To a limited extent, IMCO has begun to expand its independent activities with the appointment of regional and interregional advisers, its expanded technical assistance program, its sponsorship of symposia and training courses on pollution prevention, and its more active role in encouraging the ratification of conventions. Other areas where greater work could be done are in identifying and evaluating environmental problem areas and in proposing possible courses for legislative action. In fact, a continuous process of review and recommendation could, without being "politically biased," greatly improve the speed and efficiency with which potential problems are identified. In combination with its role in problem identification, IMCO needs a far greater capability to undertake research into scientific and technological problems of importance to its legislative sphere. At present, its independent information system is deficient. As we have seen, national studies have usually supported the preexisting policy preferences of those who have undertaken them. The absence of the *international* perspective often has been glaringly evident at IMCO meetings. One could not deny, of course, that Secretariat studies would be questioned for their underlying biases, but they would still exert great influence on those seeking to affect or evaluate the positions of member governments and hence on the decisions by the organization itself. The work of GESAMP has proven the utility of an international research capability, but its activities are quite limited. Considering the importance of scientific, technological, and economic evidence "properly interpreted," the lack of such a capability by the organization constrains its effectiveness.

From this discussion, it is apparent that an improved Secretariat also requires the abandonment of its traditional reluctance to put forward substantive proposals. This clearly implies a more *political* role, but the time has come when this is possible. It would, of course, not be wise to underestimate the problems involved in embarking on this new direction, but the opportunities must not be ignored as a result of too great a preoccupation with the strictures of the past. There is today undoubtedly much-increased opportunity for political initiatives as IMCO's substantive concerns diversify and gain in political salience, as the solidarity of the developed group erodes, and as new actors join the fray. Such action clearly requires a deft balance of independence from and cooperation with the agency's members. This is exactly what a Trilateral Commission study on international institutions has recommended: "[S]trong Secretaries-General and international staffs can help greatly in *formulating* and implementing international agreements. . . . [S]uch leaders can *propose* solutions when no country is able or willing to do so, help *galvanise support* in individual countries and *implement* decisions."[16] [Emphasis added.]

Acceptance of such a change can greatly augment the organization's contribution. If IMCO is to contribute to the real political challenges of the future, the reality of international power must not be accepted at the price of a global vision. It is perhaps trite but nevertheless accurate to observe that "it is the responsibility of global civil servants to overcome parochial myopia and to struggle with a definition of global interest. . . . A cooperative basis needs to be achieved in which global policies are articulated and then implemented by whatever organizational means . . . best serve the world's interests."[17] "As international politics becomes more interdependent, it is crucial that the systemic view be injected into issues, and international institutions possess a rather unique ability to perform this function."[18]

To date, IMCO's personality as a low-level facilitative body has reflected its subject matter and membership. Now, with a more demanding political agenda and a broadened membership, its personality should change. It will not be an easy or a sudden transition. Yet with a streamlined regulatory procedure, a larger research capability, a more activist Secretariat, and a larger world view, IMCO can enhance its contribution to international environmental regulation.

Extrinsic Developments and the Future of Environmental Protection

In addition to identifying factors and policies specific to the future of oil pollution control, our study has revealed other political factors extrinsic to this one issue but which affect it in significant ways. These factors are beyond the immediate purview of IMCO and of those involved in ocean oil pollution control, but long-term changes in them are still of central concern. Not only do these factors exert great influence on the specific problem of oil pollution, but they are central to the prospects for systemic change that may be necessary if the entire range of international environmental problems is to be effectively resolved.

16. C. Fred Bergsten et al., *The Reform of International Institutions* (New York: Trilaterial Commission, 1976), p. 7.

17. Thomas George Weiss, *International Bureaucracy* (Lexington, Mass.: Lexington Books, 1975), p. 155. On the attitudes and perspectives of scientific secretariat personnel, see Ernst B. Haas et al., *Scientists and World Order* (Berkeley and Los Angeles: University of California Press, 1977).

18. Donald W. McNemar, "The Future Role of International Institutions," in Cyril E. Black and Richard A. Falk, eds., *The Future of the International Legal Order, vol. 4: The Structure of the International Environment* (Princeton, N.J.: Princeton University Press, 1972), p. 461.

As with most other environmental issues, the ocean oil pollution problem is a spin-off of the larger global process of technological and economic growth. The environmental issue follows in the wake of growth whose side effects or externalities often have been unanticipated or ignored. This basic fact shapes not only the nature of the problem but the solution as well. IMCO members have had to adapt their solutions to existing technological and economic imperatives rather than set new standards based fully on environmental needs and in advance of commercial and technical changes. For example, it was only well after the initiation of the supertanker boom that governments began deliberations on environmental rules to "control" these behemoths. As a result, the products were reactive and only partially effective.

Some constructive changes in existing economic arrangements may be possible so as to improve this situation. At a very specific level, particular economic strategies should be developed to allow for the internalizing of environmental costs instead of emasculating environmental rules to fit existing economic objectives. As we have seen, a major impediment to higher environmental standards is fear of a loss of *competitive advantage.* Strategies must be developed to confront such economic obstacles and construct new ways to absorb costs so as to facilitate the realization of environmental and other rising social needs.[19] A second change would be to increase the prospects for planned and controlled economic growth, which considers environmental values *in advance* of such growth, not after. With the gradual development of governmental requirements for environmental impact assessments at the national level, this transition is already beginning. Internationally, UNEP, with its emphasis on such assessments as well as on planning, has been much concerned with this problem.[20] Its efforts are still highly tentative. Even so, such a development is needed if environmental considerations are to rank with, not behind, economic growth.

A third development that could ameliorate the problem of oil pollution would be a transformation within the developed states that are themselves the source of the environmental problem.[21] The introduction of alternative

19. One such strategy evident from this study was the use of "highly integrated structures" that are capable of absorbing extra costs in any particular segment of the operation. This need to devise strategies to absorb new costs is applicable to a wide range of new issues, not just that of oil pollution control. One other is, for example, the absorption in the developed countries of the costly demands of the "new international economic order."

20. UNEP has attempted to integrate proposals for environmental reform with general strategies for international and domestic economic reform. One example of this is its ill-fated Cocoyoc Declaration. For the text, see *International Organization* 29 (summer 1975), pp. 893–901. For an interesting discussion of the fate of this UNEP initiative, see John G. Ruggie, *The Structure of Planetary Politics* (forthcoming), chap. 8.

21. On the domestic aspects of many new issues of global interdependence, see Robert L. Paarlberg, "Domesticating Global Management," *Foreign Affairs* 54 (April 1976), p. 576. Paarlberg argues most persuasively that the issues are *internal* ones and that international

energy sources or a slow-down in energy demand could reduce the potential for pollution from oil as well as from other energy sources. Both of these issues are achieving increasing prominence of late. The entire future of energy development is under review in most developed countries at the same time as their "consumer ethic" is under attack. This system of unrestrained consumption entails great environmental and political costs. If the developing states continue in their quest to follow the pattern of the developed states, the potential for real environmental crisis is great indeed.[22] As a result, much needed consideration is being given to the transition from our present "cowboy" to a future "spaceman" economy. The literature on the "steady-state economy" is flourishing.[23] How this challenge is resolved will greatly affect all environmental problems, including that of oil pollution.

Just as the nature of the specific environmental problem with which IMCO is concerned has its larger context in the prevailing economic system, so too do the specific environmental policies of IMCO's member states. As we discussed earlier, the environmental issue at IMCO is really an issue for the developed, pluralist societies of the West. Clearly, therefore, two national characteristics bearing no specific relationship to IMCO have had a profound effect on the nature of IMCO's deliberations. These characteristics are the general level of *economic* and *democratic* development of IMCO's members.

To date, the modest role of the less economically developed countries at IMCO can be attributed to these factors. Although IMCO is a United Nations agency with over a hundred member states, the global structure of economic relations has meant that the interests of the vast majority of IMCO's membership have been underrepresented and tangential to the organization. Instead, deliberations have focused almost entirely on devising regulations that could be accommodated to the interests of a handful of

---

activity can accomplish little (indeed, is often a dangerous diversion) without major domestic action. "But if the entire world community is held responsible for planetary welfare, then nobody in particular will feel responsible. As it is, domestic actors are capable of blocking any interstate welfare scheme. . . . Far better would be for leadership to argue the merits of its global welfare policy at home, attacking the problem at its point of greatest resistance" (pp. 573 and 575). This is but another example of the commonly noted disintegration in the distinction between foreign and domestic policy.

22. For a first-rate treatment of the argument that a purely economic (not sociopolitical) concept of development is inadequate, see Edward J. Woodhouse, "Revisioning the Future of the Third World: An Ecological Perspective on Development," *World Politics* 25 (October 1972), p. 1. Unfortunately we cannot expect the developing countries to "revision" their future until the developed states "revision" their present.

23. On the "spaceman" economy, see Kenneth Boulding, "The Economics of the Coming Spaceship Earth," in Kenneth Boulding, ed., *Beyond Economics* (Ann Arbor: University of Michigan Press, 1968), p. 275. For an excellent discussion of the "steady-state economy," see William Ophuls, *Ecology and the Politics of Scarcity* (San Francisco: W. H. Freeman, 1977). What must be envisioned is a high technology-low consumption society. The two are not mutually exclusive but are treated as such.

wealthy nations with a major stake in the issue. As the general pattern of economic relations continues to change in the future, this pattern at IMCO will shift too. This is especially so as shipping is "high on the UNCTAD black list as an aspect of North-South exploitation."[24] IMCO's future is intimately connected with the more general question of economic development. To facilitate the orderly development of these states in a manner that will incorporate environmental considerations is a growing political challenge for IMCO as well as for the United Nations system as a whole.[25]

Again, however, developments at IMCO will depend on what happens outside its purview. To a very real extent, there is little IMCO can do to reduce the general economic imbalance that excludes the developing states from effective participation. Indeed, maritime development within these states will not proceed much faster than their general economic development. Even in the development of their maritime industries, IMCO's role will be tangential to that of the UNDP, UNCTAD, bilateral assistance, and indigenous development. IMCO's role will be limited to assisting in the development of environmentally sound and safe maritime industries and of the bureaucratic expertise needed to administer them. The larger issue of their economic progress is, however, still an urgent concern, as only states with a sufficient level of economic security will be able to develop an "ecological consciousness."

A requirement related to the need for general economic development within the developing states is the need for bureaucratic development. Only with significant bureaucratic resources and coordination can a state develop and advance meaningful policies at IMCO. Most developing states are seriously deficient in bureaucratic expertise, and their participation has suffered as a result.[26] In addition, even with a developed bureaucratic structure, the history of IMCO's environmental regulations demonstrates that generally only those government bureaucracies which are responsive and accountable to environmentalist public opinion at home will adopt environmentalist policies at IMCO.

The lack of bureaucratic accountability limits the role environmental values can play in many developed countries as well as, of course, in the

24. Susan Strange, "Who Runs World Shipping?" *International Affairs* 52 (July 1976), p. 349.

25. A recent United Nations study estimated that "the total costs of all [pollution] abatement activities" for developing states was only about "1.4–1.9 percent of their gross product." The problem is "technologically manageable and . . . [economically] within manageable limits." Maritime shipping costs would be a very minor fraction of this. Vassily Leontief et al., *The Future of the World Economy* (New York: Oxford University Press, 1977), p. 6.

26. The effect of inadequate bureaucratic expertise on the policies and influence of the developing states has been demonstrably evident at IMCO conferences. See also Cynthia Enloe, *The Politics of Pollution in a Comparative Perspective* (New York: Mackay, 1975), especially pp. 111–145.

developing and Soviet bloc states. This is another area which is not easily amenable to change on just the specific issue, as it goes to the root of governmental structures and pluralist politics. Nevertheless, it strongly affects the possible outcomes. At present, commercial interests have a continuing access to the technical bureaucracies that is denied to most "public interest" lobbies. This access could be increased were government environmental agencies to develop greater operational and political authority in the maritime field. Political pressure from within a state is clearly the major counterbalance to economic pressures. It is essential that a continuing, responsive structure of government be designed to give more access to public interest groups and legislators. What is needed is more democracy, not less. Clearly much can be learned from recent American experience in this area.[27]

Even with these developments within states, the very nature of the process of interstate negotiations imposes limits on potential achievement in this issue as in others of international complexity. The first impulse of the national representatives in such negotiations is to protect their "national interests." At one level this takes the form of economic interest; at another, national sovereignty. The concern to preserve national autonomy has time and again impeded the development of better environmental regulations at IMCO. Numerous specific proposals have been rejected because they would impinge on the sovereign legal systems of states or the inviolate "authority" of flag states. The entire legislative process at IMCO awaits the autonomous stamp of ratification while, at the Law of the Sea Conference, the very nature of the environmental problem being tackled is redefined to fit existing political, not environmental, realities. The "logic of state sovereignty" confronts "the requirements of world order"[28] nowhere

27. A trend toward more public input into government decision-making on environmental matters is already in evidence in many countries. Numerous commissions have been convened in various places to assess the environmental implications of specific proposals. This ad hoc procedure is encouraging, but it is not nearly as reliable as a legal structure that provides a right of access. Such a structure exists in its most advanced form in the United States where a powerful government committee structure, numerous independent politicians unhindered by rigid party lines, and numerous legislative instruments guarantee wide debate.

For example, one American environmentalist organization, the National Resources Defense Council, has been an omnipresent force in ensuring the implementation domestically of federal environmental legislation. Numerous court cases have served to keep pressure on the bureaucracy to fulfill its legislative mandate.

One recent commentator has argued that the very survival of our present political systems "requires above all insistence on one guiding principle—on the right and force of the general public over all sectoral and technocratic claims." Ralf Dahrendorf, *The New Liberty: Survival and Justice in a Changing World* (London: Routledge and Kegan Paul, 1975), p. 32. For a recent study on public interest participation in government decision-making, see Burton A. Weisbrod (ed.), *Public Interest Law: An Economic and Institutional Analysis* (Berkeley: University of California Press, 1977).

28. Richard Falk, "The Logic of State Sovereignty v. the Requirements of World Order," *1973 Yearbook of World Affairs* (London: Royal Institute of International Affairs), p. 1.

more clearly than in the need to protect the global environment. And the difficulties inherent in overcoming this hurdle should not be underestimated. "There is renewed emphasis throughout the world on national sovereignty," so that the tension "between the imperatives of international interdependence and the quest to retain adequate degrees of national autonomy, appears to remain the basic issue of international relationships for some time to come."[29]

It is for this reason that many commentators have put so much stock in the development of transnational organizations with a "world view." As we have argued, strengthened environmental constituencies within developed (and, ultimately, some developing) states is a matter of great priority. Considering the nature of interstate negotiations (and the high level of organization among industrial interests), their improved transnational coordination is also highly desirable. Such development and coordination are only in their earliest stages, with a few organizations (such as the International Union for the Conservation of Nature—IUCN—and the Nairobi-based Environment Liaison Center) beginning to develop sophisticated networks of interaction. With growing environmental interaction, the process of coordination cannot but increase, however slowly. Whether it will have an impact on the specific problem of oil pollution is not easy to predict. Nevertheless, in the many decades of international environmental politics looming before us, the development of an effective transnational and transgovernmental environmental network will be one of the most difficult and important challenges.[30]

Such coordination must also develop among the intergovernmental organizations concerned with environmental protection, notably between IMCO and UNEP on the marine pollution issue.[31] IMCO's responsibility in the field of vessel-source pollution is well established, despite an occasional jurisdictional squabble with other agencies. That it continue its authority here as a functionally specific body is most important. Achieving

29. Bergsten, *Reform of International Institutions*, p. 2.

30. An observer of NGOs at the Stockholm Conference concluded that "if one expects that the transnational associations on behalf of a global environmental protection policy will combine forces to mount *a concerted assault against the fortresses of archaic sovereignty*, disappointment is certain." Anne Thompson Feraru, "Transnational Political Interests and the Global Environment," *International Organization* 28 (winter 1974), p. 55.

A transnational solution is, however, highly desirable if the sources of the problem are to be adequately tackled. Indeed, one might argue for an entirely new transnational basis of legal rights and duties as a consequence of the transnational implications of the environment problem.

31. For a discussion of the nature of IMCO's relations with other agencies, see Chapter 3. These relations have had little noticeable impact on the environmental regulations studied in this book. Its relations with UNEP could become quite important in the future if UNEP is successful in developing its program.

concrete results is a most uncertain prospect for international organizations, and it is organizations under the direction of a few dominant leaders with a limited number of issues that produce tangible results. But the agency's work will be affected by and related to that of other international organizations. Not only, therefore, must there be close cooperation with specific agencies on specific problems, but there must be overall coordination.

This is the crucial function for the United Nations Environment Program. That body is best suited to establish a holistic perspective, normative frameworks, and strategies for international environmental constituencies and transnational/transgovernmental networks of environmentalists. Indeed, applying the bureaucratic politics perspective to the United Nations structure would lead one to support the organization with a basic environmental ideology as the logical focus for growth on that issue. In addition, power in such an organization is not so weighted in favor of existing commercial/industrial interests.

To date there has been some limited cooperation between IMCO and UNEP, but the relationship has also been abrasive. UNEP's representative to IMCO meetings is engaged largely in ensuring that IMCO does not unilaterally encroach on UNEP's area of authority. In turn, IMCO has certainly not looked at UNEP as the source for leadership on environmental matters. Only recently, competition between these agencies has occurred at the Law of the Sea Conference with IMCO, by all accounts, doing a very effective job of lobbying for more responsibilities. Among agencies where a "world order" perspective should be most highly developed, there is competition. If the environmental issue is to be advanced effectively, such fragmentation must be steadfastly opposed.

The long-term resolution of ocean oil pollution and other international environmental problems awaits the creation of a truly effective structure of international regulation. This will happen only if states, private interests, and international secretariats can coalesce in the international political arena to produce workable and acceptable solutions. This is itself only partially a question specific to the individual issue being addressed. To a very great extent, it is a reflection of the overarching structure of global power and the ideals to which that power is applied.[32] To ensure that the conditions for success prevail at all levels is the great challenge of the global environmental crisis.

32. See the issue of *Foreign Affairs* 56 (October 1977) which discusses the contemporary nature of power and the normative purposes to which it is applied.

'You think oil is bad? Have you tried plutonium?'

Courtesy of *New Statesman*, 12 May 1978, p. 634.

Appendix I

Tables

TABLE A

World Nontanker Registry, 1955–1975[1]

Rank	*1955*[2]	*%*	*1960*	*%*	*1965*	*%*	*1970*	*%*	*1975*	*%*
1	United States	29.82	United States	22.90	United States	16.15	Japan	12.58	LIBERIA	12.62
2	United Kingdom	19.02	United Kingdom	15.95	United Kingdom	12.90	LIBERIA	9.88	Japan	11.57
3	Norway	4.15	Japan	6.07	Japan	7.76	United Kingdom	9.76	United Kingdom	8.88
4	Japan	4.04	Norway	5.47	Norway	6.93	United States	9.75	*Soviet Union*	8.08
5	Netherlands	3.78	LIBERIA	4.64	LIBERIA	6.57	*Soviet Union*	8.10	Greece	7.41
6	France	3.74	German FR	4.50	*Soviet Union*	5.81	Norway	7.42	Norway	6.65
7	Italy	3.64	Greece	4.09	Greece	5.16	Greece	5.01	United States	4.90
8	German FR	3.19	Netherlands	3.99	German FR	4.18	German FR	4.41	PANAMA	4.24
9	*Soviet Union*	3.10	Italy	3.85	Italy	3.52	Italy	3.34	Italy	3.16
10	Sweden	2.57	France	3.34	Netherlands	3.19	Sweden	2.34	German FR	3.02
11	PANAMA	2.41	*Soviet Union*	3.10	France	2.77	Netherlands	2.28	Sweden	2.32
12	LIBERIA	2.21	Sweden	2.75	Sweden	2.75	France	2.11	Hong Kong	2.01
13	Canada	1.73	PANAMA	1.97	PANAMA	1.92	PANAMA	1.67	France	1.98
14	Spain	1.58	Denmark	1.58	Canada	1.58	Canada	1.52	Netherlands	1.58
15	Denmark	1.55	Canada	1.54	Denmark	1.51	INDIA	1.50	Spain	1.50
16	Greece	1.44	Spain	1.53	Spain	1.43	Spain	1.43	Cyprus	1.40
17			INDIA	0.95	INDIA	1.31	Denmark	1.40	*Poland*	1.31
18			BRAZIL	0.86	*Poland*	0.92	*Poland*	1.07	SINGAPORE	1.28
19			ARGENTINA	0.79	YUGOSLAVIA	0.87	YUGOSLAVIA	0.89	Denmark	1.21
20			Australia	0.70	BRAZIL	0.80	BRAZIL	0.81	Canada	1.18
Others		12.00		9.40		12.00		12.70		13.70
World Total GRT	74,114,138		88,304,398		105,345,434		141,350,011		192,105,094	

NOTE: WEO states are in standard print, developing are capitalized, and Soviet bloc states are in italics.

1. The twenty states with the largest nontanker gross registered tonnage are ranked in order of the percentage of GRT registered in each state. Nontanker tonnage was obtained by subtracting tanker tonnage from total merchant fleet tonnage, as found in *Lloyd's Register of Shipping Statistical Tables.*

2. Due to the absence of sufficient data on tanker registry for 1955 in *Lloyd's Register of Shipping Statistical Tables*, it is not possible to determine the rankings for that year beyond #16.

TABLE B

Major Exporters of Oil (Crude and Refined) by Sea, 1973

(Includes states with at least 1% of total seaborne oil exports)

	Total (thousand metric tons)	*Percentage*
1. Saudi Arabia	321,186	17.21
2. Iran	273,460	14.70
3. Venezuela	169,700	9.10
4. Kuwait	157,194	8.42
5. Libya	107,088	5.74
6. Nigeria	98,984	5.30
7. United Arab Emirates	74,085	3.97
8. USSR	69,364	3.72
9. Indonesia	57,185	3.06
10. Netherlands	48,227	2.58
11. Iraq	43,740	2.34
12. Lebanon	41,349	2.22
13. Netherlands Antilles	40,160	2.15
14. Algeria	40,010	2.14
15. Syria	33,920	1.81
16. Italy	24,010	1.29
17. Israel	20,756	1.11
18. Trinidad-Tobago	20,040	1.07
Others	224,886	12.04
Total	1,866,344	100.00

SOURCE: *U.N. Statistical Yearbook, 1975,* pp. 510–517.

TABLE C
Major Importers of Oil (Crude and Refined) by Sea, 1973
(Includes states with at least 1% of total seaborne oil exports)

	Total (thousand metric tons)	*Percentage*
1. Japan	275,666	14.82
2. U.S.	274,759	14.77
3. Italy	175,703	9.45
4. France	165,836	8.92
5. Netherlands	152,312	8.19
6. U.K.	134,634	7.24
7. West Germany	53,518	2.88
8. Netherlands Antilles	45,853	2.47
9. Spain	44,646	2.40
10. Brazil	33,880	1.82
11. Sweden	30,111	1.62
12. Singapore	27,787	1.49
13. American Virgin Islands	25,575	1.38
14. Canada	25,197	1.36
15. Israel	21,177	1.14
16. Denmark	19,560	1.05
17. Belgium	19,519	1.05
Others	334,454	17.95
Total	1,860,189	100.00

SOURCE: *U.N. Statistical Yearbook, 1975*, pp. 510–517.

TABLE D
Major Traders of Dry Cargo Goods by Sea, 1973
(Includes states which loaded and unloaded
at least 1% of world seaborne dry cargo tonnage)

	Total (thousand metric tons)	*Percentage*
1. U.S.	394,469	14.16
2. Japan	367,801	13.20
3. Australia	155,805	5.59
4. Netherlands	149,415	5.37
5. Canada	146,155	5.25
6. U.K.	128,784	4.62
7. France	83,689	3.01
8. West Germany	82,302	2.96
9. Brazil	77,103	2.77
10. Italy	74,664	2.68
11. Belgium	68,194	2.45
12. USSR	66,974	2.40
13. Sweden	58,682	2.11
14. Norway	48,305	1.73
15. India	44,447	1.60
16. Poland	42,753	1.54
17. Spain	37,174	1.34
18. Denmark	29,045	1.04
19. Venezuela	27,910	1.01
Others	701,322	25.18
Total	2,784,983	100.00

SOURCE: *U.N. Statistical Yearbook, 1975*, pp. 510–517.

TABLE E
Percentage of States' Tanker Tonnage
Owned by Oil Companies, 1973

1.	Yugoslavia	100
2.	Portugal	98
3.	Belgium	96
4.	Netherlands	76
5.	France	69
6.	U.K.	64
7.	Germany (FR)	62
8.	Spain	54
9.	Panama	48
10.	Italy	43
11.	U.S.	39
12.	Finland	26
13.	Liberia	23
14.	Turkey	17
15.	Japan	12
16.	Sweden	8
17.	Denmark	5
18.	Cyprus	3
19.	Norway	2
20.	Greece	0
21.	USSR (and E. Europe)	0

NOTE: In 1973, 31% of world tanker tonnage was owned by oil companies. This was a drop from 36% in 1960. Additional tonnage is financed by the oil companies, but figures on this "indirect ownership" are not available.

SOURCE: *B.P. Statistical Review of the World Oil Industry, 1973*, p. 14.

TABLE F
National Registry of Tankers
of the Seven Major Western Oil Companies

	Exxon	*Shell*	*Texaco*	*Mobil*	*Standard of Calif.*	*B.P.*	*Gulf*	*Total*	*% of Total*
U.K.	28	73	29	11	2	94	4	241	35
Liberia	38	3		18	43		17	119	17
Panama	36		30	4	3			73	11
U.S.	16		16	9	12		14	67	10
Netherlands		30			8		4	42	6
France	8	15		3		11		37	5
Germany	12	13		1				26	4
Antilles	4	12						16	2
Norway	1	1	9					11	2
Canada	5	1	1				1	8	1
Belgium	3						5	8	1
Italy	7							7	1
Australia	1	3		1		2		7	1
Denmark	6							6	
Venezuela		6						6	
Spain							5	5	
Argentina		3						3	4
Uruguay	2							2	
Japan		2						2	
Finland		1						1	
South Africa				1				1	
	168	163	85	48	68	107	50	688	100%

SOURCE: *The Tanker Register, 1975* (London: H. Clarkson, 1975). The figures in the cells refer to numbers of tankers and not percentage of tonnage.

Appendix II

The Insurance Industry

Insurance has been a pervasive influence in the development of international environmental controls. It is, therefore, most important to understand how the insurance industry works and how it has affected states' policies toward changes in the international environmental legal regime. Independent of government intervention, the rates set by the insurance industry discriminate against the less seaworthy vessels and less reputable shipowners and charterers. Moreover, any change in the shipowner's responsibility for pollution damage has a direct impact on the terms of insurance. Indeed, after the 1967 *Torrey Canyon* incident, one of the central issues of debate concerned the commercial ramifications for the shipowner and his insurer of any stricter system of liability.

The original insurers in the maritime field were the many private "underwriters" found at Lloyd's in London. Although in the seventeenth century Lloyd's was just a coffeehouse where these people individually transacted their business of sponsoring maritime ventures, their continued association at Lloyd's resulted in the formation of collective "syndicates" of underwriters which could more efficiently distribute the risks. Lloyd's itself still does not undertake the business of insurance but provides the facilities for the approximately 8,000 underwriters and 260 syndicates operating there.[1] The basic coverage of these underwriters has been limited in the maritime field—extending frequently to only 75 percent of the costs of the collision liability (the "running down" clause). The balance of the value as well as the many other potential liabilities of ships—loss of life, personal injury, cargo damage, dock damage, wreck removal, pollution costs, and any accompanying fines—have all been excluded from traditional insurance coverage.[2] To protect themselves from these liabilities, the shipowners themselves collaborated to form their own "mutual" Protection

1. Robert H. Brown, *Marine Insurance: The Principles* (London: Witheby, 1968), p. 9. See also John B. Ricker, Jr., "Restrictive Legislation: Its Effect on the U.S. Marine Insurance Industry," *Journal of Maritime Law and Commerce* 6 (October 1974), p. 147. Lloyd's does provide a guarantee of the undertakings of all its underwriters in order to protect its name.

2. This was first made clear by the case of *de Vaux* v. *Salvador* (1836), where damage to

and Indemnity Associations ("P & I Clubs") where they could apportion, distribute, and share the expenses incurred.[3]

There are approximately sixteen major P & I Clubs across the world. The most significant, the "London Group," consists of nine associations which, together with the three Scandinavian clubs, represent over 70 percent of the world's shipping tonnage. In comparison, the only American club, the American Steamship Owners Mutual Protection and Indemnity Association, represents just 2 percent of the world fleet.[4] The financing of the coverage offered by these clubs varies according to the range of the liability. For incidents causing damage less than about $600,000, all members of the particular club in which the offending ship is registered share the cost. Above that level to $4 million, the liability is "pooled" and apportioned on a tonnage basis among all the clubs. Above the limits of this second level (called the "retention point"), the liability is insured on the London market. With such potentially high amounts, stability can be better achieved through the payment of a premium to a Lloyd's underwriter than by carrying the liability for the large but irregular amounts directly. The coverage for this third level is called "reinsurance." The underwriters' liability has an upper limit of $50 million, however, and above that the P & I Clubs are again responsible. Until recently there was no limit on this fourth level because of the unlikelihood of its ever being reached. After the sobering experience of the *Torrey Canyon* disaster, however, the insurers were able to achieve a new limitation of $14 million for pollution liability. This is now set at $50 million and, above that amount, the shipowner cannot look to the club for assistance.

Within an individual club, a shipowner's contribution is assessed according to his proportionate share of the club's total tonnage. However, in order to ensure that the levy assessed is proportionate to the risk, it is further weighted according to regular insurance criteria—type, age and condition of vessel, quality of crew, nature of cargo and voyage, and past record. Contributions vary further according to the extra risks for which a shipowner wishes to be covered. Therefore, although the broad scheme is

---

another ship was excluded. See Victor Dover, *A Handbook to Marine Insurance*, 7th ed. (London: Witheby, 1970), and William R. A. Birch Reynardson, "The History and Development of P & I Insurance: The British Scene," *Tulane Law Review* 40 (April 1969). One particularly interesting aspect of the coverage offered by the P & I Clubs is the inclusion of fines. This must substantially reduce their deterrent effect. Indeed, one cannot but wonder whether a contract to pay for the costs of an illegal act would not itself be illegal as "contrary to public policy." See Hugh S. Meredith, "Fines, Penalties and Other Miscellaneous Liabilities . . .," *Tulane Law Review* 42 (April 1969), p. 602 (this article is a part of an Admiralty Law Institute Symposium on P & I Policy).

3. The term *mutual* means "insurance at cost since no profit element is written into the premium structure." See *Mutuality* (London: U.K. P & I Club, 1972), p. 4.

4. See "American P & I Continues to Grow," *Fairplay International Shipping Weekly*, April 15, 1976, p. 41.

one of "mutuality," individual rates of contribution can vary by many hundreds of percent.[5] Contributions for an operating year are collected through advance "calls," the amount of which is calculated on the basis of the costs shared over the previous period of five years (the average "calculation period"). If this is not enough to meet the running expenses and to cover the actual liabilities for the call period, supplementary calls are made.

Prior to 1967, oil pollution liability protection was a miniscule proportion of shipowners' overall insurance calls and was not even specified separately. Costs for *all* P & I liabilities had been as low as 15 percent of the costs of hull insurance alone,[6] and the pollution liabilities at most 5 percent of that.[7] Even after the *Torrey Canyon*, no change was predicted by the P & I Clubs. Arguing in 1969 before a U.S. Senate subcommittee which was considering raising the pollution liabilities to which ships in American waters would be subject, John Shearer of the U.K. P & I Club reported that the world's four largest P & I Associations reported only twenty-nine pollution incidents exceeding £5,000 in the seven-year period 1960–1966. Payments on these infrequent claims were only £809,652, so that any expansion in the legal regime imposing liability was "unnecessary."[8] After the *Torrey Canyon*, "it was averred that such a calamity could never happen again."[9]

While this argument was repeated time and again, it was untrue. On the one hand, the insurers argued that their experience indicated that the costs of another pollution incident would not exceed a rather small figure, while on the other hand both the shipowners and insurers adamantly opposed being made responsible now for any higher amounts.[10] Clearly the risk was not so low as they argued as Peter Miller, representing the U.K. P & I Club, admitted under questioning by Senator Muskie:

5. Annar Poulson, "Observations on the Draft Convention Articles Relating to the Establishment of an International Compensation Fund for Oil Pollution Damage," IMCO doc. LEG/CONF.2/C.1/WP.3. Mr. Poulson states (p. 2) that "we personally know of tanker insurance rates varying between 3 cents and 150 cents per g.r.t." This is a source of much contention within the P & I Clubs, as the less seaworthy operators charge they are being "discriminated" against.

6. John C. J. Shearer, *Hearings on S. 7 and S. 544 before the Senate Subcommittee on Air and Water Pollution of the U.S. Senate Committee on Public Works*, 91st Congress, 1st Session, part 1, p. 164.

7. One director of Thos. Miller and Sons, the managers of the world's largest club, the U.K. P & I Club, estimated that 50 percent of P & I costs were for loss of life and personal injury, 25 percent for cargo claims, 5 percent to cover collision damage excluded from the underwriters' coverage, and 20 percent for all other liabilities. See T. G. Coghlin, "Protection and Indemnity Insurance: The P & I Clubs," *Journal of World Trade Law* 5 (November/December 1971), pp. 591–593.

8. Shearer, *Hearings*, p. 141.

9. See "Marine Anti-Pollution and Economic Reality," *Fairplay International Shipping Weekly*, January 9, 1975, p. 125.

10. By the Convention on the Limitation of Liability (1957), ships were able to limit their liability to $67 per ton of a ship's deadweight tonnage. This ability to limit liability below the amount of damage caused is a quirk of the maritime industry (see Chapter 6).

> Well, now, to put it crudely, the *Torrey Canyon* took the wind out of underwriters. They see two things which lead them to believe that the *Torrey Canyon* is not an extraordinary incident, but one that is going to happen again and those two things are these: first, oil is being carried now, relative to 5 or 10 years ago, by ships of vastly increased size to the ship of those days, and the second point is this: the products tend to be refined at point of destination rather than at point of departure, so that the carriage of crude oil, rather than the carriage of more refined products—and I am talking in very crude terms—has been very much increased and it is the risk of pollution of crude oil rather than from refined products which is to an underwriter the grave risk of a catastrophe. They may be right or they may be wrong, but they view the future in a very black light and refuse pointblank to extrapolate the experience of the past into the future because of the two factors that I mention.[11]

Indeed, what was at issue now was not that the risks were small but, on the contrary, that they were "uninsurable." Any decision to increase the accountability of the shipowners beyond traditional limits was therefore "of little relevance because the real issue is the capacity of the marine insurance market."[12]

In fact, the risks of accidents were already becoming clear, and the insurance costs of pollution were to become obvious within a couple of years.[13] At the 1971 IMCO Fund Conference, the shipowners saw the possibility of reducing their insurance burden by shifting the costs directly onto the oil company Fund. Then the larger figures for pollution insurance were revealed. In a letter to IMCO, the director of the Norwegian P & I Association "Skuld," Annar Poulson, reported that the "final average pollution costs" for tankers in his club rose from 3.4¢ per grt for the five-year period 1960–1965 to 12.8¢ for the 1965–1970 period. For 1970 alone the average cost was 43.5¢ per grt, and Mr. Poulson projected a cost of 75¢ per grt in the near future.[14]

11. Peter Miller, director of Thos. Miller and Sons. Statement before the U.S. Senate Subcommittee on Air and Water Pollution, in Shearer, *Hearings*, p. 155.

12. Alan Mendelsohn, *Hearings before the Subcommittee on Air and Water Pollution of the U.S. Senate Committee on Public Works, July 21–22, 1970*, 91st Congress, 2nd Session, p. 44. In fact, when it looked as if some states were going to take all limitation off shipowner's liability, the P & I Clubs changed their own traditionally unlimited coverage and fixed a limitation for oil pollution that they felt they could carry. Nicholas J. Healy, "Water Pollution Liability from an Insurance Viewpoint," *Houston Law Review* 9 (1972), p. 662. For quite a period of time insurers maintained the contradictory position that higher liability was both unnecessary (because of the lack of any risk of grave damage) *and* uninsurable.

13. Noel Mostert pointed to a spate of polluting tanker accidents in 1968 (among others, the 18,000-ton *Sivella*, the 48,000-ton *Esso Essen*, the *Andron*, and a major spill from the *World Glory*) and in late 1969 (the 206,000-ton *Marpessa* and its sister ship the *Mactra*, and the 220,000-ton tanker *Kong Haakon*). *Supership* (New York: Knopf, 1974), pp. 134 and 192.

14. Poulson, "Observations," p. 4. These figures correspond to a reported average annual

But in 1969 many argued that the capacity of the market was a "specious" issue.[15] It has, on the contrary, been the single most significant factor controlling change in the field of liability for pollution damage. Both in debates at IMCO and in various national capitals, a persistent—and successful—effort was made to restrict liability to limits that the existing insurance market could accommodate.[16] Insurance, the argument goes, is at any one time a finite commodity because, as with any industry, success depends on the optimum utilization of resources. Available underwriting capital that is not being utilized in covering insurable risks to the maximum capacity is unproductive and wasteful—clearly a situation to be avoided. Any expansion of the huge underwriting pool is therefore very slow. This is exacerbated by the nature of insurance, which requires a very wide spreading of the risk in order to achieve stability. For example, the insurance cover for the *Torrey Canyon* had been placed with approximately 120 different syndicates and companies.[17] As the textbook says, the underwriter will

> not be prepared in the absence of adequate reinsurance facilities to place himself on cover for too large an amount on any one vessel . . . being impelled to bear in mind the effect of a substantial loss on his "book." His reserves may be adequate to cover a "catastrophic" loss, but his desire will be to show a reasonable profit . . . and not expose himself to violent fluctuations.[18]

The problem is aggravated by the practical nature of underwriting. Insurers traditionally require a period of actual experience with a particular

---

call for tankers in the U.K. P & I Club in 1968 of 12s 6d per ton (about $1.70). Other insurance industry officials thought that the pollution costs supplied by Mr. Poulson were "outlandishly high" (interviews).

15. At the 1969 IMCO Conference to increase the limits of liability, Donald Jamieson, Canadian Minister for Transport, said, "Any argument based on the cost of insurance and the capacity of the insurance market was specious and out of place. Figures advanced on that subject ($10 to 15 million per accident had been mentioned) were in striking contrast with the cover provided by the air insurance market, which amounts to $80 million or more per incident." *Official Records, International Legal Conference on Marine Pollution Damage, 1969* (London: Intergovernmental Maritime Consultative Organization, 1973), p. 86. What this argument fails to consider is that insurers were apprehensive about the greater risks they would be insuring with the supertankers. This is in marked contrast to the airline industry, where no increase in risk was anticipated in an already low-risk business. The total loss figures there are relatively small. On the other hand, the insurers' argument about there being no need to raise the limits was specious.

16. There was, as well, a general reluctance by insurers to become the "fall guys" of environmental activists. As the May 1971 edition of *Fire, Casualty and Surety Bulletin* commented, "The spirit of the times seems to be 'stop the polluters' by almost any means—a dangerous time for any insurer to accept the transfer of such a risk." Charles Goria, "Compensation for Oil Pollution at Sea: An Insurance Approach," *San Diego Law Review* 12 (April 1975), p. 730.

17. Ibid., p. 732.

18. Dover, *Handbook to Marine Insurance*, p. 151.

market before they will undertake a wide business in it. All insurance commentators stress over and over the importance of acquiring this unscientific but characteristic "feel": "An underwriter may not be able to give precise reasons why he accepts or declines, as the case may be, a particular risk, for the subconscious plays a large part in influencing his decisions. He must know the 'feel of the moment' and the 'sense' of a risk."[19]

This demand for higher liabilities happened to occur at a time when the world insurance market was already shrinking. Inflation was high, repair costs were escalating, and major losses were being felt in the areas of loss of life and personal injury, cargo and dock damage. Any significant increase in coverage would require a new source of underwriting cover—something that the London market opposed and the states were not willing to provide.[20] (In one area of undesirable risk, nuclear damage, commercial insurers do not provide coverage, leaving it to states to provide guarantees of the ships' liability to a limit of $100 million per incident.) Of course, insurers do not wish to lose business that is potentially lucrative. But the conservatism inherent in the insurance system often has resulted in the loss of business later found to be profitably insurable. The whole evolution of P & I cost-sharing is a prime example of such a loss by the Lloyd's underwriters, as is the creation of the International Fund.[21] It is, however, only where other interests are sufficiently motivated to provide a substantial alternative that insurers risk losing their influence and control.

In this case, although some new sources of coverage did arise, they were not themselves willing to underwrite the full risks and had therefore to turn to London for "reinsurance."[22] While reinsurance does not always provide the old market with control over the setting of conditions of insurance, it does still leave it substantial control over the amount and rate of coverage. When the U.S. passed its Water Quality Improvement Act in 1970, the Water Quality Insurance Syndicate was created to provide insurance for

19. Ibid., p. 150.

20. Another alternative was "self-insurance," but only very large, low-risk operators were capable of undertaking this—as many have done. This was also unattractive to insurers, as it meant the complete loss of some of the best business.

21. Whether such an imaginative alternative would have arisen on its own is moot, as clearly it was inspired by the lack of any economically accessible insurance for shipowners. Should it not be heavily used—as many people predict—the insurers will have sacrificed a valuable source of higher level insurance.

22. More recently, complaints have been voiced at the influx of new operators into the market, which has driven down rates worldwide. "Insurance Rates Crisis," *Fairplay International Shipping Magazine*, March 25, 1976, p. 3. Lately there has also been a sharp growth in state insurance companies, which are being used to the limits of their capacity before nationals go outside for coverage. This has "deprived the London market of much of the direct business which might otherwise have added to its growth." "The London Market and Its Future," *Fairplay International Shipping Weekly*, February 20, 1975, p. 51.

it.[23] Although originally planning to limit liability at $450 per ton, the legislation was finally passed with the limit acceptable to London insurers, $100 per ton. The State of Florida refused to follow this, and passed its own legislation with unlimited liability.[24] Despite much criticism and a major legal challenge by the P & I Clubs, Florida persisted.[25] Only after it became apparent that it would get no insurance coverage (and therefore, no ships) did it amend the act to conform with the limits of the federal legislation[26] and eventually repeal it. Similarly, in Canada, the Arctic Waters Pollution Prevention Act of 1970 was passed with provision for a far stricter regime of liability than existed internationally.[27] Yet it was not put into force until two years later when, after consultations with the London insurers, regulations were passed which conformed to the existing international regime. This allowed ships to get insurance so as to be able to sail into the area.[28]

In conclusion, both the insurance industry and the P & I Clubs wield profound influence over the control of pollution and the creation of laws governing it. Through differential rates of coverage, substandard ships are penalized. Unfortunately, insufficient positive incentives are provided for these ships to improve, as

> the attitude of the marine insurance market towards the level of the loss experience is that this is a matter for the shipowners concerned. In other words, in general, specific discounts are not extended to shipowners, e.g., in respect of the installation on board their vessels of particular fire alarm and extinguishment equipment. This represents an extension of the attitude that risks must be rated on their merits.[29]

Indeed the shipping magazine, *Fairplay*, has specifically criticized the insurance industry for this attitude.

> Yet a shipowner who takes the most effective measures to prevent water pollution [such as extra equipment, tanks or boilers] . . . gains scant recognition of his efforts by underwriters who contend that premiums cannot justifiably be cut since such measures are expensive and enhance the value of the ship to be insured.[30]

23. See Healy, "Water Pollution Liability," p. 665, and Water Quality Improvement Act, April 3, 1970, (Pub. L no. 91-224, 84 Stat. 91).
24. Florida Oil Spill and Pollution Act (1970).
25. See the case of *American Waterways Operators* v. *Askew* (1972) AMC 91, and comment in *Annual Report: U.K. P & I Club Bermuda* (1972), p. 52, and (1973), p. 16.
26. Pollutant Spill Prevention and Control Act, CH 376, July 1, 1974.
27. Arctic Waters Pollution Prevention Act (1970), c. 99.
28. Regulation SOR/72-253. The regulation restricted limits of liability to those contained in the IMCO conventions.
29. Dover, *Handbook to Marine Insurance*, p. 202.
30. "Marine Anti-Pollution and Economic Reality," *Fairplay International Shipping Weekly*, January 9, 1975, p. 125. See also S. Gibbs and D. Teeter, "Management of Marine

The effect of the insurance structure on state policy is to restrict the scope of state activity to the existing commercial framework. Even the two most vehement environmentalists at the 1969 Conference, Canada and the United States, did not achieve domestic regimes that were significantly tougher than that which the insurers would accept internationally. On the other hand, their vehemence, combined with the natural financial conservatism of the insurer and the interests of shipowners to shift the pollution burden onto other shoulders, did spark the initiative and imagination for the creation of a better and more efficient compromise solution, the International Fund. This is perhaps a net loss for the London underwriters, as experience to date indicates that the pollution risk was a profitably insurable one, despite some large spills. Indeed, on January 9, 1974, the U.K. P & I Club increased its liability coverage from $15 to $20 million, and, on February 20, 1978, to $50 million.[31] This must make governments wary of future industry projections.

Transportation in the Northeast Pacific,'' in C. L. Dawson (ed.), *The North Pacific Project* (Seattle: Institute of Marine Studies, University of Washington, 1978), p. 406. The authors complain about the total lack of consideration given by insurers to the installation of collision avoidance radar despite the important decrease in collision experience this has produced.

31. See *Report: U.K. P & I Club* (1974), p. 24. No problem now exists on spreading the risk (*Fairplay International Shipping Weekly*, November 7, 1974, p. 41). Interestingly, London insurers also found they had excess capacity as soon as the Water Quality Insurance Syndicate was created.

Index

Compositor:	Freedmen's Organization
Printer:	Publishers Press
Binder:	Mountain States Bindery
Text:	Compugraphic Baskerville
Display:	Compugraphic Baskerville
Cloth:	Holliston Roxite B53521
Paper:	55lb. P&S offset regular A69